ROUTLEDGE LIBRARY EDITIONS:
COLD WAR SECURITY STUDIES

Volume 58

SPACE WEAPONS AND U.S. STRATEGY

SPACE WEAPONS AND U.S. STRATEGY

Origins and Development

PAUL B. STARES

Routledge
Taylor & Francis Group
LONDON AND NEW YORK

First published in 1985 by Croom Helm Ltd

This edition first published in 2021
by Routledge
2 Park Square, Milton Park, Abingdon, Oxon OX14 4RN

and by Routledge
52 Vanderbilt Avenue, New York, NY 10017

Routledge is an imprint of the Taylor & Francis Group, an informa business

British Library Cataloguing in Publication Data
A catalogue record for this book is available from the British Library

ISBN: 978-0-367-56630-2 (Set)
ISBN: 978-1-00-312438-2 (Set) (ebk)
ISBN: 978-0-367-55798-0 (Volume 58) (hbk)
ISBN: 978-1-00-309520-0 (Volume 58) (ebk)

Publisher's Note
The publisher has gone to great lengths to ensure the quality of this reprint but points out that some imperfections in the original copies may be apparent.

Disclaimer
The publisher has made every effort to trace copyright holders and would welcome correspondence from those they have been unable to trace.

Space Weapons and US Strategy:

ORIGINS AND DEVELOPMENT

PAUL B. STARES

Croom Helm

LONDON & SYDNEY

Croom Helm Ltd, Provident House, Burrell Row,
Beckenham, Kent BR3 1AT
Croom Helm Australia Pty Ltd, First Floor, 139 King Street,
Sydney, NSW 2001, Australia

British Library Cataloguing in Publication Data

Stares, Paul B.
1. Artificial satellites—Government policy
—United States—History 2. Military
reconnaissance—Government policy—United
States—History
I. Title
623.74'9'0973 UG763

ISBN 0-7099-2369-4

Typeset by Words & Pictures Ltd., Thornton Heath, Surrey
Printed and bound in Great Britain

Contents

For My Parents

Acknowledgements

This book grew out of my doctoral thesis *US Policy Towards the Militarisation of Space: The Origin and Development of Antisatellite Weapons, 1945–1982* that was completed while I was a Research Fellow at the Centre for the Study of Arms Control and International Security, Lancaster University, England. I am particularly grateful to its Director, Professor Ian Bellany, who as my thesis supervisor gave me considerable support throughout the study.It was an ambitious, some argued impossible, thesis to undertake not least because of the geographical separation from the main source of information. For making it possible to conduct essential fieldwork in Scandinavia and the United States, I am indebted to the International Communications Agency, US Embassy London; the Institute for the Study of World Politics, New York; and the Social Science Research Council, London for travel grant support.

In Sweden, Milton Leitenberg of the Swedish Institute of International Affairs provided invaluable help in allowing me access to his vast collection of relevant research material. Owen Wilkes of the Stockholm International Peace Research Institute (SIPRI) and Nils Petter Gleditsch of the Peace Research Institute (PRIO) provided additional data but it was their advice on methods of political research into areas considered impenetrable that was of most help. To all three I am especially grateful. Their assistance and the generous loan(!) by Lawrence Freedman of his press cuttings on this subject provided the springboard for the two research trips to the United States.

The first of these visits took place between September and December 1980. Considerable time was spent at the John F. Kennedy Library in Boston; the Dwight D. Eisenhower Library in Abilene, Kansas and the Lyndon Baines Johnson Library in Austin, Texas. The friendly and efficient assistance given by the staff of all three libraries during my visit, and subsequently in the process of declassifying documents, was of immense benefit to me. During this and my next visit to the United States in May and June 1981, over 100 interviews were conducted. The willingness of retired and serving US government officials as well as members of its armed forces to give interviews is renowned, but their especially friendly reception for an Englishman studying this subject was particularly gratifying. On both visits to the United States, I received enormous hospitality and encouragement from a great many people, in particular, Steven Miller, Jane Sharp, Judith Reppy, the staff of the Arms Control Association (especially Bill Kincade, Jeff Porro and Marsha McGraw) who provided invaluable office space in Washington, DC., Pierce Gardner, Roger and Penny Beaumont, Robert Dove and Nick Schiele. A special thanks is due also to Gerald Steinberg.

After substantial revision, the book was finally completed while I was a Rockefeller International Relations Fellow and Guest Scholar at the Brookings Institution, Washington, DC. Without the support of Brookings and the Rockefeller Foundation this book would not have seen the light of day. For valuable comments on the manuscript, I would like to thank Raymond Garthoff, Michael MccGwire, Desmond Ball, Nicholas Johnson, John Pike, Jeffrey Richelson, Sarah Sewall and those former government officials who wish to remain anonymous. I am, of course, responsible for the final work.

As most authors will testify, the preparation of a manuscript is impossible without the help of typists. In Lancaster Anne Stubbins typed the final thesis while in Washington Tom Somuah and in particular Susan Nichols laboured long and hard over what must have seemed never ending drafts. To them, my heartfelt thanks.

Finally, various people have supported me in indirect but no less important ways. I would like to thank the staff of the North Staffordshire Polytechnic, in particular Brian White and David Dunn for an unparalleled foundation in the study of international relations. Similarly, the Armament and Disarmament Information Unit at Sussex University, especially its former Information Officer Harry Dean, gave me a welcome chance to rekindle my interest in the subject at a critical juncture. For keeping my body and soul together during my period in Lancaster, I am grateful to Tom and Pam Manion and Val Robinson who helped me in more ways than they probably realise. I am also indebted to my friend and colleague Martin Edmonds for not only considerable advice, guidance and support but also for his example of a true scholar and gentleman. In Washington, I have benefited immeasurably from the friendship of Paul Cole, Christine Helms, Joshua Epstein and Sarah Sewall. Last but not least, I wish to thank my parents and family for their love, generous support, and encouragement along the way.

In politics, what begins in fear usually ends in folly.

Samuel Taylor Coleridge

1

Introduction

> The first missile powers contemplate space with the perspective of the first oceanic naval powers, when they contemplated the globe. Their existing legal and political conceptions do not cover it, and their experience provides them only with analogies. They can have little notion of the problems to which it will give rise, or of the political, strategic and economic importance it will have for them. It is not even clear what it is, or what the human activities are that will be specially connected with it.
>
> Hedley Bull,
> *The Control of the Arms Race* (1961)

THE MILITARY IMPORTANCE OF OUTER SPACE

In recent years considerable attention has been given to the growing importance of the use of outer space for military purposes. A particular focus and cause of this interest has been the increasing likelihood that space will become an extension of the arms race between the superpowers and possibly even a future arena of combat. The plethora of books and articles on this subject can leave the impression that the military use of space is a novel activity.[1] Space, in fact, is hardly a 'new' military medium. Although weapon systems have yet to be deployed here or used in anger, space has been an integral part of the superpower arms race for over 25 years.

The extent to which space has already become militarized is indeed impressive. As will be illustrated shortly, serious attention was given to the potential military utility of artificial satellites as early as 1945, well before the launch of the first satellite in 1957. By the early 1960s, military satellites designed to perform a wide range of missions were being regularly launched. Since then, over 2,000 military payloads have been put into orbit by the United States and the Soviet Union, which accounts for approximately two-thirds of the total number of all satellites launched over the same period.[2] The investment to sustain these military space programmes has been equally impressive. While official figures are not available for the Soviet Union, the US Department of Defense between 1959 and 1984 has spent over $70 billion (current dollars). The FY 1984 military space budget alone was approximately $10.5 billion, which is over half of the total US space budget.[3]

To the public, whose appreciation of the benefits of space is generally limited to the services that satellites provide for weather forecasting and instantaneous telecommunications around the world, these facts will probably come as a surprise. This, however, would be understandable as immense secrecy surrounds the military space programmes of the two major space powers. The Soviet Union refuses even to acknowledge the existence of a military component to its space programme, while the United States chooses to release only carefully selected information on its military operations in space.

For those who have managed to keep abreast of the militarization of space through the careful collation of the limited information available, in addition to educated guesswork about the purpose of Soviet (and some US) satellites from their orbital characteristics,[4] one conclusion cannot be avoided: the United States and, to a lesser extent, the Soviet Union are now heavily dependent on the services that military satellites provide for their national security. Military satellites not only play a crucial role in the maintenance of the armed peace between the superpowers but are also vital to the planning and prosecution of warfare at almost every level. This dependency derives in large part from the unique services that satellites provide. Where they are not strictly unique they are usually more efficient and economical. This relationship has been progressively reinforced as the variety of military satellites has widened and as the reliance on equivalent terrestrial systems has diminished, often to the point of atrophy. This trend can be illustrated by a brief account of the benefits that military satellites provide. The major categories are listed below.

Photographic Reconnaissance

The information supplied by photographic reconnaissance satellites is used for a variety of purposes. Although strategic intelligence gathering continues to be the primary mission for this class of satellites, the same information is equally indispensable for monitoring compliance with arms control agreements. As the resolution of the cameras has improved and the speed of the data processing has increased, so these satellites have begun to be used for tactical purposes such as battlefield surveillance. Thus, in peacetime photo-reconnaissance satellites are invaluable for assessing the threat from a potential adversary and for enhancing confidence in their compliance with arms control treaties. In the event of a crisis, these same satellites can look for potentially escalatory actions and thus provide early warning of mobilization and an attack. Once a conflict has begun the same information can be used for locating and targeting military threats to one's own forces and afterwards measuring the success of the military engagement.

Electronic Reconnaissance

Electronic intelligence (ELINT) gathering from space by what are often referred to as 'ferret' satellites complements the photo-reconnaissance missions in a number of important ways. These satellites eavesdrop on the communications of an adversary to gauge the size, deployment and readiness of its military forces. Similarly, by listening to other electromagnetic emissions, for example that from air defence and missile early warning radars, their location and frequencies can be plotted for targeting and electronic countermeasures. Furthermore, as data on the performance of new weapon systems such as ballistic missiles are often transmitted from the test vehicle in the form of telemetry, the collection of this information can also be a valuable aid for arms control verification.

Ocean Reconnaissance

As the title implies, these satellites are used to locate and track surface shipping. They can accomplish this by using the same techniques as ELINT satellites or by the use of specially designed radars; in short, by passive and active means, respectively. In future this mission is likely to expand to include subsurface reconnaissance for submarine detection. Another area of activity where space promises to offer considerable advantages to the military is oceanographic surveillance. Such information as the height of waves; the strength and direction of ocean currents and surface winds; sea and undersea temperatures; the level of sea salinity and coastal features, which can all be obtained by space-based oceanographic sensors, are important for naval activities generally and potentially vital to future antisubmarine warfare operations.

Early Warning

While reconnaissance satellites can provide some advance indication of an attack by conventional forces, early warning satellites are designed to detect the launch of land- or submarine-based ballistic missiles. As an intercontinental ballistic missile (ICBM) is likely to take as little as 25 minutes to reach its target, the use of infrared sensors in space to detect the hot exhaust plumes of attacking missiles during their "boost" phase provides a vital margin of warning time over ground-based radars.

Nuclear Explosion Detection

A space-based worldwide nuclear explosion detection system was originally established for arms control purposes to monitor covert atmospheric testing. However, it was also recognized—certainly by the United States—that a system that could detect the location, yield and the height of nuclear bursts would also contribute to post-attack assessments during a nuclear war. As a result, nuclear explosion detection sensors have been refined and fitted to other satellites.

Communications

The use of satellite relays for communicating rapidly and reliably over long distances was recognized from an early date and they were therefore among the first group of satellites to be developed. Not only have they reduced the reliance on expensive and vulnerable land lines, undersea cables and radio relay stations (which are often on foreign soil), but they have increased markedly the level of control, and with it, efficient usage of military forces. This trend will increase as the number of military users at the tactical level multiplies with the miniaturization of satellite ground terminals.

Navigation

Navigation satellites have a similar "force multiplier" effect. In addition to providing very accurate information to submarines, ships, aircraft and land-based forces, these same satellites can also be used to guide weapon systems such as ballistic missiles and even "dumb" conventional munitions to their target with great precision.

Meteorological

Apart from allowing more accurate weather forcasting for general-purpose military operations than was possible before by traditional methods, meteorological satellites also provide invaluable assistance to other military missions. For example, near real-time weather surveillance over specific areas allows for the more efficient usage of photoreconnaissance satellites, whose film may otherwise be wasted in taking pictures of cloud tops. Furthermore, as the amount of wind and precipitation over a target area can affect the accuracy of strategic warheads, timely weather information may allow final targeting adjustments to be made in the event of a nuclear war.

Geodetic

Geodetic satellites provide vital information on the size and shape of the earth's surface as well as its shifting gravitational fields. This permits more precise mapping of the earth's surface and also improves predictions of the effect of gravitational forces on the accuracy of ballistic missiles.

From this brief summary of the many uses of military satellites one can see that they have, as Walter Clemens, Jr observed, a "janus-like" quality to them.[5] Just as they can help reduce fears of a surprise attack and provide invaluable assistance to the monitoring of arms control treaties, they can also facilitate a whole range of warlike operations on earth. This is a characteristic that has caused endless debates about the "peaceful" nature and with it the legal status of military activities in space. Leaving this question aside, the distinctive character of the military use of space has none the less been essentially *supportive* or *ancillary* to terrestrial military missions. Space has not yet become a theatre of military operations in which weapon systems *per se* have been deployed or actually used.

However, the chances of space remaining a "sanctuary" like this into the twenty-first century appear today to be remote.[6] This prognosis stems from two relatively recent developments. First, both superpowers are actively developing antisatellite (ASAT) systems.[7] In 1977 the United States declared that the Soviet Union had an "operational" satellite interceptor. This has in part stimulated the United States to develop its own ASAT system, which is scheduled to become operational by the end of this decade. Second, both the United States and the Soviet Union are developing more "exotic" technologies to create laser and particle beam weapons. While the military applications of these Directed Energy Weapons (DEWs) are manifold, the two most commonly cited missions are for antisatellite operations and ballistic missile defence (BMD). It was in regard to the latter that, on 23 March 1983, President Reagan called for a major US research effort. While he did not explicitly state that space-based directed energy weapons would be the primary method of defence, this is likely to be the main avenue of development.[8]

In this respect the recent public interest in the military use of space does reflect something new; the trend towards the introduction of weapons for use in or from space does represent a qualitative departure from the dominant trend over the past 25 years of the "space age". In short, we do appear to be entering a new phase in the militarization of space.

A Sense of Déjà Vu

For those familiar with the history of the US military space programme, there must be a strong sense of *déjà vu*. The very same weapon systems that are currently being developed were all proposed in a remarkably similar way during the late 1950s and early 1960s. Even the arguments that were used to promote these space weapon proposals are similar. For example, the use of such concepts as "space denial", "space control", and the need to take the "high ground", which are familiar today, were just as common in the early years of the space age.

This early interest in space weapons was a direct result of the fears engendered by the launch of Sputnik. To many the Soviet achievement posed a challenge to the technological supremacy and with it the political credibility of the United States around the world. Others believed it presaged a more direct threat to US national security. In addition to demonstrating that the US homeland was now vulnerable to Soviet intercontinental ballistic missiles (ICBMs), it also raised the less tangible fear that the United States would somehow become "blackmailed" or "dominated" from space. For example, an editorial of the *New York Times* declared that the United States was in "a race for survival" and US House of Representatives Speaker John McCormack even went so far as to say that the United States faced the prospect of "national extinction".[9] To many in the United States, to use Tom Wolfe's vivid description, "Nothing less than *control of the heavens* was at stake".[10]

After Sputnik, a succession of Soviet space achievements reinforced the feelings of inferiority within the United States and heightened the concern about Soviet military intentions in space. In anticipation that the United States would respond militarily to this "threat", a variety of space weapon concepts were proposed and studied by all three of the armed services as well as numerous aerospace companies. These included antisatellite weapons, orbital bombardment systems, and space-based ballistic missile defences.[11] Of the three, the most extensive development work was carried out on antisatellite systems. Contrary to most people's understanding of the history of military space activities, the United States became the first nation to test an antisatellite system—well before the first identifiable Soviet ASAT test. Moreover, two ground-based ASAT systems were eventually deployed by the United States in the Pacific during the 1960s. Although the testing of both programmes had ceased by 1970, one remained nominally operational until 1975.

In hindsight, however, the level of the US effort in this area seems remarkably restrained when set against the public anxieties of the late 1950s; the interest of the services and industry in space weapons; and, later, the widespread concern that the Soviet Union would shoot down US reconnaissance satellites or deploy nuclear bombs in orbit. Even when the Soviet Union began testing a satellite interceptor in 1968, the United States seemed surprisingly unconcerned and continued to phase down its own limited ASAT capability.

The level of Soviet interest in ASAT weapons is just as puzzling as, if anything, there were more incentives for the Soviet Union to develop these weapons. First, the Soviets would have been aware before 1957 that the United States was developing reconnaissance satellites to spy on their homeland which had remained virtually closed to foreign inspection. Second, the United States also declared its interest in developing satellites for other military missions—such as for communication purposes—that arguably benefited the United

States more than the Soviet Union. Thus, a vigorous programme aimed at denying the United States use of space in wartime, if not during peacetime, was to be expected. While the Soviets did eventually begin to test a satellite interceptor, it has a limited range and questionable operational reliability. The tests that began in 1968 ceased three years later and were only resumed in 1976, which suggests that this has not been a high-priority development project.

Thus, contrary to what was expected at the dawn of the space age, an arms race in the true meaning of the term has been surprisingly absent from space. This observation raises some important questions which provide both the motives and focus of this study:

* Why were space weapons never extensively deployed by the United States and the Soviet Union when all the conditions were apparently ripe?
* What changed in the late 1970s to now make an arms race in space appear inevitable?
* What are the likely implications of the development and use of antisatellite weapons?

Due to the limited amount of "hard" information on the motives behind the Soviet space programme, the analyst is left to examine the behaviour of the United States for reliable answers to the first two of these questions. As a result, this study concentrates on explaining the evolution of US military space policy from 1945 to 1984 and in particular its interest and involvement in the development of antisatellite weapons. However, where Soviet space activities and US perceptions of Soviet intentions have affected US policymaking, they are obviously discussed here. Moreover, on the basis of the information that is available on the Soviet military space programmes, an explanation of Soviet actions over the same period will be attempted.

Although this study is primarily an historical narrative based on the policies of successive presidencies, it is divided implicitly into two parts. After an opening chapter on the origins of the US military space programme, the first part of the study spans the period from the launch of Sputnik to the end of the Johnson administration. This was a period of rapid expansion for the US military space programme and, as noted earlier, considerable uncertainty over the intentions of the Soviet Union towards outer space. By the time the Outer Space Treaty was signed in 1967, the organization and direction of the US space effort had become established and many of the earlier fears of a Soviet space threat had largely subsided. A separate chapter on the developmental histories of the major US antisatellite projects from 1957 to 1970 concludes this section.

With the first full test of the Soviet satellite interceptor in 1968, in addition to growing evidence throughout the 1970s that certain Soviet military satellites could pose an indirect threat to US forces in wartime, Soviet space activities once again became a cause for concern among US decision-makers. The character of these Soviet space activities is discussed in Chapter 7 and acts as a backdrop to the second part of the study: the evolution of US policy from the Nixon to the Carter administration. This section also concludes with a separate chapter on US ASAT and other "space defense" research and development from 1971 to 1981.

Although the origins of the new phase in the militarization of space can be traced back before the Reagan administration, its policy none the less represents a break from the past. The commitment to deploy the US ASAT system and the rejection of Soviet arms control initiatives represent a qualitative departure from the policy of previous administrations. The Strategic Defense Initiative, as the new US BMD research effort is called, further underlines this trend. The chapter on the Reagan administration is therefore deliberately separated from the main body of the text to emphasize that the current period is the beginning of a new era in US–Soviet space activities.

An historical account of US policy towards the militarization of space should provide more than a record of how and why decisions were taken. As both superpowers avoided an arms race in space for over 25 years, this case study raises some more fundamental questions about the militarization of space. In particular are there any lessons to be learned from this experience, especially for arms control? The concluding chapter to this study looks back to address these questions and also looks forward with a brief assessment of the likely implications of an arms race in space.

2

The Origins of the US Military Space Programme, 1945–1957

In making the decision as to whether or not to undertake construction of such a [space] craft now, it is not inappropriate to view our present situation as similar to that in airplanes prior to the flight of the Wright brothers. We can see no more clearly all the utility and implications of spaceships than the Wright brothers could see fleets of B29s bombing Japan and air transports circling the globe.

"Preliminary Design of an Experimental World Circling Spaceship",
Rand Report No. SE11827, 2 May 1946

Introduction

The United States did not have a dedicated military space programme until the Air Force was given permission to begin development of the WS-117L reconnaissance satellite in 1954. Even so, it was not until after the launch of Sputnik in 1957 that this project was pursued with any sense of urgency. The origins of the US military space programme, however, can be traced back well before 1954 to the feasibility studies conducted by the armed services at the end of the Second World War. These were remarkably sophisticated in demonstrating the viability of satellite development. But, like the early ballistic missile proposals that the embryonic space programme would depend on for launch vehicles, further progress was hindered by a combination of factors, principally political indifference, military conservatism, interservice rivalry and the austerity of the post-war defence budgets.

In May 1955, the United States also declared its intention to develop a "scientific satellite" as part of the preparations for the forthcoming International Geophysical Year (IGY). Although the early military and scientific space research was almost indistinguishable, the decision to develop in great secrecy a military satellite alongside an "open" scientific space programme clearly marked the beginning of two avenues of development. Despite a deliberate partition, the two could never be entirely separated for technical, financial, and organizational reasons. As a result, there were problems from the outset over the relative priorities, goals and public profiles of the civil and military space programmes. They are still evident today. However, these problems were barely surfacing when the Soviet Union became the first nation to launch an artificial satellite. It was only with the shock of this Soviet achievement that the United States became sufficiently motivated to support a substantial space effort. This in turn necessitated a re-evaluation of the priorities of the national space programme.

Comparatively little is known of US space research prior to 1957 and even less of its military component. This chapter traces the development of the US space programme during this period in three parts: the events leading up to the decision to develop a reconnaissance satellite in 1954; its progress prior to Sputnik; and a more general description of the origins of the scientific space programme.

EARLY DAYS

The early US interest in the development of artificial satellites was inevitably linked to the potentialities of the rocket as a launch vehicle. With the influx into the United States of German rocket scientists, their technical information and moreover their surplus V-2 rockets which had all been rounded up as part of Operation Paperclip, there was every reason to believe that rapid progress would be made towards building the longer-range ballistic missiles necessary to launch a satellite.[1] An Army Air Force Scientific Advisory group study commissioned by the Commanding General H. H. Arnold and led by Theodore von Karman reported in 1945 that long-range rockets were not only feasible but also that artificial satellites were a "definite possibility".[2] This was echoed by General Arnold's own "War Report" in which he recommended that the Army Air Force (AAF) pursue the further development of such missiles. He even went a stage further in anticipating the development of missile defences and the role that space systems could play in overcoming them:

> We must be ready to launch them [rockets] from unexpected directions. This can be done from true space ships capable of operating outside the earth's atmosphere. The design of such a ship is all but praticable today; research will unquestionably bring it into being within the foreseeable future.[3]

Others foresaw more near-term military benefits of space. In an interview in 1945 at Wright Field, Ohio, Wernher von Braun, one of the leading members of the group of German scientists, stated that:

> The whole of the Earth's surface could be continually observed from such a rocket [in Earth orbit]. The crew could be equipped with very powerful telescopes and be able to observe even small objects, such as ships, icebergs, troop movements, construction work, etc. . . .[4]

Unfortunately for the proponents of space research, the various proposals for long-range ballistic missiles did not get the necessary support from either the War Department or the services. The same fate also befell the early satellite projects proposed by small service research groups. Dr Vannevar Bush, who had already made known his scepticism of the feasibility of long-range rockets in the now legendary hearings before the Senate Committee on Atomic Research in December 1945, played a similar role in the progress of the early space research. As Chairman of the newly created Joint Research and Development Board (JRDB), he was responsible for advising the Secretaries of War and the Navy on the direction of military research. His position and influence on the JRDB were enough to prevent any "unification of interest or joint project for orbital developments between the various interested groups. . . ."[5]

Discouragement of joint space projects was also manifest within the services, though for reasons other than scepticism of their military worth. For nearly ten years, inter-service rivalry prevented any collaborative work on space research, with the result that their respective research groups worked largely in isolation from one another and with minimal funds. A combined effort would have undoubtedly helped overcome the prevailing inertia and accelerated satellite development in the United States.

In most of the existing accounts of the origins of the US military space programme, credit for the earliest research is given to the Army Air Force and their contract with the Project RAND group at the

Douglas Aircraft Company. This group eventually produced the now-famous report "Preliminary Design of an Experimental World Circling Spaceship" in 1946.[6] Although this was undeniably the single most important piece of research at the time, it was almost certainly the US Navy that conducted the earliest studies on the development of artificial satellites by any of the armed services.

As early as 1945, the Naval Research Laboratory (NRL) had carried out a review of the technical feasibility of satellites but had concluded that it would be too ambitious an enterprise for their limited resources. The main thrust of the Navy's early space research, came instead from the Bureau of Aeronautics (Bu/Aer) and in particular the work of Commander Harvey Hall of the Bureau's Electronics Division.[7] The Bu/Aer's studies of the German rocket programme in 1943 and 1944 and later, from captured records, led to the formation of the Committee for Evaluating the Feasibility of Space Rocketry (CEFSR) on 3 October 1945.

By the end of October, CEFSR had submitted to Bu/Aer a satellite development proposal which resulted in contracts for further feasibility studies with the Guggenheim Aeronautical Laboratory, the Glen C. Martin Company, North American Aviation and also the Douglas Aircraft Company. It soon became apparent, however, that while the Navy was prepared to fund research at this level, further development, which was estimated to cost between $5 and $8 million (at 1946 prices), was beyond their independent financial resources. Faced with this stumbling block, Hall and his associates saw the only way forward to be a co-operative venture with the Army Air Force.

As a result, on 7 March 1946, the Aeronautical Board, which was responsible for securing co-ordination in the research and development of Army and Navy projects, met to review the Navy's proposal and consider future collaboration. The Army Air Force appears to have been caught off balance by the Navy's proposal as it delayed giving an official response until after it had given the matter further study. The prospects for further development did not appear bright as the memorandum summarizing the meeting acknowledged the general advantages of pursuing satellite development, but stated that there were "no obvious military, or purely naval applications to warrant the expenditure . . ."[8] A formal meeting of the Aeronautical Board's Research and Development Committee was set aside in May for the Army Air Force to provide its response.

In the meantime, the issue was brought to the attention of the office of the Commanding General Army Air Forces which, according to

Robert Perry's account:

> ... decided that the position of the air forces in an interservice conference would be compromised unless its representatives could produce a paper demonstrating equal competence with the Navy—and equal interest—in space research. Air staff authorities also felt that the Army Air Forces should have primary responsibility for any military satellite vehicle, considering such activity to be essentially an extension of strategic air power.[9]

This was probably the first formal statement that space was an extension of the air medium and therefore the responsibility of the Air Force. It was an argument that would be constantly reiterated by the Air Force in its interservice battles for the next 15 years.

In a crash effort to demonstrate "equal competence" and avoid being excluded from future military space research, General Curtis LeMay, then director of research and development for the Army Air Force, turned to the Douglas Aircraft Company and its Project RAND group to undertake studies on their behalf. These were barely completed in time for the May 14th meeting of the Research and Development Committee.

Given the constraints on time (a three-week deadline!), the final RAND report was a remarkable achievement as it encompassed not only a thorough technical evaluation of the feasibility of an artificial satellite but also an accurate prophecy of the likely implications and utility of space vehicles. The vast majority of the 321-page study deals with the technical problems of building and launching a satellite vehicle. However, Chapter 2 by L. Ridenour on the "Significance of a Satellite Vehicle" does examine possible military applications, albeit with an understandable bias to aerial warfare. For example, it states that the "military importance of establishing vehicles in satellite orbits arises largely from the circumstances that defenses against airborne attack are rapidly improving".[10] Due to advances in radar, guided missiles and proximity fuses, the "air offensive of the future will be carried out largely or altogether by high speed pilotless missiles".[11] The missiles would need to use outer space for a "minimum energy trajectory" which would enable satellites to play a vital guidance role "to bring the missile down on its intended target" or "alternatively be considered as the missile itself".[12] Apart from these possible missions, the report also noted that a "satellite offers an observation aircraft [*sic*] which cannot be brought down by an enemy who has not mastered similar techniques".[13] The "spotting of the points of impact of bombs launched by us, and the observation of weather conditions over enemy territory" were identified as the two most important goals for observation.[14] Furthermore, the benefits of satellites for communications relay at relatively low and high (geosynchronous) orbits were also recognised.[15]

Notwithstanding its modest conclusion that the "full military usefulness of this technique cannot be evaluated today", this very first RAND report not only predicted the major military benefits that satellites would offer by way of reconnaissance, communication, missile guidance, attack assessment and weather forecasting but also questioned—possibly for the first time—the continued utility of the manned penetrating bomber.[16]

Despite the Army Air Force's efforts to be ready with their own proposal, the May 1946 meeting of the Research and Development Committee proved to be inconclusive; no agreement was reached between the two services. As a result the Aeronautical Board, to which the Committee reported, postponed deciding service responsibility for satellite development until it had received higher-level guidance. In the meantime, both the Navy and Army Air Force continued with their preliminary investigations without much enthusiasm or much funding.[17]

Further efforts by the Navy's Bureau of Aeronautics and sections of the Army Air Force to gain approval for their own satellite proposals became caught up in the organizational changes that followed the creation of an independent Air Force and Department of Defense with the passage of the National Security Act in 1947. The JRDB, which changed its title to the Research and Development Board (RDB), became a policymaking body with more authority than it had previously possessed as a co-ordinating committee. To the dismay of the proponents of long-range missiles and rocket boosters, Dr Vannevar Bush carried on as chairman and continued to oppose their plans. The guided missile budget for FY 1947 was reduced from $29 million to about $13 million with the result that the original 28 programmes were cut to eight, including the last of the long-range missile projects—Convair's MX-774.[18]

The RDB's Committee on Guided Missiles, which had assumed "responsibility for the coordination of the Earth Satellite Vehicle" in September 1947, exhibited a similar attitude to the various satellite proposals that came before it. Perry even goes so as to state that "such proposals did not even receive serious consideration".[19] The justification was summed up in a March 1948 report of the Committee's Technical Evaluation Group, which stated after reviewing the latest CEFSR and RAND studies, that "neither the Navy nor the USAF has as yet established either a military or a scientific utility commensurate with the presently expected cost of a satellite vehicle. However the question of utility deserves further study and examination."[20]

The Air Force had in fact come to the same conclusion following their own independent assessment of the technical feasibility and operational utility of satellites.[21] While this study acknowledged that the construction of a satellite was not beyond available technical resources, especially with the appropriate level of support, it also recognized that the development of guided missiles and particularly jet aircraft took precedence in the allocation of scarce funding and the use of limited scientific expertise. With this in mind the official position of the Air Force was stated in January 1948 by Vice Chief of Staff, General H. S. Vandenberg:

> The USAF, as the service dealing primarily with air weapons—especially strategic—has logical responsibility for the satellite.
>
> Research and Development will be pursued as rapidly as progress in the guided missiles art justifies and requirements dictate. To this end, the program will be continually studied with a view to keeping an optimum design abreast of the art, to determine the military worth of the vehicle—considering its utility and probable cost—to ensure development in critical components if indicated, and to recommend initiation of the development phases of the project *at the proper time*.[22]

This did little more than encourage the continued support of the RAND studies. The situation in 1948 was summed up in a paragraph of Defense Secretary James Forrestal's *First Annual Report*:

> The Earth Satellite Vehicle Programme which was being carried out independently by each military service was assigned to the Committee on Guided Missiles for coordination. To provide an integrated program with resultant elimination of duplication, the committee recommended that current efforts in this field be limited to studies and competent designs; well-defined areas of such research have been allocated to each of the three military departments.[23]

Although the RDB Committee on Guided Missiles had allowed the Army to pursue preliminary development of its Hermes (later Redstone) missile project, thereby bringing all three services into space-related research, the Navy's attempts to go beyond preliminary investigations with its satellite proposals continued to be frustrated. A second attempt at a joint project in mid-1948 was rejected by the Air Force, as was a further proposal to the RDB.[24] With such an unencouraging environment the Navy abandoned its satellite research—at least for the time being—in late 1948. With the blessing of the RDB the Air Force through Project RAND eventually became "the only service authorized to expend defense department funds on studies of satellite vehicles".[25]

With the prevailing funding constraints and general scepticism of the military utility of satellite vehicles, RAND set about demonstrating a clear military requirement to overcome the inertia. A RAND-sponsored conference on the utility of satellites held in 1949 noted that, while satellites could be expected to involve a costly undertaking, possess small payloads and have a limited capacity for destruction compared to existing delivery systems, there were none the less some redeeming characteristics. In particular, they could be equipped with "photographic and television equipment", which could pass over "selected areas of the earth" and, significantly, they "could not be brought down by present weapons or devices".[26]

Of the potential military applications discussed, it was the reconnaissance mission that received the most attention. With the intensification of the cold war and the Soviets' capacity for technological and strategic surprises (witness their atom bomb test in 1949, the Sino-Soviet Pact and later the Korean invasion in 1950), the concomitant requirement for better and more timely intelligence meant that satellites had at last found their military *raison d'être.*

In November 1950, RAND submitted definitive recommendations to Air Force headquarters that it pursue advanced research into specific aspects of the satellite reconnaissance mission. This resulted in a further two RAND reports completed in April 1951. These represented the most comprehensive analysis to date of using satellites for reconnaissance purposes.[27] With the growing appreciation of the need for better strategic intelligence, considerable enthusiasm greeted the two April 1951 RAND reports within the Air Force and parts of the Pentagon. Further support from the RDB and the Air Force followed for RAND to continue its studies. More importantly, the newly established Air Research and Development Command (ARDC) authorised RAND to make specific recommendations for the start of developmental work on a reconnaissance satellite system. This project was given the title of "Feedback" and became the basis of the first US military satellite programme.

THE US RECONNAISSANCE SATELLITE PROGRAMME

Project Feedback generated an impressive collection of studies on every aspect of a satellite reconnaissance system. At its peak, the project involved many hundreds of scientists and engineers from RAND and various industrial subcontractors. By 1 March 1954 the final two-volume report, which condensed the findings of the studies, was complete. This recommended, *inter alia*, that "the Air Force undertake the earliest possible completion and use of an efficient satellite reconnaissance vehicle" as a matter of "vital strategic interest to the United States".[28] It also recommended that the project, which was estimated to cost $165 million and take seven years to complete, should be conducted in the strictest secrecy.

The recommendations of the Feedback study received high-level support within the Air Force, including the acting chairman of the Scientific Advisory Board, J. A. Doolittle; the Air Force Chief of Staff, General N. F. Twining; and the heads of Strategic Air Command and ARDC—Generals LeMay and Power, respectively.

By May 1954, ARDC had been given responsibility to proceed with the Feedback recommendation, and received final approval from the Office of the Secretary of Defense (OSD) Co-ordinating Committee on Guided Missiles) in July. Referred to in its early stages as Project 1115 by ARDC, the satellite reconnaissance programme was given the more familiar weapon system designation WS-117L and the unclassified title "Advanced Reconnaissance System". By 27 November, a System Requirement (No. 5) had been issued, with the General Operational Requirement (No. 80) coming in March of the following year. The latter defined the Air Force's operational objective to be the provision of continuous surveillance of "pre-selected areas of the earth" in order "to determine the status of a potential enemy's warmaking capability".[29] Prime reconnaissance targets were airfield runways and future intercontinental missile launch sites. Electronic intelligence gathering and weather forecasting were also added as secondary functions.

Shortly afterwards, design proposals were solicited from Lockheed, Martin and Bell. While these individual companies were vying for the contract, the organizational responsibility for the project was shifted from the Wright Air Development Division to the Western Development Division of ARDC, which had been recently established under General Bernard Schriever's command to oversee the Air Force's ballistic missile programme.[30] The move also corresponded with the publication of the WS-117L Preliminary Development Plan within the ARDC. In March 1956, the contractor evaluation board met to assess the WS-117L design studies carried out by the various companies. On the basis of their findings, Lockheed was awarded the prime contract on 29 October, 1956.[31] This award was virtually a foregone conclusion as Lockheed monopolized the existing pool of satellite research personnel and was also extensively involved with the U-2 high altitude reconnaissance aircraft programme.[32] Although the subsequent progress of WS-117 was slow, due principally to continuing funding constraints, a military satellite programme had at last been approved.

Various factors had helped gain approval for the satellite reconnaissance programme. Foremost among them, as noted above, was the growing requirement for better strategic intelligence. This had gained widespread acceptance by the beginning of the 1950s and was given official sanction following the influential report of the Technology Capabilities Panel (TCP). Completed in 1955 and entitled "Meeting the Threat of Surprise Attack", the report called for the development of advanced reconnaissance methods, including the use of satellites. As its Chairman James Killian later stated "... the TCP reception by President Eisenhower and the NSC encouraged the CIA and Air Force plans to accelerate the development of reconnaissance satellites".[33]

The deliberations of another panel also undoubtedly helped the fledgeling space programme. The decision to accelerate the ballistic missile programme in the spring of 1954 following the recommendations of the Strategic Missile Evaluation Committee (the "Teapot Committee"), under the leadership of Professor John von Neumann, also meant that a sufficient launch capability would be available for satellites. The integration of the missile and satellite programmes, however, was met initially with some resistance. The managers of the IRBM and ICBM projects were worried that the satellite programme would affect the design of the rocket boosters and ultimately dictate the flow of skilled manpower and finances. Only with the influx of further funding were these fears alleviated.[34]

Another important factor, though more difficult to evaluate, is the role that the Central Intelligence Agency played in promoting satellite reconnaissance. Created in 1947 out of the wartime Office of Strategic Services (OSS), the CIA inevitably became interested in strategic reconnaissance. Among the early programmes they supported was WS-119L or Project Genetrix, consisting of high altitude balloons with cameras that were released to drift over the Soviet Union and be recovered afterwards. This programme, which ran between November 1955 and the spring of 1956, had evolved from a similar system, known as Project Moby Dick, for acquiring meteorological information over the United States.[35] By the time WS-119L had become operational, President Eisenhower had made the CIA responsible for the development of the U-2 high altitude reconnaissance aircraft. After the U-2's first flight over Soviet territory in 1956, the need for further balloon operations had ceased.[36]

The extent to which the early satellite reseach projects were Air Force "fronts" to CIA initiatives cannot be gauged from the available information. While the official Air Force histories make no reference to CIA involvement, it is now generally accepted that the Agency did provide considerable technical and financial support for the early US satellite reconnaissance studies, including Project Feedback. These became the basis for the CIA's own reconnaissance programme, known as Project Corona.[37] A key figure in the CIA programme was Richard Bissell, who as Deputy Director of Plans was also responsible for U-2 development. As Victor Marchetti and John Marks maintain:

> ... Bissell had been a driving force behind the development of space satellites for intelligence purposes— at times to the embarrassment of the Air Force. He had quickly grasped the espionage potential of placing high-resolution cameras in orbit around the globe to photograph secret installations in the Soviet Union and China. And due in great part to the technical advances made by scientists and engineers working under Bissell, the CIA largely dominated the US government's satellite reconnaissance programs in the late 1950s and well into the 1960s.[38]

Important though these factors were in influencing the *adoption* of a satellite reconnaissance programme, they were not sufficient to ensure that the necessary financial support was allocated to the project. For instance, the FY 1957 budget was restricted to just \$3 million, which the ARDC acknowledged as being "inadequate initial funding".[39] As General Schriever later reflected, "I can recall pounding the halls of the Pentagon in 1957 trying to get \$10 million approved for our [USAF] space program. We finally got the \$10 million, but it was spelled out that it would be just for component development. No system whatsoever".[40]

In fact, DoD's support for satellite development was only marginally better than it had been in the late 1940s. Part of the reason, ironically, was the decision to develop a scientific satellite known as Project Vanguard in 1955. As Perry states:

> Through the whole of the period when the supporters of WS-117L were seeking program approval and adequate funding, the general attitude of the Department of Defense remained hostile towards satellites. Although not openly proclaimed, it was departmental opinion that satellite vehicles were not feasible and further, that until Vanguard experiments confirmed feasibility itself, the WS-117L program should be funded at the "study level".[41]

The Scientific Satellite Programme

The US scientific space programme has virtually the same origins as its military counterpart. In fact, the Bu/Aer and RAND feasibility studies of artificial satellites were arguably more scientific than military in nature.[42] However, it was not until October 1948, when the *Journal of Applied Physics* published the "Grimminger Report" based on the unclassified portions of the RAND studies, that this research was exposed to the wider scientific community. This stirred considerable interest among the small national rocket societies as well as the scientists who had been carrying out atmospheric and meteorological research using sounding rockets.[43] Another group that began to play a more active role was the space flight and astronautics enthusiasts. As it was still a relatively small research community at this stage, informal contacts developed between the various interested groups. A notable example was the contact between Wernher von Braun (then chief of the Guided Missile Development Division, Ordnance Missile Laboratory, at Redstone Arsenal, Huntsville, Alabama) and S. Fred Singer (a leading proponent of scientific satellites) that led to the MOUSE proposal (Minimum Orbital Unmanned Satellite, Earth), which received serious consideration during the early 1950s.[44]

At the same time, von Braun, who had been canvassing support within the Army Ballistic Missile Agency (ABMA) for an experimental satellite project, formally recommended that the Army develop a "minimum satellite vehicle based upon components available from missile development of the Army Ordnance Corps".[45] Like Commander Hall before him, von Braun believed that with the prevailing fiscal constraints it was essential to secure the support of the other services. Although the Air Force later declined participation because of its prime involvement in the reconnaissance satellite programme, the Navy responded favourably. By the spring of 1955. both the Army and the Navy had worked out the details for a joint satellite proposal—dubbed Project Orbiter.

At this point, the future progress of these embryonic satellite projects became caught up in the proposal by a group of eminent scientists that a specific year, to begin in July 1957, should be dedicated to worldwide scientific endeavour. As the centrepiece and climax of what would become known as the International Geophysical Year (IGY), the goal of launching a scientific satellite gained considerable support. In August 1954 the Congress sanctioned US participation in these activities, and later the IGY co-ordinating committee (CSAGI—*Comité Spécial à l'Année Geophysique Internationale*) accepted the US proposal "... that thought be given to the launching of small satellite vehicles ..."[46]

As the development of a scientific satellite encroached on the existing US satellite and ballistic missile programmes, the National Security Council (NSC) met in May 1955 to discuss guidelines for US participation in this effort. The resultant directive—NSC 5520, entitled "U.S. Scientific Satellite Program" and signed on 26 May—decreed that the US satellite could not employ a launch vehicle currently intended for military purposes. While this decision was in part the result of a desire to enhance the peaceful image of the US space effort, it was primarily designed to protect the ballistic missile programme from diversion and disruption.[47] After further discussions, the White House publicly announced on 29 July 1955 that the United States would launch a satellite as part of the IGY activities.[48]

The task of selecting the satellite launch vehicle was assigned to Donald A. Quarles, then Assistant Secretary of Defense for Research and Development. Quarles established an "Ad Hoc Advisory Group on Special Capabilities" under the chairmanship of Dr H. J. Stewart, which became known as the Stewart Committee. All three of the services eventually put forward IGY proposals: the Army with Project Orbiter using the Redstone (later Jupiter C) missile; Project Vanguard from the Naval Research Laboratory using the Viking sounding rocket; and the Air Force with its World Series proposal based on a combination of Atlas and Aerobee-Hi rockets. With NSC 5520, however, the Stewart Committee's choice was limited to the Vanguard project, as this was the only one based on a nonmilitary launch vehicle. Given the advanced development of the Redstone missile and the expertise behind it, the final choice was a great blow to the Army's Ballistic Missile Agency led by von Braun and General Medaris. Their Orbiter proposal would undoubtedly have been the better choice, as was illustrated when a Jupiter C missile launched America's first artificial satellite in January 1958.

Although the IGY satellite project was designed to be primarily a scientific enterprise, it was recognised from the start that certain military benefits might accrue from it. For example, NSC 5520 noted that as scientific satellites promised to return data on the content of the atmosphere these "significant findings will find ready application in defense communication and missile research". Furthermore, "Antimissile missile research will be aided by the experience gained in finding and tracking artificial satellites".[49] These observations in addition to the relatively low cost of the whole Vanguard project (estimated to be in the order of $20 million) help explain the DoD's attitude to WS-117L.

NSC 5520 is also significant in that it represents the Eisenhower administration's earliest statement of attitudes and guidelines to the exploration of space. For example, on the subject of whether satellites could be used for offensive purposes it states:

> It should be emphasised that a satellite would constitute no active military offensive threat to any country over which it might pass. Although a large satellite might conceivably serve to launch a guided missile at a ground target it will always be a poor choice for the purpose. A bomb could not be dropped from a satellite on a target below, because anything dropped from a satellite would simply continue alongside in orbit.[50]

Furthermore, with the embryonic satellite reconnaissance programme almost certainly in mind, US participation in the IGY scientific satellite project was to be conducted in a manner that "Preserves U.S. freedom of action in the field of satellites and related programs" and moreover "Does not involve actions which imply a requirement for prior consent by any nation over which the satellite might pass in its orbit, and thereby does not jeopardize the concept of 'Freedom of Space'".[51]

The administration's concern about preserving the principle of "Freedom of Space" was based on the assumption that the United States would be the first to launch an artificial satellite. Although this problem was alleviated by the precedent set with Sputnik, it was lost in the repercussions of that event.

Conclusion

By October 1957 the United States was proceeding with two separate space programmes; one military and the other "scientific". Both had evolved from common origins and both had faced similar problems in gaining acceptance. Foremost among these was the persuasive argument that technology had not matured sufficiently to warrant spending a large amount of money for what appeared to be limited returns. Many satellite proposals were dismissed as being either fanciful or mere dreams. As a Congressional Report on Military Astronautics would later put it, "for a long time it was considered not quite respectable to talk in terms of space flight".[52]

This kind of conservatism was manifest among the civilian defence advisors, most notably Vannevar Bush, and within the armed services. Despite the clear appraisal of the military benefits of satellites outlined in the early RAND proposals, the armed services concentrated on the procurement of military systems more directly related to their individual service roles and missions—an attitude encouraged by existing financial constraints. As Laurence Kavanau stated in a speech in 1963: "the sponsors and authors of these studies were unable to justify them on the basis of their military utility. Since they were not weapons, satellites could not compete for War Department funds with other projects which would develop weapons."[53] This became a recurring problem to the supporters of "passive" space systems.

Just as the individual services were not prepared to support advanced R & D either separately or jointly, they were also not prepared to see another service gain the sole rights to the "space mission". Again, as Kavanau points out:

> The early Navy and Army Air Force space studies were, to some extent, undertaken to convince the Joint Research and Development Board of the War Department that one or both of these services had a claim to weapons developments in the space medium.[54]

This early inter-service rivalry was a recurring hindrance to the development of US space systems, as it produced a duplication of research and ultimately a dilution of the national effort.

By the beginning of the 1950s, attitudes were beginning to change. York and Greb maintain that the increased support for satellites and other reconnaissance projects was a result of technology evolving to make such developments more attractive. They "were approved at precisely the time that the underlying technology matured and became available and that the technology reached the state of readiness for reasons that had nothing to do with reconnaissance objectives".[55] While advances in such areas as chemical propellants, ballistics, rocket engine design, radar, communications, computers and optical surveillance equipment undoubtedly created a "critical mass" of basic building blocks which helped undermine the prevailing inertia, York and Greb underestimate the role that individuals played in initially hindering and then later promoting those same technologies. There was never much doubt about the technical feasibility of building ballistic missiles and satellites; rather, the question was whether it was worth spending the money to do it. They also undervalue the role of the requirement for strategic intelligence, which promoted satellite development and with it further advances in critical technologies. A major consequence of this was that the CIA became an important proponent and innovator of satellite technology which, as in the case of the U-2, allowed development to continue without some of the constraints associated with service procurement.

However, the promotional aspects of the strategic intelligence requirement have to be matched against the effect that the resultant secrecy and obvious sensitivity of the satellite reconnaissance programme had on the subsequent evolution of the US space effort. Apart from the constraints of technology transfers, it undoubtedly contributed to the forced division of the US space effort and the numerous problems that this created for successive administrations.

3

Eisenhower and the Space Challenge

Perhaps the starkest facts which confront the United States in the immediate and foreseeable future are (1) the USSR has surpassed the United States and the free world in scientific and technological accomplishments in outer space, which have captured the imagination and admiration of the world; (2) the USSR, if it maintains its present superiority in the exploration of outer space, will be able to use that superiority as a means of undermining the prestige and leadership of the United States; and (3) the USSR, if it should be the first to achieve a significantly superior military capability in outer space, could create an imbalance of power in favor of the Sino-Soviet Bloc and pose a direct military threat to U.S. security.

The security of the United States requires that we meet these challenges with resourcefulness and vigor.

Introductory Note to NSC 5814/1,
"U.S. Policy on Outer Space",
20 June 1958

INTRODUCTION

The period immediately following the launch of Sputnik was a challenging time for the Eisenhower administration. Despite repeated warnings, it was still caught unprepared for the crisis that Sputnik would cause. Public and congressional concern at the implications of the Soviet achievement for the prestige and security of the United States immediately created pressures for an accelerated and expanded space effort. What Eisenhower had hoped would be a relatively leisurely and orderly US entry in space was transformed overnight into a national obsession to wrest the lead from the Soviet Union.

This imperative posed additional challenges. New organizational structures and procedures had to be formed to manage the expanded space programme. A complete review of US interests, priorities and goals in the exploration of space was also necessary. Moreover, the exaggerated fears of Soviet intentions after Sputnik spurred a whole range of space weapon proposals to counter the perceived threat. These had to be assessed not only for their military utility and technical feasibility but also for their long-term impact on the future use of space. These, then, were among the most important problems that faced the Eisenhower administration from 1957 to 1960.

The Reaction to Sputnik and the Reorganization of the US Space Programme

The launch of Sputnik I in October 1957 first shocked and then galvanized the American people into committing vast resources for the missile and space programmes. The impact of the Soviet achievement was immense and went beyond even the wildest predictions of those who had been alerting the country to this very possibility for many years before. The effect was aptly described by James Killian:

> As it beeped in the sky, *Sputnik I* created a crisis of confidence that swept the country like a windblown forest fire. Overnight there developed a widespread fear that the country lay at the mercy of the Russian military machine and that our own government and its military arm had abruptly lost the power to defend the mainland itself, much less to maintain U.S. prestige and leadership in the international arena. Confidence in American science, technology, and education suddenly evaporated.[1]

While the Soviet achievement came as a great surprise to many—especially the general public—Eisenhower and his senior advisors had known for some time that the Soviet Union was on the verge of launching an artificial satellite.[2] Eisenhower not only misjudged the impact that Sputnik would make but also underestimated the force of congressional and public opinion after the event had occurrred. This swamped the administration's confident assurances that Sputnik did not represent a threat and that the United States was not "behind" or in a "race" with the Soviet Union. Eisenhower could do little else as the whole affair placed the administration in a difficult dilemma. To belittle the Soviet achievement would not only undermine the current American effort but also the whole rationale of the IGY exercise, while to praise it would risk fuelling the paranoia still further. Moreover, Eisenhower could not be seen to be unduly worried for similar reasons which in turn exposed his administration to criticisms of complacency and negligence. Furthermore, he could not even reveal the source of his confidence in America's relative position, namely the highly secret U-2 reconnaissance missions, for fear that this would lead to their demise.[3]

After the initial attempts to dampen the public alarm, the NSC conducted its own post-mortem and assessed ways to deflect the inevitable recriminations.[4] During the cabinet meetings held on 8 October, Eisenhower reaffirmed the deliberate separation of the military and scientific space effort and emphasized the "peaceful character" of the US programme. He also insisted that, as there had never been a "race" with the Russians to get a satellite into orbit, the United States had not "lost".[5] Moreover the Soviet Union had probably done the United States a favour as Sputnik had set the precedent of freedom of passage in outer space. This was viewed with some significance given the possible international repercussions to US satellite reconnaissance.

Although the White House press release issued the following day reiterated the President's sentiments, it soon became clear that explanations of past actions and calm assurances of America's present situation would not be enough to assuage public concern.[6] The administration was quite clearly going to have to review national space policy, the organization of the programme, and the relative priorities of the projects within it. This, however, would take time. The launch of the much larger Sputnik II with its canine passenger on 2 November demonstrated that the earlier Soviet achievement had been no fluke and intensified public fears still further. The demand for immediate remedial action to redress the crisis in national confidence could no longer be ignored by the administration.

On 7 November Eisenhower announced that he had created a Special Assistant to the President for Science and Technology and appointed Dr James Killian to fill the post. The Special Assistant would also chair a Presidential Science Advisory Committee, or PSAC as it was widely referred to, which would be available to provide independent science advice to the President. On the next day the Department of Defense authorised the Army's satellite programme—now known as Explorer—to be a back-up to Project Vanguard, thus revising the guidelines set out in NSC 5520. After a disastrous and embarrassing Vanguard test failure in December, Explorer I using a modified Redstone missile (Jupiter C) eventually became the first successfully launched US satellite on 31 January 1958.

The Army's authorisation to launch satellites highlighted the unresolved question of which agency or service would have ultimate responsibility for the space programme. Since Sputnik, the White House had been assailed with competing demands for projects and missions from the Defense Department and all three services. As Schoettle observed: "each political participant sought to convince the administration of its own special capability in space by calling loudly for recognition of its skills and resources. It was a veritable 'Anvil Chorus'."[7]

With the administration's declared intent to accelerate and expand the space programme, Eisenhower's immediate preoccupation was the need to avoid duplication and rivalry among the services. His solution was to concentrate authority within the Department of Defense. On 27 November Secretary of Defense McElroy announced plans to establish a special projects agency within the Pentagon to oversee all advanced research and development of space technology. By the time the congressional authorization hearings were held in January of the following year, this had become formally known as the Advanced Research Projects Agency (ARPA) with its own director (Dr Roy Johnson) and staff.[8] Responsibility for all space projects—scientific and military—was given to ARPA although operationally many were still assigned to the individual services. This new creation was bitterly resented and opposed by the services, particularly the Air Force, who continued to argue for *sole* development rights on the basis that space represented an extension of the air medium.[9]

At the same time there was strong pressure to establish a separate civilian space agency by members of PSAC and Congress. Their reasons were that civilian control would enhance the US image of favouring the peaceful exploitation of space as well as prevent the compromise of purely scientific programmes by the military. Eisenhower, however, resisted the creation of a separate space agency in the belief that this would foster duplication.

At a Legislative Leadership meeting on 4 February 1958, the proposal for a new independent space organization was raised. According to the meeting's minutes, Eisenhower argued that for "nonmilitary" research "Defense (DoD) could be the operational agent, taking orders from some nonmilitary scientific group".[10] Later that day, Eisenhower met with Killian, George Kistiakowsky and Herbert York (both members of PSAC) to discuss further the organization of the missile and space programmes. Again, Eisenhower opposed any dissolution of DoD's overall responsibility and "interjected a caution not to put the satellite job in any service" because as he had stated earlier, he had "come to regret deeply that the missile program was not set up in OSD (Office of the Secretary of Defense) rather than in any of the services".[11] Later, as the minutes of the meeting record:

> The President said that space objectives relating to Defense are those to which the highest priority attaches because they bear on our immediate safety... He did not think that large operating activities should be put in another organization, because of duplication, and did not feel that we should put talent etc. into crash programs outside the Defense establishment.[12]

By the beginning of March, however, Eisenhower had bowed to the growing pressure to set up an independent civilian organization based on a reconstituted National Advisory Committee for Aeronautics (NACA) to be called the National Aeronautics and Space Agency (NASA). It appears that the President and the Defense Department agreed to this separation in the belief that NASA would handle just *basic* rather than applied space research.[13] The differing interpretations of the role of NASA became apparent in the subsequent congressional hearings following the submission of the National Aeronautics and Space Act on 2 April 1958.[14]

Since Sputnik, Congress had become an important forum for discussing the US space programme, particularly after the creation of two congressional Ad Hoc Space Committees, which would later become the basis for the House and Senate Select and Special Committees, respectively. As Logsdon recounts:

> The primary foci of congressional concern during the hearings were the interaction between civilian and military space efforts; the freedom of DoD to engage in military-oriented space research; and the lack of a mechanism for overall policy direction for the space effort which would recognize all aspects, military and civilian, political and scientific, of the program.[15]

After much debate and redrafting, compromises were reached on the issues of military control and overall responsibility. The Act finally stated in Section 102(b) that space activities:

> ... shall be the responsibility of, and shall be directed by, a civilian agency... except that activities peculiar to or primarily associated with the development of weapon systems, military operations, or the defense of the United States (including research and development...) shall be the responsibility of, and shall be directed by, the Department of Defense.[16]

A National Aeronautics and Space Council (NASC) was created to advise the President on all aspects of space policy. Congress also added a provision for the establishment of a Civilian-Military Liaison Committee where NASA and DoD could consult and advise one another.

The results of the reorganization of the space programme were mixed. Neither of the two co-ordinating bodies worked well as Eisenhower preferred the NSC to the Space Council while the Civilian-Military Liaison Committee consisted of relatively junior officials without real authority. Furthermore, friction and rivalry continued to permeate the military space programme. ARPA and the services never established a working relationship, while the co-ordination and transfer of projects between the DoD and NASA created a series of bitter interdepartmental battles.[17]

By August 1959, moves were again under way to reorganize the burgeoning military space programme. The initial impetus for the reorganization came from the Director of Defense Research and Engineering (DDR&E), Dr Herbert York, who made his plans clear to the new Special Assistant for Science and Technology, Dr George Kistiakowsky, on 15 August and 26 August 1959.[18] In short, York proposed to transfer the responsibility for the development of the military space projects back to the services as he believed the creation of ARPA had, if anything, increased the amount of duplication and rivalry. This proposal was put to the President by Kistiakowsky on 15 September, in a memorandum on the "Coordination of Satellite and Space Vehicle Operations".[19] Given Eisenhower's earlier opposition he was initially wary of this recommendation as he scribbled on the attached memorandum for the Joint Chiefs of Staff: "I think this needs a lot of study. It *appears* to be going in the wrong direction."[20] However, after consultation with Secretary of Defense McElroy, he eventually concurred. As a result, all of the space projects under ARPA's control were transferred to the services and ARPA was left to conduct just basic research on advanced military technology.

In parallel with these deliberations on the future role of ARPA, the services were also discussing a proposal put forward by Admiral Arleigh Burke, Chief of Naval Operations, that a unified space command be created. Although support was forthcoming from the Army, the Air Force was adamantly opposed to the idea in the belief that this would pre-empt their campaign for the space mission. McElroy was also not enamoured with the proposal and took the opportunity at the official announcement of the transfer of projects from ARPA on 18 September 1959, to state that the "establishment of a joint military organization with control over operational space systems does not appear to be desirable at this time".[21] This was reiterated by his successor, Thomas Gates, on 16 June 1960.[22]

The various civil-military oversight and co-ordinating bodies also underwent reorganization. In January 1960, Eisenhower recommended to Congress that it disband the NASC and the Civilian-Military Liaison Committee as he considered both to be superfluous and ineffective. The House of Representatives accepted but on the condition that the CMLC be replaced by a more senior co-ordinating group. As a result, the Aeronautics and Astronautics Co-ordinating Committee Board was created on 1 July 1960. The NASC survived, however, after Lyndon Johnson (then Majority Leader and Chairman of the Committee on Astronautical and Space Sciences) blocked the House bill in the Senate. Although it existed in name, it virtually ceased to function after January 1960.[23]

Concern over the poor management and progress of the satellite reconnaissance programme also led to the creation of the highly classified National Reconnaissance Office (NRO) in August 1960. The origins of this initiative stem directly from the growing involvement of the CIA in the *development* of reconnaissance satellites and particularly the technical problems encountered in transmitting data from space. Following a review of the satellite reconnaissance programme by the President's Board of Consultants on Foreign Intelligence Activities in February 1958, Eisenhower directed the CIA to develop a reconnaissance satellite of its own. While this decision ran counter to his intention to avoid duplication, the PBCFIA had reported that the WS-117L Advanced Reconnaissance System would not be able to meet the near-term requirement for a reconnaissance satellite as it relied on the Atlas booster then still a long way from operational readiness. Instead, it recommended that the CIA develop a reconnaissance satellite—in a similar way to the U-2—that could be launched by the Thor IRBM and be available by the spring of 1959.[24]

The resultant CIA project, known as Corona, was directed by Deputy Director of Plans Richard Bissell who, as noted earlier, had a similar responsibility for the U-2 programme. Although the CIA had participated in the Air Force's reconnaissance satellite programme from its inception, Project Corona involved very different surveillance and recovery methods to the Air Force system. According to Thomas Powers' account:

> Bissell modified the Air Force plan in two important ways. First, he switched

> the satellite from one in which the camera was fixed while the satellite spun, to one in which the satellite was stabilized and the camera scanned. Second, he abandoned the Air Force plan for televising the satellite's photographs and developed in its place a more complex system whereby the actual film would be jettisoned from the satellite and recovered in mid-air.[25]

The project was funded in part by the CIA's own funds and indirectly by the appropriations for the Air Force's Discoverer programme, which acted as its "cover" in the development stage. The advertised purpose of the Discoverer satellites was to provide support to the space programme by way of advanced experimental research, including *inter alia* biomedical experiments using mice and, more truthfully, the "development of capsule recovery techniques".[26]

As a result of Eisenhower's decision to proceed with the CIA project, the Air Force reconnaissance programme, now known as Sentry but later changed to SAMOS (for Satellite and Missile Observation System), did not receive the priority status and with it funds that were expected. This, when combined with continuing doubts about the technical efficacy of its transmission system, further eroded support for the SAMOS programme. In fact, by the summer of 1959 the US satellite reconnaissance programme was in a state of crisis. The Corona/Discoverer tests had begun without a single success and the SAMOS development schedule was slipping at an alarming rate due to technical problems.

On 19 August 1959, the new Presidential Scientific Advisor, George Kistiakowsky, received a call from Bruce Billings, Director of Special Projects at DDR&E. According to Kistiakowsky's diary, Billings "complained about the unbelievable chaos among the highly classified projects—the piling up of one project on top of another without any effective mechanism for evaluating even the potential usefulness of each".[27] The SAMOS project was singled out as being particularly mismanaged and overly ambitious.

It was almost certainly a result of Kistiakowsky's doubts about the SAMOS project that PSAC recommended, in the autumn of 1959, that the Air Force switch to the "recovery" type method. The Air Force, having abandoned this option in the spring of 1959, was understandably angry.[28]

On 8 December 1959 Kistiakowsky received a briefing from the Air Force at the Ballistic Missile Division (formally WDD) in Los Angeles. According to his diary, the briefing:

> ... emphasized that BMD is suffering greatly from contradictory orders which they receive almost on a weekly schedule from Washington and

> which keep changing their program. BMD acts as if they were at the point of despair. They will clearly object to the PSAC recommendations because they believe that "readout" SAMOS is much more promising than "recovery" SAMOS and also feels that the polar communications satellites is the system on which to concentrate in that field. In both cases our recommendations go diametrically opposite to these feelings.[29]

Although Kistiakowsky continued to oppose the "readout" method the Air Force finally got its way, mainly because the downing of Gary Powers' U-2 on 1 May 1960 created enough concern for the Congress to inject funds into the SAMOS project far in excess of the administration's requests for FY 1961.[30] However, the battle over SAMOS was enough to cause a complete re-evaluation of the management of the satellite reconnaissance programme.

On 10 June Eisenhower formally asked Secretary of Defense Gates to re-evaluate the programme and brief the NSC on US intelligence requirements, the technical feasibility of meeting those requirements with SAMOS, and DoD's plans for the system.[31] This seems to have been part of a review panel organized by Kistiakowsky but co-chaired by James Killian and Dr Edwin Land of the Polaroid Corporation, which was informally referred to as the "SAMOS panel".[32] Its primary objective was to recommend changes to the management of the reconnaissance programmes. After high-level discussions with the CIA and the Air Force, the panel eventually reported its conclusions to the President in a special NSC meeting on 25 August 1960.[33] They recommended *inter alia* the creation of a separate organization—the NRO—which would be led by a civilian from the Department of the Air Force, with direct command to the officers in charge of the various projects. The Defense Department and the President agreed to this and an NSC directive was issued. Although the NRO was established on 31 August within the Office of the Secretary of the Air Force, under the cover of the Office of Missile and Satellite Systems, its purpose was to co-ordinate both the Air Force's, CIA's and later Navy and NSA reconnaissance activities. Brig. Gen. Robert E. Greer was put in charge of its "Development Office" at BMD in Los Angeles, which reported directly to the NRO.[34] The NRO was also made subordinate to an interdepartmental panel known as the Executive Committee (EXCOM), composed of the Secretary of the Air Force, the Director of the CIA and the DDR&E.[35] So secret were the operations of the NRO that its very existence was only inadvertently revealed by the Senate Select Intelligence Committee in 1973.

US Policy Towards the Military Use of Space, 1957–1960

As the organization of the US space programme evolved so too did the policy to guide it. Although Eisenhower stated after Sputnik that he considered the US satellite programme "well designed and properly scheduled to achieve the scientific purposes for which it was initiated" the subsequent "anvil chorus" calling for a variety of military space programmes made a review of national space policy unavoidable.[36]

At the beginning of 1958, Eisenhower requested that the PSAC recommend the outlines of a national space programme with the result that, on 4 February, Killian announced that a special panel led by Edward Purcell had been established to carry this out. The Purcell Panel's report, which received presidential approval on 26 March 1958, both reinforced Eisenhower's own views on the exploitation of space and endorsed much of what had been stated in NSC 5520.[37] Although the report's emphasis was very much on the scientific benefits of space exploration it also indicated the range of military applications, most notably reconnaissance, communication, and weather forecasting, that held the most promise. Moreover, in keeping with NSC 5520 the report stated:

> Much has been written about space as a future theater of war, raising such suggestions of satellite bombers, military bases on the moon and so on. For the most part, even the more sober proposals do not hold up well on close examination or appear to be achievable at an early date. . . In short, the earth would appear to be after all, the best weapons carrier.[38]

The report's endorsement of the passive military benefits of space and its unequivocal rejection of the utility of space weapons established the basic guidelines for the US military exploitation of space. This was later reaffirmed and refined by a succession of NSC and NASC directives during the remainder of the Eisenhower administration. The most important of these were NSC 5814/1, entitled "Preliminary U.S. Policy in Outer Space" and approved on 18 August 1958; and an unnumbered NASC paper entitled "U.S. Policy on Outer Space" which was given the President's assent on 26 January 1960.[39]

In contrast to the basic inclination of Eisenhower's space policy, various sections of the armed forces and the aerospace press waged a highly vocal campaign in support of a variety of space weapon projects for orbital bombardment, ballistic missile defence and antisatellite purposes. In much the same way as bomber and missile "gaps" were projected at the time so the absence of a US space weapons programme was also frequently identified as a serious deficiency in the US defence posture.

In fact, even before the early Soviet space feats encouraged many Americans into thinking about a "space threat", some US officers had already made known their support for such weapons in a form that would become all too familiar in subsequent years. Their statements generally emphasized the US requirement to gain "space superiority" or take the "high ground" in space or "deny" its use to an adversary. As General Schriever stated in an address at San Diego in February 1957:

> In the long haul our safety as a nation may depend upon our achieving "space superiority". Several decades from now the important battles may not be sea battles or air battles, but space battles, and we should be spending a certain fraction of our national resources to ensure that we do not lag in obtaining space superiority.[40]

This theme received further attention after it became more acceptable to talk about space matters following Sputnik. Air Force Chief of Staff General White, in a speech on 29 November 1957, stated:

> Whoever has the capability to control the air is in a position to exert control over the land and seas beneath. I feel that in the future whoever has the capability to control space will likewise possess the capability to exert control of the surface of the earth. . . In speaking of the control of air and the control of space, I want to stress that there is no division, *per se*, between air and space. Air and space are indivisible fields of operations. . .[41]

The latter point on the continuum of air and space had now become a popular Air Force argument for reasons other than the belief that the national security interests of the United States were at stake in space. As noted above, it had already been used by the Air Force to claim its inalienable rights to the "space mission" and it would later become the basis for its campaign to gain a manned military capability in space.[42]

Following the launch of Sputnik, the number of speeches calling for a vigorous US space programme increased dramatically. The services and aerospace companies both sensed that "rich pickings" would be available in the post-Sputnik budgets and any proposal that contained the word "space" was bound to be attractive. Moreover, they thought it inevitable that the White House would support proposals to meet what they perceived to be a Soviet threat in space.

As US Army General James Gavin stated in his book *War and Peace in the Space Age*:

> It is inconceivable to me that we would indefinitely tolerate Soviet reconnaissance of the United States without protest, for clearly such reconnaissance has an association with an ICBM program. It is necessary, therefore, and I believe urgently necessary, that we acquire at least a capability of denying Soviet overflight—that we develop a satellite interceptor.[43]

As early as June 1957 Gavin requested his deputy to direct the Army's Redstone Arsenal at Huntsville to look into the feasibility of satellite interceptors for this very purpose. As a result, by 19 November 1957—just six weeks after Sputnik—the Army was ready with its own space programme recommendations, which stated among other things that:

> . . . the United States has an equally urgent national requirement for a satellite defense system. Sooner or later, in the interest of survival, the United States will have to be able to defend itself against satellite intrusion, otherwise it will be helpless before any aggressor equipped with armed reconnaissance satellites. . . A program to provide a weapon system capable of destroying a satellite in space has been under study by the Army for the past six months. At the moment, early preliminary indications from these studies point toward two possible solutions to this problem. . .[44]

The "two possible solutions" were either a modified Nike Zeus antiballistic missile or a "homing satellite" carrying a nuclear or a "shaped charge" warhead. At much the same time the Air Force and the Navy put forward their own antisatellite proposals.

In order to both placate the proponents of space weapon systems as well as provide insurance in case circumstances dictated a change in US policy, ARPA and all three of the armed services were allowed to pursue research on a variety of space weapon concepts but to a level no further than preliminary investigations (see Chapter 5 for further details). Research of this nature was approved in the programme goals for military R&D discussed at the 13 February 1958 meeting of the NSC. This was later embodied in NSC 5802/1 entitled "U.S. Policy on Continental Defense". Here "Defense against Satellites and Space vehicles" was included in the list of areas of "particular importance" where a ". . . vigorous research and development program should be maintained in order to develop new weapons and needed improvements in the continental defense system and to counter improving Soviet technological capabilities for attack against the United States".[45]

As many of the early investigatory projects reached fruition in the latter stages of the Eisenhower administration, so the services once more requested that they be allowed to embark on advanced development and deployment. However, in keeping with its early stated policy goals, the Eisenhower administration successfully resisted this pressure for the following reasons.

First, they considered the potential threat from Soviet reconnaissance satellites and orbital bombardment systems to be insufficient to warrant a US antisatellite programme. While NSC 5814/1 estimated that the Soviet Union "could probably orbit surveillance satellites capable of low resolution (approximately 100-200 feet) at any time within the next year [1959] to obtain weather and perhaps some additional data of military intelligence value",[46] the development of a weapon system to counter Soviet satellite reconnaissance offered marginal benefits to the United States, especially as the Soviet Union could gain virtually all the information that satellites provided from "open" US sources. Also, in contrast to the widespread concern that the Soviet Union would threaten the United States with space weapons, the NSC and the NASC directives make no reference to this possibility in their annexes on projected Soviet space capabilities. This appreciation of the situation was echoed later by the Defense Department after Gordon Gray, the Special Assistant to the President for National Security Affairs, requested its views on the requirement for a satellite interceptor. In its reply of 23 May 1960 Deputy Secretary of Defense James Douglas stated:

> Quite some time ago you raised a question with regard to statements in the public press concerning intercept and destruction of satellites. The Defense position is that there is no urgent requirement for a capability to intercept satellites. There is no clear indication that the Soviets are expending effort on reconnaissance satellites or on weapon-carrying satellites. Such reconnaissance would seem to offer little attraction to them, and the utility of an offensive satellite weapon in the near future is very questionable. It should be noted that until the Soviet satellite launching a few days ago, they had apparently not launched one for two years.
>
> Although we are not presently concerned with the interception of satellites, we have embarked on studies to develop a capability to inspect satellites at close range in the interest of our own satellite operations.[47]

Second, if the need for a satellite interceptor was marginal, many of the other space weapon proposals were even less attractive. As

NSC 5520 and the Purcell Report had first acknowledged, there was little if any military rationale to deploying weapons in space for use against targets on earth. In every way, IRBMs and ICBMs represented superior delivery systems to orbital bombs.

Third, where a military rationale could be demonstrated, such as with space-based ballistic missile defence systems, severe technical problems made further development not only questionable in the near term but also prohibitively expensive. Even the post-Sputnik boom in defence spending could not accommodate the envisaged funding levels for such programmes.

These factors, though important, were none the less secondary to the most influential determinant of Eisenhower's space policy: the need to preserve the principle and later practice of satellite reconnaissance was the overriding US concern. Given the "closed" nature of Soviet society, reconnaissance satellites were seen to have greater value to the United States than the Soviet Union. Indeed, even before the first US reconnaissance satellites were launched in 1959, NSC 5814/1 stated: "Reconnaissance satellites are of *critical* importance to U.S. national security".[48] In addition to emphasizing their early warning and intelligence gathering functions, both NSC 5814/1 and the 1960 NASC directive also foresaw with remarkable prescience the future importance of reconnaissance satellites for arms control verification.

The international acceptance of satellite overflight for reconnaissance purposes, however, was not expected to be straightforward. Despite the precedent set by Sputnik it was by no means clear that the Soviet Union would let reconnaissance from space go unchallenged. On the contrary, there was every likelihood that the Soviet Union would exhibit the same hostility to satellites as it had to aerial reconnaissance. On at least two occasions in 1958 the Soviet Union had shot down US aircraft on intelligence gathering missions over its territory.[49] Furthermore, they were also desperately trying to shoot down the U-2 aircraft that periodically violated Soviet air space. It was only a matter of time before these efforts would prove successful and bring to an end aerial overflight of the Soviet Union. As Robert Amory, Jr, Deputy Director of the CIA, recollected, "everybody knew it [the U-2] had a limited life. The Russian radar would improve; their fighters and intercepts and other things like their surface to air missiles would improve. And a precisely accurate prediction was made of about a four year life."[50]

With the resultant dependancy on reconnaissance satellites that this would create, senior US officials believed that the Soviet Union already had a powerful incentive to develop antisatellite weapons without the added encouragement of the United States taking the lead in ASAT development and deployment. It is from the recollections of these officials that we can gauge the impact of this argument on US space weapons policy.

In his book *Race to Oblivion,* Herbert York while he was DDR&E recalls that:

> The Air Force . . . strongly promoted the idea that we should undertake on an urgent basis the development of a "satellite interceptor", to be known as SAINT. The President himself, in recognition of the fact that we didn't want anybody else interfering with out satellites, limited this program to "study only" status and ordered that no publicity be given either the idea or the study of it. The other two military departments independently promoted the same idea and volunteered their services for its accomplishment.[51]

Furthermore, as one former official at the Department of Defense remembers about the ongoing ASAT research:

> During the Eisenhower administration, there was one point when the program had reached such a size or where the commitments were getting to be sufficiently serious and important that they had to be brought to the attention of the President and then he said he didn't want an antisatellite program. He was opposed to antisatellites because he felt that satellites were more to our benefit than to the Soviets and he did not want us to do anything that would initiate antisatellite warfare.[52]

George Kistiakowsky is rather more explicit in his diary of the instances when antisatellite weapons were discussed. On 15 January 1960 he recalls replying to a memorandum from Richard Morse (Assistant Secretary of the Army for Research and Development) concerning a proposal to demonstrate the use of a Nike Zeus missile against a satellite. He states:

> I emphasized my unalterable opposition to the project, which he outlined in a memo to me, of shooting down a satellie, because once we had downed our own satellite, and of course made much to-do about it, the Soviets could easily shoot down one of their own over Soviet territory, accuse us of doing it and make a big public issue of it, which would then give them an excuse for shooting down our reconnaissance satellites.[53]

Kistiakowsky took this issue up later that day with Gordon Gray, to whom he complained "about the total lack of policy directives regarding public statements or the peaceful nature of reconnaissance satellites, and cited this proposal of Morse as an instance of how we could easily damage what I consider an extremely important measure of national security".[54] Gray apparently "agreed" with Kistiakowsky and decided to arrange a special meeting of the NSC to discuss the matter.[55] The meeting was duly held three weeks later, by which time the subject had become particularly topical following the discovery of an "unknown satellite" by US tracking facilities. This prompted further calls for a US capability to inspect and destroy enemy satellites.

At the NSC meeting on 5 February 1960, Kistiakowsky "raised the question of the dangerous statements about the destruction of enemy satellites if they overfly the United States" with the argument that this would later "prejudice the use of our reconnaissance satellites". Later in the meeting Dr Joseph Charyk, Assistant Secretary of the Air Force:

> ... presented the plans for a satellite interceptor [SAINT], emphasizing the inspection rather than the destruction aspect of this plan. This was followed by a general discussion in which I [Kistiakowsky] presented my arguments against any demonstrations of our ability to destroy satellites, etc. The President rather unemphatically agreed with my position, but it didn't sound as if it was a directive.[56]

This unemphatic support none the less seems to have prevailed as the Eisenhower administration did not permit at any time the advanced development of an antisatellite system. However, the period of uncertainty surrounding the unidentifiable object appears to have contributed to the eventual decision to proceed with the further development of the *inspection* variant of the SAINT proposal in November 1960 (see Chapter 6 for details).[57]

While the Eisenhower administration rejected the advanced development of antisatellite weapons for fear of encouraging the Soviets to do the same and even worse to use them, there was still the question of how the United States should protect its reconnaissance satellites once they became operational. After all, as noted earlier, there were strong incentives for the Soviets to take countermeasures regardless of a US ASAT research and development programme.

In fact, the need to protect US reconnaissance satellites posed a considerable problem to the Eisenhower administration. Although direct countermeasures against satellites were likely to prove more difficult than with reconnaissance aircraft—given their greater altitude and speed—the physical security of satellites could not be guaranteed by defensive measures against dedicated opposition over the long run. The limited launch capacity of US boosters placed severe constraints on anything other than essential payload items. Furthermore, as satellites are extremely delicate instruments, it is difficult to see what protective measures could have been employed. For example, as early as May 1954 *Time* magazine reported on a meeting of the American Rocket Society, where one of the participants observed that the operation of military satellites could be easily undermined by relatively cheap countermeasures. One of the methods mentioned was to launch clouds of "lead or steel particles" into the path of the satellite which, due to its high orbital velocity, would become "shredded" as it passed through them.[58] Furthermore, though this would be argued more often later, US officials believed they could not deter the Soviet Union from taking action against US satellites by the threat of retaliation in kind as the Soviet Union could arguably afford to live without its military satellites.

Faced with a situation in which the United States could not turn to its usual methods of securing a vital asset, namely defence or deterrence, US officials deliberately sought what can be described as a politico-legal solution to the vulnerability of US reconnaissance satellites. This amounted to the proposal of international agreements to sanction or legitimize those military activities in outer space that were considered beneficial to the United States and outlaw those that were not.

Much of the early US thinking on arms control in outer space stemmed from the perceived lessons of the attempt to control atomic energy after the Second World War. Many believed, particularly in the State Department, that because effective action had not been taken at an early date, the further military exploitation and proliferation of atomic power had gone beyond international control.[59] As a result, the earliest of the official US statements and policy guidelines on the international control of activities in space stressed that this new medium should be used exclusively for *nonmilitary* purposes. Thus one of the first NSC directives on this matter—NSC Action No. 1553 of 21 November 1956 stated:

> It is the purpose of the United States, as part of an armaments control system, to seek to assure that the sending of objects into outer space shall be exclusively for peaceful and scientific purposes and that under effective control the production of objects designed for travel in or projection through outer space for military purposes shall be prohibited.[60]

This directive was reflected in a number of subsequent disarmament proposals during 1957 and early 1958, culminating in two letters from President Eisenhower to Soviet Premier Nikolai Bulganin in January and February 1958. These proposed a cessation of all military activities in space including the testing of long-range missiles.[61] However, these proposals were made conditional on the creation of an international system of observation or verification to control international space activities. It is difficult to assess how sincere Eisenhower and his administration were with these proposals as they would have ultimately banned ICBMs as well as military satellites. They were certainly viewed by the Soviet Union as US ploys to hinder the more advanced Soviet missile and satellite programmes.

By the spring of 1958 the expected opposition to US satellite reconnaissance and the need to secure its international acceptance caused a subtle but important change in the new US proposals on the regulation of military activities in space. This requirement was recognized in the Policy Guidance section of NSC 5814/1, which stated that the United States should:

> In anticipation of the availability of reconnaissance satellites, seek urgently a political framework which will place the uses of U.S. reconnaissance satellites in a political and psychological context most favorable to the United States. . .[62]

In seeking a favourable "political framework" the United States now promoted the view that space could and should be used only for "peaceful" rather than "nonmilitary" purposes, thus permitting the deployment of military satellites that were not in themselves weapons systems. As a result, many of the subsequent discussions of arms control in space were linked to questions of sovereignty (where does the jurisdictional control over air space end and outer space begin?) and rights of passage (what is the legitimate and illegitimate use of space?).

In the NASC document that superseded NSC 5814/1 on 26 January 1960, these questions were given further consideration. It states:

> It is possible that certain military applications of space vehicles may be accepted as peaceful or acquiesced in as noninterfering. On the other hand, it may be anticipated that states will not willingly acquiesce in unrestricted use of outer space for activities which may jeopardise or interfere with their national interests.

In response to this problem it proposed that the United States should:

> Continue to support the principle that, insofar as peaceful exploration and use of outer space are concerned, outer space is freely available for exploration and use by all, and in this connection: (a) consider as a possible U.S. position the right of transit through outer space of orbital space vehicles or objects not equipped to inflict injury or damage; (b) where the U.S. contemplates military applications of space vehicles and significant adverse international reaction is anticipated, seek to develop measures designed to minimize or counteract such reaction; and (c) consider the usefulness of international arrangements respecting celestial bodies.[63]

A natural corollary to the drive to seek the legitimacy of "peaceful" uses of space was to propose that "nonpeaceful" activities be prohibited. Thus by 1960 US arms control proposals not only stressed the international verification of space activities generally, but also the prohibition of "weapons of mass destruction" in orbit. This came first with the submission on 16 March 1960 to the Ten Nation Committee on Disarmament meeting in Geneva (with the UK, Canada, France, and Italy) of a "Plan for General and Complete Disarmament in a Free and Peaceful World", and later in a separate "Program for General and Complete Disarmament Under Effective International Control". The latter proposed that the "... placing into orbit or stationing in outer space of vehicles carrying weapons capable of mass destruction shall be prohibited".[64] President Eisenhower repeated this proposal in his address to the 14th General Assembly on 22 September 1960.

With the exception of the creation of the UN Committee on the Peaceful Uses of Outer Space by UN General Assembly Resolution 1472 (XIV) in December 1959, very little progress was made towards regulating activities in outer space. US hopes of securing international recognition of the legitimacy of military reconnaissance satellites, however, were temporarily given a boost with the much quoted remarks by Khrushchev at the Paris summit meeting in May 1960.

According to Eisenhower's account of the summit, Khrushchev complained about the U-2 incident to a select gathering at the conference, but when President de Gaulle observed that a Soviet satellite had recently passed over France and had probably taken photographs, "Krushchev broke in to say he was talking about airplanes, not about satellites. He said any nation in the world who wanted to photograph Soviet areas by satellite was completely free to do so."[65] Kistiakowsky also noted this incident in his diary and in a footnote stated: "The remark of Khrushchev that he did not object to intelligence satellites became the foundation of a consistent policy of both superpowers."[66] However, this was probably an unguarded statement by Khrushchev as the Soviet Union later launched a diplomatic offensive in the UN aimed at outlawing US "espionage" satellites. As Chapter 4 will illustrate, Kistiakowsky's observation overlooks the fact that it would take a number of years before Soviet officials and commentators ceased to object to US reconnaissance satellites.

CONCLUSION

In just three short years, from the autumn of 1957 to the end of 1960, the US space programme was transformed from a small and struggling effort to a large and multifaceted enterprise with considerable public and congressional support. The key to that change was, of course, the psychological and political impact of Sputnik.

As we have seen, this transformation was not without its teething troubles. The orderly development of space technology was often sacrificed for the imperatives of the space race, with the result that there were many early test failures. There was also considerable trial and error with the organizational procedures that were set up to manage the sudden expansion of the space programme. However, by the end of the Eisenhower administration the most important and most tangible product of the US space programme, namely the intelligence gathered from reconnaissance satellites, had begun to be delivered.[67] Although this intelligence would be reaped more effectively by later administrations, the high priority invested in the satellite reconnaissance programme had already paid off as U-2 flights over the Soviet Union ceased abruptly in May 1960.[68]

The need to safeguard this vital asset largely dictated the US approach to the militarization of space. In order to reduce the anticipated international opposition to US satellite reconnaissance, the Eisenhower administration increasingly emphasized that this was a "peaceful" use of space and therefore legitimate. Conversely, "nonpeaceful" uses of space should be prohibited. This approach inevitably affected the further development of a variety of space weapons proposals put forward at the time. While Eisenhower always doubted the military rationale for these proposals, the foremost concern was to avoid projecting an aggressive image to the US space programme and stimulating Soviet countermeasures.

Some commentators have seen a contradiction between this stated policy and the considerable research that was carried out on space weaponry culminating in the decision to develop SAINT. However, this was not inconsistent as research was kept to a strict ceiling of development while SAINT was primarily an *inspection* system. The SAINT programme would have aided or complemented ground-based space surveillance systems and would only have become a platform for antisatellite activities if it proved necessary. This policy of providing a minimum research base to "hedge one's bets" became the main rationale for US ASAT research and development from 1957 to 1981. If there was any inconsistency, it was in the publicity that SAINT and other projects received as "antisatellite weapons systems" which undoubtedly contributed to Soviet suspicions of US intentions in space. It was only in the Kennedy administration that rules regarding disclosure of information on US military space programmes were rigorously applied.

Thus, in the final years of his presidency, Eisenhower left a considerable legacy: a substantial military and space programme that would reach fruition under later administrations, and also the foundations of a policy on outer space that would guide future US presidents for many years.

4

Kennedy and the Years of Uncertainty

I do not say that we should or will go unprotected against the hostile misuse of space any more than we go unprotected against the hostile use of land or sea, but I do say that space can be explored and mastered without feeding the fires of war, without repeating the mistakes that man has made in extending his writ around this globe of ours.

President John F. Kennedy
Address at Rice University in Houston, Texas,
12 September 1962

INTRODUCTION

In hindsight, the years spanned by the Kennedy administration represent a watershed in US-Soviet relations not least for their impact on the future exploitation of space. It was not until the last weeks of Kennedy's foreshortened presidency that many of the problems that his predecessor had foreseen were finally resolved. In particular, the anticipated clash over the right to conduct reconnaissance from space reached its climax during this period. In 1962 the Soviet Union began a diplomatic offensive at the United Nations to prohibit reconnaissance from space. This was carried out against a backdrop of veiled threats that military action might also be taken if the diplomatic effort failed. Moreover, a series of Soviet space "firsts" not only appeared to add weight to this possibility but also to growing fears that the Soviet Union would deploy orbital bombs in space. This concern increased the pressure on Kennedy and his advisors to depart from the guidelines established by the Eisenhower administration for the military space programme. In short, this was a period of considerable uncertainty: uncertainty over whether the principle and practice of satellite reconnaissance could be secured; uncertainty over the future direction of the US space programme; and, in particular, uncertainty over whether a nuclear arms race would develop in space.

Review and Reorganization

Although Senator John Kennedy had made some passing comments about the US space programme during his presidential campaign, it was only after his electoral triumph that he paid serious attention to this issue. During the transition period, Kennedy appointed Jerome Wiesner of MIT, who had been his principal campaign science advisor, to establish what was formally referred to as the "Ad Hoc Committee on Space" but more generally known as the Wiesner Committee. Its purpose was to examine the national space programme and recommend policy for the future. Apart from Wiesner, the committee included Trevor Gardner, Edwin Purcell, Donald Hornig, Bruno Rossi, Edwin Land, Harry Watters, Kenneth BeLieu, and Max Lehrer. Kennedy met with the Committee on 10 January to discuss their recommendations and an unclassified report, which omitted their detailed military recommendations, was released the next day.[1]

The report proceeded from the belief that the United States was lagging behind the Soviet Union in missiles and space technology. It attributed this position to inadequate management of the national space programme, particularly the duplication and poor co-ordination between NASA, the Defense Department, and the three services. As Logsdon states, the report "deplored the tendency of each military service to create an independent space program, and it called for the establishment of single responsibility for space programs among the military services".[2] Thus, the problem that the Eisenhower administration had considered settled was once again the subject of criticism and alternative recommendations.

The Wiesner Report recommended that the National Aeronautics and Space Council be revitalized by amending the existing legislation to establish the Council in the Executive Office of the President. The Vice President would serve as Chairman with the Secretary of State, the Secretary of Defense, the NASA Administrator and the Chairman of the Atomic Energy Commission as its other members. The Aeronautics and Astronautics Co-ordinating Board was also to be revamped and used to facilitate planning and avoid duplication between NASA and the DoD.

In response to the Report's criticisms of the military space programme, the new Secretary of Defense, Robert McNamara, immediately directed the Office of Organizational and Management Planning Studies to review the responsibility for military space research and development. As a result a new Defense Directive (No. 5160.32) on "Development of Space Systems" was issued on 6 March 1961. Under the new Directive each of the military services was permitted to conduct preliminary research to develop new ways of using space technology within guidelines prescribed by the Director of Defense Research and Engineering (DDR&E). All proposals for advanced research and engineering would be reviewed by the DDR&E and sent to the Secretary of Defense for final approval. However, except in certain circumstances, the further development of these proposals would become the responsibility of the Department of the Air Force.[3] Furthermore, on 28 March McNamara assigned the Air Force the responsibility for research, development and operation of all Department of Defense reconnaissance satellite systems and for research and development of all instrumentation and equipment for processing reconnaissance data from satellite sources.[4] This, however, did not exclude the CIA and increasingly the National Security Agency (NSA) from developing and operating their own reconnaissance satellites. Moreover, the Army was also allowed—at least for the time being—to continue with its Advent satellite communications programme and the Navy with its Transit navigation satellite. Thus, while the Air Force was now the *primary* service involved in military space R&D, it still did not have *sole* responsibility —a vital difference.

Even more frustrating to the Air Force was the growing influence of NASA, which caused a great deal of acrimony between the two organizations. This was particularly galling because since its inception NASA had become virtually dependent on Air Force personnel and facilities. While this was largely inevitable given the Air Force's prevailing expertise, Nieburg maintains that the Air Force deliberately penetrated NASA from infancy to dominate its activities. Furthermore, when the Air Force felt threatened by the introduction of McNamara's more stringent guidelines for military research, it saw NASA "as a means of sidestepping Defense Department authority".[5] This inevitably brought into question the status of NASA as a truly independent civilian space agency.

While the Air Force's participation in NASA activities was consolidated during the Kennedy administration, its influence actually declined. The tenfold leap in NASA's appropriations stemming from President Kennedy's commitment in May 1961 to land a man on the moon by the end of the decade not only increased NASA's political constituency but also sealed the primacy of NASA's manned space flight programme over the Air Force's. Also, McNamara's policy directive of February 1962 that assigned the Air Force responsibility "for the research, development, test and engineering of satellites, boosters, space probes and associated systems necessary to support specific NASA projects and programs . . ." (and other agreements) was really only a reflection of the administration's drive to prevent unnecessary duplication in the national space programme.[6] As we shall see, it was also a reflection of McNamara's and Harold Brown's (then DDR&E) scepticism of the utility of manned military operations in space that together doomed the Air Force's own manned space aspirations. However, NASA's programmes were regularly used to conduct Defense Department experiments. While McNamara used this to deflect some of the Air Force criticism, it did not help in the projection of NASA's "civilian" status—a fact that did not escape Soviet attention.

Although ties between the Defense Department, the Air Force and NASA were not very harmonious, a working relationship did evolve in succeeding years. A similar transition occurred between the various government agencies involved in executing national space policy. Despite the Wiesner Report's recommendations, the execution of US space policy, particularly in the diplomatic arena, was not successfully co-ordinated until after a special high-level interagency committee had been established in May 1962. This had become imperative to meet the Soviet challenge to US satellite reconnaissance activities.

The Legitimization of US Satellite Reconnaissance

By the autumn of 1960, US photoreconnaissance satellites were beginning to return much-needed if still relatively crude intelligence on the Soviet Union. With the exception of the highly classified CIA involvement, the existence of a US satellite reconnaissance programme had been openly admitted in Congress. Moreover, the aerospace press carried extensive descriptions of the various projects. During the Eisenhower administration this openness was designed to support the "peaceful" image of the US space programme and also exhibit the growing US prowess in space exploration. By the time Kennedy entered office this policy had already begun to be re-evaluated in light of the demise of U-2 flights over the Soviet Union and the consequent shift to satellite reconnaissance.[7] It now seemed just a matter of time before the Soviets turned their attention to opposing satellite reconnaissance overflight.

As a result, a division of opinion developed in the new administration on the visibility of future US satellite reconnaissance activities. Both sides used the U-2 incident to support their case. To some, particularly in the State Department, the political and diplomatic "fallout" following the downing of Gary Powers' aircraft illustrated the dangers of too much secrecy. They argued that the US reconnaissance programme should remain as open as possible to gain international approval. In contrast, the intelligence community, particularly those involved with the NRO, believed that potential international opposition could be reduced by enforcing strict secrecy in all further satellite reconnaissance operations. Abram Chayes, Legal Advisor to the Department of State, recalls the basic problem and the respective schools of thought:

> The question was this: How was one to protect our satellite reconnaissance operations politically? Everybody had the U-2 incident in mind. Although we had a beautiful technical instrument for reconnaissance, when the thing was disclosed, there was no political defense for it. And in the State Department on the whole we thought that you gained something politically by being somewhat more open about our operations and *developing a climate of legitimacy* about them instead of trying to keep them completely secret. The opposite view was let's keep them very, very secret. First of all, nobody knows about them. And what they don't know won't hurt them. And anyway nobody knows how good they are. And secondly, to the extent that this gets out in public, it forces the Russians to make a challenge of some kind because they can't accept the fact that they are being observed and, therefore, they would have to make some form of political challenge.
>
> And that we could never defend against a political challenge because there was so much power in this idea of peaceful uses of outer space, and this might not be regarded as a peaceful use and so on.[8]

While the White House favoured from the outset a clampdown on the amount of information released on US reconnaissance activities, it took over a year before the administration and armed services could be co-ordinated and the divergent views reconciled. The first indication of this poor co-ordination came after the introduction of new guidelines on the release of information about military space launches.

As soon as Kennedy entered office, he directed McNamara to eliminate the advance notice of US military space flights, particularly those involving the launch of reconnaissance satellites. As a result, in January 1961 the amount of information released on the upcoming SAMOS II launch was sharply reduced. Henceforth, no prelaunch notification or briefing was to be given to the press. Instead, short stylized statements were to be provided *after* the launch. Arthur Sylvester, the Assistant Secretary of Defense for Public Affairs, later notified the President of the new restrictions. In the memorandum he stated:

> In summary this readjustment is a big step towards the gradual reduction of volunteering information on our intelligence acquisition systems which Mr. McNamara has informed me is your desire. As might be expected, there may be criticism from the media, the contractors and the Congress as this change in procedure becomes known but we are prepared for it.[9]

Sylvester had been advised on the amount of information to impart by Dr Charyk, now Under Secretary of the Air Force and Director of the NRO. Both were obviously aware of making too great a change in policy as Sylvester later stated in the memorandum:

> Dr. Charyk has reviewed these changes and is satisfied that they meet all his security requirements and those of his SAMOS Project Director, Brigadier General Greer. Dr. Charyk agreed to make public this amount of data in view of the great volume of previous news stories which have been written already about SAMOS.[10]

The launching of SAMOS IV and V in November and December also went unannounced. The absence of any reference to SAMOS was also conspicuous in the President's Annual Report to Congress for 1961.

Although the new procedures reduced the level of publicly available information on US reconnaissance activities, the services resisted any attempt to extend this to a general clampdown on all aspects of the US military space programme. In particular, the Air Force continued to give great publicity to its space activities and plans. The situation remained like this for most of 1961, causing President Kennedy considerable frustration and anger on more than one occasion. As one observer recalls, Kennedy would often get on the telephone to Arthur Sylvester and demand to know why he had "let those bastards talk". Apparently, "everytime the Air Force put up a space shot and any publicity was given to it, he just went through the roof . . ."[11] Eugene Zuckert, the new Secretary of the Air Force, also learned from McNamara that Kennedy had been very unhappy with a speech given by General Schriever in the autumn of 1961 that called for an expanded "military oriented space program". He admitted later that "I guess it was then that I started cracking down a little harder on what our people said in their speeches on space".[12]

It was almost certainly the result of the uneven exposure of US military space activities that led on 23 March 1962 to the "blackout" directive from the Department of Defense. This directive, although signed by Deputy Secretary of Defense Roswell Gilpatric, had been drafted by Charyk.[13] It prohibited advance announcement and press coverage of *all* military space launchings at Cape Canaveral and Vandenberg AFB. It also forbade the use of the names of such space projects as Discoverer, MIDAS and SAMOS. Military payloads on space vehicles would no longer be identified, while the programme names would be replaced by numbers. *Aviation Week* reported at the time that "defense officials are justifying the secrecy on the grounds that it will lessen the chances of provoking attacks on the U.S. space program by Russia and other foreign countries... These officials contend that announcing U-2 flights over Russia in advance would have had the provoking effect they hope to avoid under the new secrecy policy."[14]

By the time the directive became known to the press, it had already received considerable criticism from NASA, Congress and some industrial contractors. Many believed that this ruling would stifle the free flow of information and hinder the development of the space programme. Others criticized it on the ground that it would compromise the US position of advocating the peaceful use of space and draw attention to the clandestine nature of the US military space programme. As Solis Horwitz, the Director of Organizational and Management Planning, argued:

> My objection was that in the effort to hide it, we were actually creating more publicity. The easier way would have been to say "Yes, we're doing it. Everybody knows the Air Force has a space program, but the shot [space launch] was for the purposes of measuring this weather condition and doing that." We were not fooling anybody except our own people. The Russians knew we had an Air Force program, and they knew we were not limiting it certainly just to NASA type shots.[15]

Despite the opposition, the directive remained in force. Although the restrictions on public references to other parts of the military space programme were gradually relaxed during later administrations, any acknowledgment of reconnaissance from space remained strictly taboo. It was only marginally broken when President Carter admitted in 1978 that the US operated satellites for this purpose.

At the same time that the White House and Defense Department were trying to reduce and finally eliminate the amount of information being made available to the public, the State Department was working in a different—sometimes contradictory—way to reduce international opposition to satellite reconnaissance. As noted earlier, officials in the State Department favoured not only an "open" US space programme but also a diplomatic solution to the anticipated clash over satellite reconnaissance with the Soviet Union. As early as 2 February 1961, Secretary of State Dean Rusk, in a memorandum for the President, sounded out the White House views on utilizing the embryonic UN Committee on the Peaceful Uses of Outer Space to discuss matters of co-operation on space matters with the Soviet Union.[16] Although this Committee had been established in December 1959, it had not met due to Soviet insistence that its arrangements for voting be made more equitable. Although Rusk was optimistic that a solution could be reached between their respective UN ambassadors, Stevenson and Zorin, the President's Special Assistant for National Security Affairs, McGeorge Bundy, replied in a memorandum to Rusk that the Soviet attitudes to the Committee:

> ... have raised a question over here [the White House] as to whether we really want to take active steps in that particular forum on this particular issue, with the Soviet Union, at this time. My own feeling is that the President would be reluctant to see us move in this direction now.[17]

None the less, during the summer of 1961 the State Department continued to work on a group of proposals which were designed to be submitted in the form of a draft UN resolution at the 16th Session of the UN General Assembly. In line with their desire to foster a climate of legitimacy, the State Department proposed that all space activities be covered by international law, including the Charter of the UN. Furthermore:

> ... states launching objects into orbit or sustained space transit should furnish data such as orbital or transit characteristics as soon as these have been determined to the Secretary General for the purpose of regularization of launchings.[18]

These proposals were discussed in detail at the White House on 29 August 1961, and were then taken up with the Defense Department. On 14 September 1961 John McNaughton, the Deputy Assistant Secretary of Defense for International Security Affairs, wrote to Under Secretary George Ball informing him of the DoD's concurrence with the State Department's proposals but with the proviso that "we should avoid any attempt in the UN to define the limits of outer space or to limit the military use of space".[19] After receiving clearance from the other interested departments, the proposal was introduced to the United Nations' First Committee and adopted unanimously by the General Assembly on 20 December 1961 as UNGA Resolution 1721 (XVI).

The UN registration agreement came as a complete surprise to the NRO and its Director, Joseph Charyk. Although the State Department had consulted with the DoD, the satellite reconnaissance community had now become "an entirely separate structure" with the result that "the normal staff processes did not disclose certain potential difficulties with the registry proposal".[20] The poor co-ordination of policy continued further when the bureaucracy set about determining what information should be supplied to the UN registry. On 29 December the State Department informed NASA and DoD of its proposals on what the registry should include. By 12 January, 1962 the DoD had replied that the United States should take the lead on this matter—presumably to set a precedent. It suggested that only objects in "sustained orbit" be registered "in order to protect US freedom of action, if in future it should prove necessary to launch satellites for only two or three orbits *so as to minimize vulnerability to hostile counteraction*".[21] The DoD also requested that only minimal information be provided—international designation, the booster, date and time of launch, approximate orbital parameters—and rejected the State Department's suggestion that the vehicle's *purpose* be supplied to the UN. In response to this, the State Department outlined the proposed format for registration in a memorandum to NASA and the DoD on 23 January, but only received acknowledgment from the Pentagon on 6 March, one day *after* the United States had submitted its first registration of US space launchings to the United Nations. To the probable dismay of the DoD, the general purpose of the satellite was included in the US submission.[22]

The need for co-ordination was becoming ever more pressing as the Soviets had finally accepted the composition and voting procedures of the UN Committee on Outer Space. The first meeting, due in March, was set to discuss further legal principles in the exploration of space. In preparation, an interdepartmental group consisting of representatives from NASA, DoD and the State Department was established to co-ordinate US planning for these discussions. It was agreed at the first meeting of this "Special Group" on 26 April that U. Alexis Johnson (the State Department's representative) would meet with Charyk or Gilpatric "to reconcile the existing differences between State and Defense on this" (that is, registration). Charyk met with Gilpatric on 8 May but was still unhappy with the arrangements, and wrote to Johnson on 25 May expressing his concern about the amount of information to be released voluntarily.[23] As a result of this intervention, the State Department cabled the US delegation in Geneva on the following day "cautioning it that in its statements on this subject it must make clear that any information to be provided would be on a voluntary basis and at the discretion of the reporting State".[24] Although this note would affect further reporting, the United States had already submitted its second registration report on the previous day (25 May).

The distinct lack of co-ordination shown over the registration of space launchings in particular, and the projection of US space policy in general, naturally concerned the White House. As one observer recounts: "it became clear that we could not afford to go to the Peaceful Uses of Outer Space Committee of the UN, the Legal Subcommittee, the Technical Subcommittee—and all these organizations vary—without a rather coherent Federal policy on space". The Air Force was also apparently concerned about the UN restricting the United States' freedom of action in space and asked Dr Killian to bring this to the attention of the President.[25] Moreover, it was becoming increasingly obvious that the Soviets were about to use these forums to mount a diplomatic offensive to outlaw observation satellites, which necessitated a coherent and co-ordinated policy to meet the challenge.

As a result, Kennedy issued National Security Action Memorandum (NSAM) 156 on 26 May 1962, which directed the Secretary of State "to establish an inter-agency committee to review political aspects of U.S. policy on satellite reconnaissance".[26] The existence of this committee, which had no official title but was known to its members as the "NSAM 156 Committee", has only recently (1980) been made public. All reference to its activities and deliberations received the highest security classification. The Committee was chaired by Deputy Under Secretary for Political Affairs Ambassador U. Alexis Johnson, with Raymond Garthoff as its Executive Secretary. The other agencies represented were the DoD (Assistant Secretary for International Security Affairs [ISA] Paul Nitze, usually represented by his Deputy, John McNaughton); the CIA (Deputy Director Herbert P. Scoville" Jr); NRO (Joseph Charyk); NASA (Deputy Director Robert Seamans, for James Webb, the Director); and ACDA (Deputy Director Adrian Fisher). The White House representatives were Carl Kaysen (McGeorge Bundy's Deputy at the NSC) and

Dr Jerome Wiesner.[27]

As expected, the Soviet Union began its diplomatic offensive to ban US satellite reconnaissance at the meeting of the Legal Subcommittee in June 1962 with the proposal that the United Nations adopt its "Draft Declaration of the Basic Principles Governing the Activities of States Pertaining to the Exploration and Use of Outer Space". Paragraph Eight of this document declared: "The use of artificial satellites for the collection of intelligence information in the territory of foreign states is incompatible with the objectives of mankind in its conquest of outer space".[28]

While the formulation of US strategy to meet the Soviet objections to reconnaissance satellites dominated the deliberations of the NSAM 156 Committee, this inevitably overlapped with other related issues, principally: sovereignty and rights of passage in space; disclosure of information/photos to the public and to US allies; interpretation of current agreements relating to outer space; and also possible arms control initiatives. The Committee's initial report, entitled "Report on Political and Informational Aspects of Satellite Reconnaissance Policy", was submitted to Secretary Rusk on 1 July 1962 and then passed to the President. The report, which was apparently unanimous on all but two of its 19 recommendations, remains classified. However, with regard to satellite reconnaissance Abram Chayes recalls:

> . . . it was decided to embark on a series of briefings of our closest allies. They were briefed quite fully, sometimes by the President on state visits, more often by special teams who talked only to the head of state, foreign minister, or defense minister and gave them a sense of what the scope of our program was, how good it was, and what its relation was to our overall strategic picture which was very intimate. Thus, when issues of this kind arose in the United Nations for example, these people understood the implications of the political issue and were prepared to give us support. This was really quite successful.
>
> At the same time we were doing these briefings on the technical side, we began to state the legal position which was that outer space was not under the sovereignty of any nation. It is like the high seas, and there just is nothing illegal about observing another nation from the high seas. . . Here was a completely non-territorial regime and, therefore, anything other than an aggressive act from that medium was consistent with international law . . . and so on.[29]

Furthermore, the Report endorsed the March 1961 "blackout" directive prohibiting the disclosure of information on US reconnaissance satellite operations and the advanced notification of satellite launchings. This was designed to reduce international censure and the likelihood of Soviet countermeasures. The Committee members, however, were not unanimous in their support of these measures. In a separate memorandum to the President, ACDA Director William Foster argued:

> Recommendation 18 of the Satellite Reconnaissance Report would preclude the United States from making or endorsing proposals for advanced notification of space vehicle launching. In the case of space vehicle launchings, the recommendation is based on the contention that advance notification would facilitate passive and active countermeasures against reconnaissance satellites...
>
> ACDA does not wish to underestimate the countermeasures problem, but we have difficulty in assigning it overriding significance. Past and continuing publicity respecting reconnaissance satellites is sufficient to have aroused Soviet interest in the possibility of countermeasures. If passive countermeasures are practical, it is not unreasonable to suppose that the Soviet Union may undertake them whether or not advance notification is provided. Insofar as active countermeasures are concerned, it is not evident that advance notification need be so precise to facilitate in the pin-pointing of targets in outer space...

Also

> By carrying out concern regarding our space vehicles to the point where we are unwilling to discuss advance notification of missile launchings, we may encourage the conclusion that we are attempting to shield activities which we ourselves regard as suspect. It is difficult to see how such an approach could contribute to the political and legal defense of satellite reconnaissance.[30]

Foster's views did not prevail, as on 10 July, following the 502nd meeting of the NSC that reviewed the Committee's findings, the President issued NSC Action 2454. This approved all but "Recommendation 19", which was referred back to the Committee for "further study".[31]

The first major US response to the Soviet Draft Proposal came with Ambassador Gore's speech to the First Committee of the United Nations on 3 December 1962. This had been drafted by members of the NSAM 156 Committee and included the key statement:

> It is the view of the United States that Outer Space should be used for peaceful—that is, non-aggressive and beneficial—purposes. The question of military activities in space cannot be divorced from the question of military activities on earth.
>
> There is, in any event, no workable dividing line between military and

> non-military uses of space. One of the consequences of these factors is that any nation may use space satellites for such purposes as observation and information gathering. Observation from space is consistent with international law, just as observation from the high seas.[32]

On the afternoon of the same day, Soviet Representative Morozov responded with the Soviet view:

> We cannot agree with the claim that all observation from space including observation for the purpose of collecting intelligence data is in conformity with international law... The object to which such illegal surveillance is directed constitutes a secret guarded by a Sovereign State, and regardless of the means by which such an operation is carried out, it is in all cases an intrusion into something guarded by a Sovereign State in conformity with its sovereign prerogative.[33]

The Soviet Union continued its opposition to satellite reconnaissance when the Legal Subcommittee reconvened in April 1963. However, by July there were indications that the Soviets were prepared to remove their objections with the news that Khrushchev had offered to show satellite photos to the Belgian Foreign Minister, Paul Henri Spaak, at a picnic along the Dnieper River! This indication was confirmed when the new Soviet delegate to the UN Outer Space Committee, Dr Nikolai Fedorenko, omitted the customary objections to espionage satellites in a speech on 9 September 1963.[34]

The timing of this change in position was no coincidence. The Soviet Kosmos reconnaissance pogramme had by now begun regular intelligence gathering operations. Moreover, progress was being made at the Test Ban negotiations, in which satellite reconnaissance appeared to offer a way around the interminable problem of on-site inspection. Last but not least, the prospects for a ban on nuclear weapons and weapons of mass destruction in space looked promising and any further Soviet obstruction would have reduced the chances of success. Although various commentators in the Soviet press continued to persist with what had been the official line, Soviet diplomatic opposition to US satellite reconnaissance effectively ceased in September 1963.

The Pressure to Expand the Military Space Programme

After the disappointments of the Eisenhower administration, the services—particularly the Air Force—believed that a change in government would bring a change in attitudes towards the militarization of space and with it the fulfilment of their desire for an expanded military space programme. Indeed, at the height of his campaign on 10 October 1960, Kennedy had stated:

> We are in a strategic space race with the Russians and we have been losing. . . Control of space will be decided in the next decade. If the Soviets control space they can control earth, as in the past centuries the nations that controlled the seas dominated the continents.[35]

In anticipation of a change in policy—as the Republican candidate Richard Nixon had also been making similar statements—Lt. Gen. Bernard Shriever requested Trevor Gardner (former Air Force Assistant Secretary for Research and Development) on 11 October to assemble a group of experts similar to the von Neumann Committee to recommend a future military space development plan. Later, on 1 December, the Office of the Secretary of the Air Force sent a memorandum to all Air Force Commanders and contractors to ready themselves for an increased emphasis on military space projects as Kennedy's campaign statements had indicated "a realization at the highest levels . . . that military supremacy in space is as essential to our security as military supremacy at altitudes near Earth".[36] The Air Force followed this up in the transition period by preparing briefings in support of their case, for incoming DoD officials and key congressional members. The consistent theme of these briefings was that the military space programme should receive greater emphasis and that the Air Force was the service to carry it out. Later the Gardner Report would declare that: "National security considerations alone justify a major increase in the Department of Defense space effort."[37]

At first, it appeared as if Air Force expectations would be realized. Kennedy did increase the budget for military space systems and, as noted earlier, gave the Air Force primary responsibility for their development. However, as the directive still allowed the other services development rights in certain circumstances, both the Army and the Navy lost no time in outlining their own plans for further space projects including antisatellite weapons. During congressional hearings on the new directive, the Army repeated its earlier proposal to convert its Nike Zeus antiballistic missile to the antisatellite role while the Navy suggested a similar modification to its Polaris missile under the code name Early Spring. The Air Force was well aware of the threat posed by these proposals to their own plans for the SAINT programme and they immediately argued against them. With regard to the Army's proposal, the Air Force stated:

> It is believed critically important that a foreign satellite be inspected at close range to determine the purpose and mission of the satellite. Only after inspection has been accomplished should we make a decision to destroy the satellite or allow it to continue on its mission. Further, it is felt that an antisatellite system should have the capability to inspect multiple targets for the purpose of economy and operational flexibilty. Since the envisioned antisatellite Nike-Zeus would not include these capabilities, the Air Force would not seek this development program.[38]

As for the Navy's Early Spring proposal:

> The development of Early Spring by the Air Force would be dependent on its evaluation as a solution for the antisatellite problem... If such were approved by DoD the Air Force should have the primary role for its development using the resources of the Navy where applicable. It appears that sea launching of antisatellite systems offers no unique advantages to an antisatellite capability.[39]

Although the Army's and Navy's ASAT proposals were not accepted at this time, the Air Force was also denied its request to transform the SAINT programme from a satellite inspector to a true antisatellite system. Moreover, while SAINT received some additional funding it was still not the increase the Air Force had anticipated. The expansion of the military space budget was also misleading as it was for existing programmes, not for the projects and missions that had been denied by the previous administration.

In fact, this turned into a frustrating time for the Air Force; in short, a period of false hopes and false starts, as on a number of occasions it appeared as though Kennedy would acquiesce to the kind of expansion that it wanted. What made it doubly frustrating was that the priority of NASA's manned space programme undermined the Air Force claims to this mission and the new regulations de-emphasizing US military space activities robbed it of the chance to appeal in the public arena. This was especially galling as they believed that many of the arguments that the previous administration had used to curb the Air Force's space plans no longer seemed valid. In particular, the Air Force now believed there was clear evidence of a Soviet threat developing in space. After a lull in the Soviet space flight programme at the end of Eisenhower's term in office, a succession of Soviet space achievements at the beginning of the Kennedy administration again raised fears of Soviet military intentions in space. Moreover, Soviet diplomatic opposition to reconnaissance satellites and the often bellicose statements by Soviet officials threatening to use space for offensive purposes added weight to the Air Force's assertions.

The first test in what appeared to be a new phase in the Soviet space programme came soon after Kennedy's inauguration, when an interplanetary probe was launched towards Venus from Sputnik IV on 4 February 1961. While this feat gained the world's admiration it also had the double effect of raising the fear that a similar orbital platform could be used for launching nuclear missiles against targets on earth. Similarly, the first manned orbital flight by Major Yuri Gagarin in his five-ton Vostok I spacecraft on 12 April 1961, to be followed on 6 August by Major Titov in Vostok II, also raised forebodings about the massive Soviet booster capability and with it apprehensions of future Soviet military operations in space.

Partly in response to these space activities and partly as a post-mortem for the budgetary setback in the FY '62 debates, the Air Staff commissioned a ten-year space plan in May 1961. This was completed by September 1961 with the predictable recommendation for an expansion of the military space programme to meet the emerging Soviet threat. In addition to proposing passive satellite systems which had already been justified, the plan presented an urgent requirement for a satellite interception system, space-based ballistic missile defences and fast-reaction space bombers that could re-enter the atmosphere. Above all else, however, the Air Force wanted to demonstrate a manned military capability in space. As General Ferguson stated when describing the plan in congressional testimony: "Man has certain qualitative capabilities which machines cannot duplicate. He is unique in his ability to make on the spot judgments... Thus by including man in military space systems, we significantly increase the flexibility of the systems, as well as increase the probability of mission success."[40]

The latest Soviet space activities failed to change the administration's attitude towards a manned military space programme, but they did cause it to re-evaluate the requirement for an antisatellite weapon. In particular, it was the direct Soviet references to space vehicles as potential nuclear delivery systems that most concerned the administration. On 9 August 1961 at a Kremlin reception honouring Titov's space flight, Khrushchev stated:

> You do not have 50 and 100 megaton bombs. We have bombs stronger than 100 megatons. We placed Gagarin and Titov in space and we can replace them with other loads that can be directed to any place on earth.[41]

The significance of this statement became all the more ominous after the Soviet Union announced that it would break the moratorium on nuclear testing. In August 1961 they informed the world that they had "worked out designs for creating a series of superpowerful nuclear bombs of 20, 30, and 50 and 100 million tons of TNT".[42] In the event, tests of 30 and 58 megatons were recorded in the autumn of 1961. Although the resumption of testing was correctly perceived as being designed to increase the psychological pressure on the West during the latest Berlin Crisis, the prospect of Soviet "superpowerful" orbital bombs grew with further—albeit veiled—statements by Khrushchev in December 1961. On 9 December, Khrushchev referred to 50 and 100 megaton Soviet bombs as "a sword of Damocles" that would "hang over the heads of the imperialists when they decide the question whether or not they should unleash war".[43] On the following day in a speech to the Fifth World Congress of Trade Unions, Khrushchev remarked: "If we could bring the spaceships of Yuri Gagarin and Gherman Titov to land at a prearranged spot we could, of course, send up 'other payloads' and 'land' them wherever we wanted."[44]

For all their crude bellicosity, Khrushchev's statements could not be taken lightly and were enough to worry the White House and the Defense Department. Although the limitations of orbital bombs in comparison to intercontinental rockets were widely accepted, the threat was still apparent. Not only was it a potential source of blackmail in the event of an international crisis but failure to heed these types of warnings could embarrass and undermine the administration domestically.

Kennedy was particularly wary of both these possibilities and directed McNamara to take remedial action. As a result McNamara cabled the Aerospace Corporation in Los Angeles over Christmas 1961, stating it was of the "highest priority" to get a system that could respond to this threat. He requested that the Aerospace Corporation immediately begin work on this. Within a month, however, the request had been cancelled and the original telegram was later attributed to being a "heat of the moment decision".[45] Although the administration seems to have held back—for the time being at least—from committing itself to an antisatellite weapons programme, the need to take precautions against possible Soviet space activities was especially evident in the congressional budget hearings of the following year.

In his Fiscal Year 1963 testimony, McNamara stated with reference to US space research:

> . . . our program is directed at: (a) achieving a technology which will permit us to engage in military operations in outer space if the requirement does develop in the future, and (b) developing certain of the basic equipment required for such military operations—specifically boosters for launch vehicles sufficiently large to place into outer space equipment of the size we might possibly require.[46]

Similarly, his Deputy, Roswell Gilpatric, stated: "We are very conscious of the need for taking out . . . certain technological insurance. . . We don't want to be caught by surprise if any hostile use of space should occur."[47] Later, in a statement that would characterize the US approach to the research and development of antisatellite weapons for the next 20 years, Dr Harold Brown outlined what would be referred to as the "building block" approach:

> At this stage of development, it is difficult to define accurately the specific characteristics that future military operational systems of many kinds ought to have. We must, therefore, engage in a broad program covering basic building blocks which will develop technological capabilities to meet many possible contingencies. In this way, we will provide necessary insurance against military surprise in space by advancing our knowledge as a systematic basis so as to permit the shortest possible time lag in undertaking full-scale development programs as specific needs are identified.[48]

One of the contingencies that Brown referred to was quite obviously the antisatellite or antiorbital bomb mission. The Air Force was still receiving funding for the SAINT programme and it was widely expected that if such a contingency did arise, the Air Force would be allowed to develop a more advanced SAINT vehicle that could disable "hostile" objects in space. However, it was to the US Army and not the Air Force that McNamara and Brown would turn to provide a "quick fix" antisatellite system.

In a highly secret decision, McNamara instructed the Army to proceed with the development of a modified Nike Zeus system in May 1962. This received the code name MUDFLAP in its developmental stages at Bell Laboratories but later became generally known as Program 505. It completed its first test a year later in May 1963 and became "operational" on 1 August 1963[49] (see Chapter 6 for further details). It is not known whether this decision was made at the behest of the White House, although in light of the later presidential memorandum authorizing an antisatellite system in May 1963, it seems likely that McNamara's request to the Army was for advanced *development* rather than full deployment. It is possible that McNamara had already decided to turn to the Army when he cancelled the request to the Aerospace Corporation in January 1962. After the existence of the US ASAT programmes had been officially disclosed in September 1964 McNamara did state that:

> . . . when I issued the first instruction in May of 1962, instructing the Army to proceed, we weren't entirely clear that the Air Force had a capability with the Thor to develop such a system, and moreover, it wasn't entirely clear that the Army system by itself would not be adequate.[50]

At the same time that McNamara privately authorized the Army to begin development of the MUDFLAP variant of its Nike Zeus programme, there were public indications that the administration was about to change its policy towards antisatellite weapons. On 13 May 1962 the *New York Times* quoted a speech by Roswell Gilpatric in which he stated that the United States needs:

> . . . to be ready to anticipate the ability of the Soviets at some time to use space offensively. We may want to develop satellites or other space systems which could be used to defend the peaceful or other defensive satellites now in operation.[51]

Whether this speech was made before Gilpatric knew of McNamara's May 1962 decision is uncertain, as he also stated that the "Defense Department had decided to develop the technology of manned orbital systems able to rendezvous with satellites and then land at preset locations on earth". The report noted that these same manned systems would also have the capability of "neutralizing" or "destroying" hostile satellites. The Air Force certainly took this as a cue to proceed with studies of a manned SAINT interceptor in the expectation that support would be available.[52] Later, on 12 June, the *New York Times* printed another story stating that the Defense Department was "embarking upon a man-in-space program to prevent foreign military control of space as well as its exploitation".[53]

The furore that greeted these reports was immense; not only did it appear to signal a reversal of the administration's position on the peaceful exploitation of space, but it also seemed to members of Congress that the Defense Department was competing with NASA for the manned space mission.[54] Kennedy was particularly incensed over this apparently unauthorized statement and it is likely that it was another contributory factor in his NSAM 156 directive of 26 May.

After the 12 June *New York Times* article, administration officials went out of their way to deny any change in policy. Defense

Department officials stated that no "hard" decisions had been taken on manned military systems for space, while Gilpatric and Brown appeared before the Senate Committee on Aeronautical and Space Sciences in June and denied that the Defense Department was pre-empting NASA's role. Brown also stated, with reference to manned military space systems, that "I cannot define a military requirement for them. I think there may, in the end, turn out not to be any."[55] After this latest disappointment, the Air Force received another blow to their hopes when the Defense Department took steps to control the Air Force's proliferating "study projects" on offensive space warfare systems.[56]

The Air Force campaign to change the administration policy continued, however, and reached a peak following the public disclosure of another Soviet space first with the "rendezvous" of Vostok III and IV between 12 and 15 August. Although there were contradictory reports about the separation distance—ranging from three to 100 miles—it inevitably fuelled the speculation about possible Soviet satellite inspection and interception capabilities.[57] Administration officials would not comment on the possible significance of this latest test but let it be known that the US military space programme would be reassessed. Due to the vociferous criticism that the administration received over this incident, Kennedy personally telephoned Secretary of the Air Force Zuckert on 21 August 1962, to request a full appraisal of the possible shortcomings of the military space programme.[58] Zuckert replied the next day, quoting Kennedy's campaign statements to imply an unfulfilled promise, and requested in general terms that funding for military space projects be increased.[59] Kennedy wrote back on 27 August admonishing Zuckert for quoting his previous speeches—which he had no interest in—and also for not providing a full breakdown of Air Force projects that had been requested but rejected in the FY 1963 budget process.[60] Zuckert replied on 4 September that:

> My chief concern ... is with the future. Space technology is reaching the point, I believe, where specific military and psychological warfare possibilities are emerging. Although the Soviets reassert the peaceful intent of their space program, they frequently couple their assertions with allusion to the military capabilities inherent in a specific accomplishment. We have no choice now, I believe, but to increase the vigor of our exploitation of space technology for military uses.
>
> The present planning of the defense space program basically does not envision the necessity or feasibility of an expanded major military operational role in space. . .[61]

Although he did not recommend a "crash program", Zuckert did state that "our military space thinking and programming are in need of a stimulant".[62] Among the systems that he proposed was a satellite interception capability as well as the now customary request for a manned military programme. Zuckert also referred to the forthcoming Air Force Five Year Space Plan that had originally been requested by McNamara in June 1962. In his reply on the next day Kennedy wrote that he would wait for this to be completed before any further discussions could take place.

The Air Force Space Plan was finally submitted to McNamara on 5 November but by the autumn of 1962 the administration was already emphasizing a new "hard-nosed" approach to military space proposals. In a key speech given on 9 October 1962 John H. Rubel, Deputy Director of Defense Research and Engineering, stated that the level of defence spending on space systems was:

> ... as close to the optimum size as we can make it in the light of all the uncertainties that must accompany such a program. In fact we probably err on the side of allowing too generous a margin of safety for the effects of these uncertainties. Henceforth the DoD would emphasize hard military requirements and that proposals which served abstract doctrines about the military role in space would not be entertained.[63]

Although "proposals which served abstract doctrines" was in part a reference to the more futuristic Air Force space weapons concepts, it was especially aimed at its manned space aspirations. For the FY 1964 budget request the Air Force proposed procuring some of NASA's Gemini spacecraft under its own project heading of "Blue Gemini". The Air Force also proposed a separate programme to develop a space station called MODS, for Military Orbital Development System. However, in line with administration policy, both were deleted from the budget in January 1963.[64] The DoD's attitude to manned military space projects was re-emphasized at the budget hearings, where McNamara stated: "We do not today see clearly a military requirement for men in space, in contrast to unmanned satellites in space utilized for military purposes."[65] The Air Force did get some compensation however in that it was allowed to conduct experiments aboard the later NASA Gemini flights.

As a result of the deliberations of the NSAM 156 Committee, administration officials had also begun in the autumn of 1962 to emphasize not only the peaceful intent of the US space programme but also the desirability of banning weapons of mass destruction in space.[66] This did not, however, preclude ASAT research and development from going forward in great secrecy as a precaution against the deployment of Soviet orbital bombs. In addition to McNamara's May 1962 decision to develop the ASAT variant of the Nike Zeus ABM missile, the Air Force was also given permission in December 1962 to begin testing some Thor IRBMs converted to the ASAT role as a second option.

The SAINT programme, which the Air Force had originally hoped would eventually become the main US antisatellite system was, however, cancelled on 3 December 1962.[67] Although there was speculation that the SAINT programme was a victim of the administration's heightened interest in projecting a peaceful image of the military space programme, the cancellation was "made by the Air Force entirely on its own without any prompting from the Director of Defense Research and Engineering".[68] Contrary to subsequent interpretations that it was a deliberate signal to the Soviets, SAINT was cancelled for technical and financial reasons. In fact, it was announced at the time that the Air Force would continue with basic research in this area and also participate in NASA's Gemini programme.[69]

By February 1963 the Air Force was told to prepare for an "operational standby capability" after it had completed the initial testing of the Thor ASAT missile.[70] While it is clear that the decision to develop an additional ASAT system was a continuation of the general insurance policy of hedging against possible Soviet orbital bomb deployments, it is not clear whether it was in response to a specific event. In January 1963 McNamara stated in congressional testimony that "the Soviet Union may now have or soon achieve the capability to place in orbit bomb-carrying satellites" and that while he was sceptical of their utility as weapon delivery systems "we must make the necessary preparations now to counter it if it does develop".[71] Further indications that the Soviets might actually do this came in February when Marshall Biriuzov, then Chief of the Soviet Strategic Rocket Forces, declared in a broadcast interview: "It has now become possible at a command from earth to launch missiles from satellites at any desired time and at any point in the satellite trajectory."[72]

Final approval for the Thor ASAT system—now known as Program 437—was given by the President on 8 May 1963. Kennedy approved the recommendation drafted by the State Department (with the concurrence of the Secretary of Defense, Director of the CIA and the Director of ACDA), which stated that as a contingency against the possibility of Soviet orbital bombs the United States should "develop an active antisatellite capability at the earliest practicable time, nuclear and non-nuclear".[73] Such was the urgency to develop an operational capability that on 6 July 1963 NSAM 258, entitled "Assignment of Highest National Priority to Program 437", was also issued.[74] By February 1964 the first test of Program 437 had been successfully completed on Johnston Island in the Pacific (for details see Chapter 6). Later McNamara justified the possession of two ASAT systems:

> . . . the Air Force proceeded further in their study of the Thor application, determined that it would be possible to develop an antisatellite capability based on it, and not only determined that, but indicated ways in which it could provide capabilities that the Army system could not.
>
> Now, the next question of course, why would we keep the Army system then after you get the Air Force system? And the answer to that is that the Army system has capabilities the Air Force system doesn't have, so we consider it desirable to keep both.[75]

The respective advantages of the two systems were that the Nike Zeus could react more quickly due to its solid propellant, while the Thor missile could be fired against targets at higher altitudes. Both systems had their limitations; their fixed base sites meant that they would have to wait for the target to be in a position overhead before they could be launched, and their nuclear warheads, if used, would have threatened US satellites in the vicinity of the explosion. Their use would also have contravened the Partial Test Ban Treaty, which, ironically, had been signed only the day before NSAM 258 was issued. Despite these operational constraints, both systems remained in place for a considerable time—Project 505 to 1967 and Project 437 to 1975. Although both programmes were designed to be "secret" precautionary measures, their existence became known in leaks to the press after Congress had been privately informed by the White House in 1983.[76] It was not until 18 September 1964, however, that the United States officially admitted the existence of the two systems.

On the face of it the decision to develop an antisatellite capability appears to represent not only a triumph for the services but also a departure from the policy established by Eisenhower. This interpretation, however, would be wrong. While the services proposed both systems, they did not represent the level of commitment that they had expected from the Kennedy administration. The two systems were procured from existing programmes with a minimum of effort and cost. Moreover, they ultimately amounted to a limited capability with severe operational constraints. The projects that the Air Force had hoped would reach operational status such as SAINT and the various manned military space system proposals did not get the requisite support. In fact, the ASAT programme complemented rather than contradicted the administration's underlying military space policy. Its primary goal—like its predecessors—was to maintain space for the passive use of military systems primarily by example and diplomatic action. Yet if this failed, the administration wanted to be in the position to respond to any threatening developments in space. Thus in many respects the Kennedy administration was implicitly following a "twin track" policy before that term became fashionable during the Carter administration.

Arms Control and the Orbital Bomb Threat

As noted in the preceding chapter, the Eisenhower administration had included prohibiting the orbiting or stationing of weapons of mass destruction in outer space as part of their proposals for General and Complete Disarmament (GCD). A similar proposal was outlined again by President Kennedy in "The United States Program for General and Complete Disarmament in a Peaceful World", which was presented to the 16th General Assembly on 25 September 1961.[77] Later on 18 April 1962, the US Delegation to the Eighteen-Nation Disarmament Conference at Geneva submitted a similar scheme in their "Outline of Basic Provisions of a Treaty on General and Complete Disarmament in a Peaceful World", which again put disarmament measures in space as part of an overall GCD package.[78] The US position was that progress towards disarmament in space was conditional on the acceptance of GCD. The Soviet position was virtually the same.

However, in March 1962, the US delegation at Geneva had been taken by surprise with a proposal from the Canadian Secretary of State for Foreign Affairs, Howard Green, for a *separate* ban on weapons of mass destruction in space. This was unusual in another aspect as it did not contain the standard Western demand for inspection. It had also come without advance consultation and against the wishes of the US officials present.[79] Although the US representatives immediately spoke out against any "declaratory measures" that did not contain certain provisions for "inspection and control", Green's proposal was brought to the attention of the NSAM 156 Committee as a possible option.[80]

On 19 June 1962 Raymond Garthoff, the Executive Secretary of the

NSAM 156 Committee, drafted a paper on the pros and cons of a separate ban. Although the State Department and ACDA supported such an agreement (with the further recommendation that the matter be transferred to the Committee of Principals, the formal interagency co-ordinating committee for arms control in the Kennedy/Johnson era), the representatives from the Defense Department, Joint Chiefs of Staff (JCS), the NRO and CIA all opposed a separate ban. Apart from the JCS, who wished to retain this military option, their arguments were almost entirely based on the belief that inspection was a vital prerequisite to any agreement. In addition, as Garthoff points out, there was also the fear of the resultant "negotiations somehow impeding our other highly important military uses of space, especially satellite reconnaissance".[81] The fact that there was not a unanimous recommendation on this issue was entered in their final report on the "Political and Informational Aspects of Satellite Reconnaissance Policy", which was submitted to the President via Secretary Rusk on 2 July.[82]

Prior to the review of the Report's findings at the 10 July meeting of the NSC, William Foster of ACDA wrote to the President on 6 July to add weight to the State Department and ACDA views on a separate ban. The memorandum stated:

> ACDA believes that if an agreement could be reached to prohibit the placing in orbit of weapons of mass destruction, such an agreement would be in the interest of national security. To prevent the extension of the arms race to outer space should be an important objective of arms control, and hence national security policy. This conclusion appears valid even if such an agreement were fully effective only against very large and hence very high yield weapons such as those tested by the Soviet Union in its last test series. The fact that the Soviet Union would gain increased knowledge of our satellite reconnaissance capabilities does not seem a compelling argument against our acceptance of inspection procedures particularly in view of the present Soviet ability to estimate these capabilities with a substantial degree of accuracy, a fact that is noted in the Satellite Reconnaissance Report.[83]

As regards the argument that a separate agreement could provoke "a public airing of the satellite reconnaissance issue", Foster argued that the Soviets were already doing this and therefore the best tactical ploy for the United States would be to shift the debate away from the Outer Space Committee and the General Assembly to the Eighteen-Nation Conference in Geneva, where the United States would have "reasonable procedural grounds for ignoring, if not defeating, condemnatory principles of space law or hortatory resolutions which might embarrass the satellite reconnaissance program".[84]

As a result of the differing opinions on this matter, President Kennedy issued NSC Action 2454 on 10 July 1962 which, as noted earlier, approved all the report's recommendations but referred Recommendation 19 back to the Committee "for further study". The State Department had in fact anticipated further discussion, and on the next day Garthoff drafted a memorandum from Ambassador Johnson (the Committee's chairman) to the other members of the NSAM 156 Committee. Entitled "Recommended U.S. Position on a Separate Ban on Weapons of Mass Destruction in Outer Space", it received their concurrence and was eventually submitted to the President by Secretary Rusk on the following day, 12 July.[85] The rapidity of their decision-making was an almost unprecedented feat considering the number of bureaucratic actors involved. Briefly, the memorandum recommended the following: opposing a declaratory ban; not taking an American initiative in raising a separate ban; attempting to place the onus for rejecting such a ban if it were proposed on the Soviet Union; and, finally, noting that ACDA would "urgently" study the problem "to determine the inspection requirement for a separate ban".[86]

However, Kennedy, after consulting McGeorge Bundy and his assistant, Carl Kaysen, questioned the Committee's recommendations, particularly its belief that unilateral verification of a declaratory ban would not be adequate. About a week after the 12 July report had been submitted, Carl Kaysen telephoned Garthoff to question whether a declaratory ban had really been given serious attention, saying that President Kennedy was ready to accept such a judgment if it were a carefully considered one, but otherwise not.[87] On 23 July in a memorandum from McGeorge Bundy to Secretary Rusk, the President formally requested a review and further study. It stated that the President was not prepared to accept the "total rejection of any possible declaratory ban" in the belief that a ban of this type may be better than nothing. He also requested that a study of national means of verification be completed. Furthermore, as Garthoff states, the President "questioned the general position calling for inspection in all cases regardless of possible U.S. interest in declaratory agreements in special cases".[88] In short, the Committee was asked to reconsider its recommendations.

As this was now primarily an arms control issue, the interagency debate was formally shifted from the NSAM 156 Committee to the Committee of Deputies (the working group of the Committee of Principals). However, the NSAM 156 Committee continued to take an interest due to its overlapping membership and its periodic need to cover related issues. For example, on 27 August 1962 NSAM 183 was issued.[89] Entitled "Space Program for the United States", it requested the NSAM 156 Committee to translate the recommendations of NSC Action 2454 (that is, their original report's recommendations) into practical contingency papers. On 21 September 1962 Ambassador Johnson enclosed a number of contingency papers in a memorandum to the White House ("Position and Contingency Papers Pursuant to NSAM 183" TOP SECRET).[90]

Meanwhile, the Committee of Deputies, having discussed a draft paper on "Recommendations on a Separate Arms Control Measure for Outer Space" on 5 September, in turn recommended in a memorandum to Secretary Rusk that the United States support a declaratory ban. On the same day Deputy Secretary of Defense Roswell Gilpatric delivered a keynote speech intended to signal to the Soviets the US desire to prevent an arms race in space with a proposal for such a ban on weapons of mass destruction. This initiative was the result of a separate and less formal grouping of representatives from the NSC and the Departments of State and Defense which met at the now famous "Tuesday Lunches". The speech had been carefully drafted by Adam Yarmolinksy—Special Assistant to McNamara—with additional assistance from Garthoff and Kaysen. In the speech Gilpatric stated:

> The United States believes that it is highly desirable for its own security and for the security of the world that the arms race should not be extended into outer space, and we are seeking in every feasible way to achieve that purpose. Today there is no doubt that either the United States or the Soviet Union could place thermonuclear weapons in orbit, but such an action is just not a rational military strategy for either side for the foreseeable future. We have no program to place any weapons of mass destruction into orbit. An arms race in space will not contribute to our security. *I can think of no greater stimulus for a Soviet thermonuclear arms effort in space than a United States commitment to such a program. This we will not do.*

Later in the speech Gilpatric stated "we will of course take such steps as necessary to defend ourselves and our allies, *if the Soviet Union forces us to do so*".[91]

This approach reflected the belief of McNamara, and probably members of the White House National Security Staff as well, that an implicit invitation for an informal agreement with the Soviet Union was tactically the best solution to this question. It also reflected their growing perception of the dynamics of the arms race and its corollary that unnecessary actions and reactions could be controlled by unilateral restraint. Both these views would become articulated in a more explicit manner at a later date.

On the other front, ACDA Director William Foster submitted to the "Principals" on 15 September the draft "U.S. Approach to a Separate Arms Control Measure for Outer Space", which supported a declaratory ban to be monitored by national means of verification. The Committee of Principals met on 19 September to discuss the draft. At this meeting the differences of opinion emerged on both the principle of a unilaterally verified ban and also the question of what tactics to adopt with the Soviet Union. The JCS, rather than appear too negative towards the proposal, made their support conditional on it being part of a GCD package, which was just an indirect way of trying to kill the idea.[92] Gilpatric, who was representing the Defense Department, reported that he and McNamara favoured a private approach to reach an informal agreement with the Soviet Union. Foster, who was chairing the meeting in Rusk's absence, agreed to try this method.

Although agreement was not reached at the meeting, Foster gave the impression in his report to the President on 17 September that a general consensus on the desirability of reaching a ban had been attained, but that opinions varied on how to achieve it. Indeed, a separate JCS submission (JCSM 719-62) giving their own views was attached to Foster's report.[93] This was submitted to the President on 1 October via Carl Kaysen, who also added his own recommendations. As a result on the following day NSAM 192 was issued.[94] This stated that:

> The President has reviewed the memorandum of September 1962, submitted to him by the Director of the United States Arms Control and Disarmament Agency reporting the views of the Committee of Principals on 'A Separate Arms Control Measure for Outer Space' and the comments of the Joint Chiefs of Staff. He has approved the recommendations contained in the memorandum, which was a final response to the directives of NSAM 156 and NSC Action 2524.
>
> The President wishes to be informed before we make an initiative in the United Nations General Assembly or elsewhere to put forward the proposal for a ban on weapons of mass destruction in outer space.
> (signed by Carl Kaysen)[95]

NSAM 192 represented, possibly for the first time, the willingness of the US government to conclude an arms control agreement with the Soviet Union that did not make inspection or verification a necessary prerequisite. This change of attitude was possibly the result of Kaysen's influence and, according to one source, it was not based on "anything we had in terms of actually inspecting vehicles in space" but rather it was a "prime case of the balance between the degree of verification risks versus the benefits of a treaty".[96]

In response to NSAM 192, Foster approached Soviet Foreign Minister Gromyko and Ambassador Dobrynin privately in New York on 17 October with the proposal for a separate ban on stationing weapons of mass destruction in outer space. According to Garthoff, Gromyko assumed this to be a renewal of the 1957 Western Proposal, which covered all objects "sent through" space (see Chapter 3) and which also called for inspection. Gromyko responded with the standard Soviet reply that this would only be considered if it was tied to the removal of American "Forward Based Systems" in Europe.[97] Foster's New York proposal proved to be the last time that the Soviets were approached directly on this matter until the autumn of 1963. The drama of the Cuban Missile Crisis intervened to interrupt the US initiative.[98]

After the Cuban crisis, the thread was taken up again; first, in a policy planning paper on "Post-Cuba Negotiations with the USSR" dated 9 November, which included a proposal for a "declaratory commitment not to station weapons of mass destruction in outer space"; and second, in a similar speech to the one given by Gilpatric on 5 September.[99] This time Ambassador Gore stated at the United Nations on 3 December 1962:

> ... it is especially important that we do everything now that can be done to avoid an arms race in outer space—for certainly it should be easier to agree now not to arm a part of the environment that has never been armed than later to agree to disarm parts that have been armed. My Government earnestly hopes that the Soviet Union will likewise refrain from taking steps which will extend the arms race into outer space.[100]

This sentiment was later repeated in a speech by Rusk on 3 January 1963.

The early part of 1963, as we have seen, brought a distinct increase in the amount of Soviet propaganda on the militarization of space. In addition to the Defense Department's preparations described earlier, the State Department also began to consider US policy in the event the Soviet Union deployed orbital bombs. As one former State Department official recalled: "We had a number of ambiguous indications that the Soviets might be working along these lines but we weren't sure whether these were indications that the Soviets were suspicious of U.S. intentions or a Soviet interest in them."[101] On 7 March 1963, U. Alexis Johnson submitted to Secretary Rusk a memorandum on "Further Initiatives on a Ban on Weapons in Space", which Garthoff had drafted. This reviewed the progress of US attempts to obtain a ban and also noted that a "contingency plan for U.S. reaction to Soviet placing of a nuclear weapon in space" was being transmitted separately.[102] However, on 29 April Rusk decided for "tactical reasons" to "suspend for the present" any further approaches to the Soviets even though a memorandum to the President recommending the opposite had been drafted and agreed by the other Principals.[103]

The precautions against a possible Soviet orbital bomb continued, however, and on 8 May 1963 Rusk submitted to the President the "Contingency Plan for U.S. Reaction to Soviet Placıng of a Nuclear Weapon in Space", which had also been drafted by Garthoff. This, as noted earlier, contained the recommendation that the United States "develop an active antisatellite capability at the earliest practicable time, nuclear and non-nuclear".[104] The President approved. This recommendation, according to one source, was "based on the possibility that the Soviets might proceed with deployment before an agreement. After that it did not cover or affect the decision".[105] In other words, the decision to go ahead with the development of an antisatellite system would be a precaution against the possible Soviet abrogation of an agreement. Moreover, it would provide insurance for possible domestic criticism that the administration had not taken the necessary precautions.

Garthoff's central role in the progress towards an agreement was further demonstrated when he visited Moscow while on a tour of Eastern European capitals in early September 1963. In a private exchange of views with Usachev, a senior Soviet diplomat, Garthoff explained the purpose of Foster's approach in New York. He emphasized that the US proposal did not concern missiles transmitting *through* space but only weapons of mass destruciton stationed *in* space. Furthermore, the United States would not demand inspection with this agreement. As Garthoff recalls, "he [Usachev] was very interested and grateful for my clarification of the U.S. proposal, its objective and the position on inspection".[106]

On 19 September, five days after Fedorenko had omitted the Soviet objections to US reconnaissance satellites from its latest proposal to the Committee on the Peaceful Uses of Outer Space, Gromyko addressed the General Assembly in New York with the statement that:

> . . . the Soviet Government deems it necessary to reach agreement with the United States Government to ban the placing into orbit of objects with nuclear weapons on board. We are aware that the United States Government also takes a positive view of the solution of this question. We assume also that an exchange of views on the banning of the placing into orbit of nuclear weapons will be continued between the Governments of the USSR and the U.S. on a bilateral basis. It would be a very good thing if an understanding could be reached and an accord concluded on this vital question. The Soviet Government is ready.[107]

Kennedy was not slow to respond to this offer, for on the following day he acknowledged the Soviet proposal and stated: "Let us get our negotiators back to the negotiating table to work out a practicable arrangement to this end."[108] Rusk took this up with Gromyko and confirmed that the Soviets wanted to proceed with a treaty or some other formal undertaking. This left the US administration with the question of how to react to the Soviet request. On 1 October Fisher of ACDA submitted a paper to the Committee of Deputies, entitled "Proposed U.S.-Soviet Arrangement Concerning the Placing in Orbit of Weapons of Mass Destruction". Three alternative "arrangements" or options were outlined to Rusk in a memorandum from Foster after the Deputies meeting: (1) unilateral statements of intention; (2) a US-USSR executive agreement; or (3) a UNGA Resolution. The State Department representatives favoured a UN resolution with unilateral statements in support. They believed this would preclude multilateral negotiations as well as pre-empt potential domestic opposition. The JCS opposed this in a paper issued on the same day (7 October), and recommended instead a joint declaration of intent. Having failed to prevent an agreement, the JCS objective was to minimize the significance of whatever resulted from the discussions with the Soviet Union. They recommended that the term "weapons of mass destruction" rather than "nuclear weapons" be used and also that a "withdrawal provision" be included. Both would have made the agreement virtually meaningless.

At the meeting of Principals on 8 October, the opposing views were aired by both Rusk and the JCS representative. Rusk informed the meeting that the term "weapons of mass destruction" would be used but that this would be held to include nuclear weapons. At that juncture, Under Secretary Ball joined the meeting to bring the views of the President. Ball informed them that, for domestic reasons, Kennedy did not want an executive agreement "or anything looking like an executive agreement". Instead he favoured an UNGA Resolution with supporting statements by the United States and

USSR.[109] This course of action was adopted by the meeting.

On 15 October, Mexico, which had earlier proposed a draft treaty on banning of weapons of mass destruction from space (21 June 1963), retabled its draft resolution after it had been approved by the United States and the Soviet Union. This was passed two days later as Resolution 1884 (XVIII). It called upon all states to refrain from placing in orbit or stationing in space "nuclear weapons or any other kinds of weapons of mass destruction".[110] Thus, for the time being at least, a thermonuclear arms race in space had been averted.

Conclusion

By the autumn of 1963 the Kennedy administration had not only headed off the Soviet attempt to prohibit reconnaissance satellites but had also reached an agreement with the Soviet Union to ban nuclear weapons from space. These two achievements must rank in importance alongside the "Hot Line" and Partial Test Ban Treaty that were signed in the summer of 1963. They are all the more creditable when one considers the uncertainty over the future exploitation of space when Kennedy entered office. Far from departing from Eisenhower's policy, Kennedy quickly grasped the importance of reconnaissance satellites to US national security and the logic of his predecessor's efforts to secure their international acceptance by emphasizing the "peaceful" nature of the US space programme. However, Kennedy found the projection of this peaceful image increasingly jeopardized by the uncoordinated actions and statements of his officials. Following a reappraisal of policy and the creation of a new decision-making body (the NSAM 156 Committee), US policy became more coherent and organized. In addition to emphasizing the peaceful efforts of the US space programme, the United States also deliberately de-emphasized its military component to reduce potential foreign opposition. Moreover, while the necessary precautions were taken to insure against potential threats to the security of the United States, the military space programme remained primarily "passive" in nature. By also canvassing support for the US position within the legal subcommittee of the UN Committee on the Peaceful Uses of Space, the Soviet diplomatic offensive against "espionage satellites" was progressively undermined.

In the long drawn out deliberations over banning weapons of mass destruction from space, Kennedy, as Garthoff observes, also showed great tactical skill: first, in challenging the idea that a declaratory ban without inspection was unacceptable; and then using "his executive power to gain a *de facto* moratorium until the time was auspicious for a treaty".[111] It was this legacy that his successor, President Johnson, converted into treaty form in 1967 with unanimous support from Congress—something that would have been inconceivable in 1963.

5

The Johnson Years: The Consolidation of Policy

I am convinced that we should do what we can—not only for our generation, but for future generations—to see to it that serious political conflicts do not arise as a result of space activities. I believe that the time is ripe for action. We should not lose time.

President Johnson,
Statement on the Need for a Treaty Governing
Exploration of Celestial Bodies,
7 May 1966

INTRODUCTION

President Johnson inherited a military space programme and associated policy that was now well-established and reasonably co-ordinated. After the period of uncertainty over whether space would become a new arena of superpower arms competition and confrontation, the United States and the Soviet Union had apparently reached a *modus vivendi* in their military exploitation of space. It was left to Johnson to further the broad policy guidelines of his predecessors which had contributed to this beneficial arrangement and also bring to fruition many of the projects started in the preceding years. It would be wrong, however, to suggest that the evolution of US military space policy during the Johnson administration was without incident or devoid of his own individual contribution.

The earlier concern over Soviet space activities and intentions did not recede entirely after 1963. It resurfaced on a number of occasions, particularly after the Soviet Union began testing a Fractional Orbital Bombardment System (FOBS) in 1967. As a result, the requirement for additional or improved US ASAT systems was discussed periodically, though it was never a major issue. A more controversial issue was the decision to allow the Air Force to develop a manned space system after numerous rejections. Furthermore, Johnson, in one of the major achievements of his administration, successfully converted the UN resolution banning weapons of mass destruction in space into a more satisfactory international treaty. While it did not go much beyond the 1963 agreement, it was none the less a significant accomplishment.

US Policy: Conformity and Contradiction

Just as President Johnson entered office, an interagency study group chaired by Raymond Garthoff and organized under the auspices of the State Department's Policy Planning Council was putting the final touches on a study document entitled "Planning Implications for National Security of Outer Space in the 1970s".[1] This had been authorized during the Kennedy administration as one of a series of studies under the general rubric "Basic National Security Planning Tasks". Although this study appears to have made little impact, largely because it reaffirmed the existing direction of US space policy, it is none the less interesting as further evidence of official US thinking at that time.[2]

Included in the study's recommendations were the now standard US policy objectives:

> We should continue to stand on the general principle of freedom of space. We should actively seek arms control arrangements which enhance national security. We should pursue vigorously the development and use of appropriate and necessary military activities in space, while seeking to prevent extension of the arms race into space.[3]

As had been the case with previous administrations, preventing an arms race in space meant in this context avoiding the deployment of nuclear and other weapons of mass destruction *in* space. It did not preclude the right of the United States to take precautions against this possibility. Thus, as the report states: "An anti-satellite capability (probably earth to space) will be needed for the defense of the United States... Current high priority efforts should be continued and extended as necessary in the future."[4] The primary justification for the US ASAT programme continued to be a hedge against technological surprise and Soviet abrogation of the UN resolution prohibiting nuclear weapons in orbit. However, two other rationales were also provided in this study. These were "to deal with other potential space targets in time of war whether or not the [Soviet] orbital nuclear delivery vehicles were introduced" and, secondly, "to 'enforce' the principle of noninterference" in space.[5]

Other parts of the study, however, are less emphatic, to the point of almost contradicting these secondary rationales for a US ASAT programme. The Soviet space threat was certainly not considered to be great and, if anything, was downplayed throughout the study. For example:

> Much . . . is alleged about Soviet intentions in space, based largely on their aspirations on earth and assumption of maximum capable effort in space. Here, too, it is quite enough to say they *may* try, and *may* succeed, in pursuing *some* promising avenue of development with important military implications. It is too much to say they will do so; it is not a foregone conclusion.[6]

With regard to the near-term Soviet orbital bomb threat, the report states that "no more than token deployment is likely, and such a move would be regarded by the United States as being primarily of psychological significance".[7]

The report also acknowledges doubts about the ability of ASAT weapons to enforce the principle of noninterference in space by the threat of reciprocal action. In addition to recognizing that there may be ways to interfere with satellites without the source being identified, the study notes that the party threatening reciprocal action would also in turn face the likelihood of retaliation.[8] Given its greater dependence on space systems in general, the United States could become self-deterred from retaliating in kind. As the report later states: "The usefulness to the United States of observation [satellites] . . . as a means of penetrating Soviet secretiveness is obvious. The value to the USSR may be less clear; indeed, the value is probably much lower."[9] While the study notes that the principle of non-interference is only relevant in peacetime, and that the benefits of space to the Soviet Union may have grown by the 1970s, it none the less reflects a general ambivalence towards the utility of ASAT weapons.

Although US fears about Soviet orbital bombs gradually receded, it was still considered prudent to continue with the two programmes authorized by Kennedy, not the least for domestic political reasons.[10] Apart from various leaks to the press, there had as yet been no public acknowledgement of the existence of these two systems. But, in quite a remarkable departure from his predecessor's policy of maintaining a low profile in such matters, Johnson decided to state openly that the United States possessed an antisatellite or, more correctly, an antiorbital bomb capability in the autumn of 1964.

During the run up to the 1964 presidential election, the Johnson administration was criticized for neglecting US national security interests by Barry Goldwater, the Republican candidate. Johnson was well aware that this kind of election ploy had proved particularly fruitful for Kennedy in the 1960 campaign. He decided, therefore, to announce the existence of some highly classified projects to pre-empt or deflect further Republican criticism. Johnson had already disclosed the existence of the SR-71 high-altitude military reconnaissance aircraft when—possibly in response to reports that Khrushchev had boasted on the previous day that the Soviet Union possessed a "terrible" new weapon (albeit hastily retracted soon afterwards)—he decided to announce publicly the existence of the two US ASAT systems. As the *Los Angeles Times* leader commented:

> This [Khrushchev's] clumsy bragging about supposed might could not have been far from the President's mind when he made his quiet revelation from the Capitol steps in Sacramento. The chance to contrast Soviet development of a weapon allegedly capable of wiping out all life with U.S. development of a weapon designed solely for defense may simply have been too good to ignore.[11]

In the speech President Johnson stated:

> Seven years ago America awakened one morning to find a Soviet satellite orbiting the skies. We found that our adversaries had acquired new capabilities for the use, or misuse, of space.
>
> This administration moved to meet that challenge. We sought and we supported a resolution unanimously approved in the United Nations banning the use of weapons of mass destruction in outer space. We have stated that we have no intention of putting warheads into orbit. We have no reason to believe that any nation now plans to put nuclear warheads into orbit. We have more effective systems today.
>
> At the same time, we recognize the danger that an aggressor might some day use armed satellites to try to terrorize the entire population of the world, and we have acted to meet that threat. To insure that no nation will be tempted to use the reaches of space as a platform for weapons of mass destruction we began in 1962 and 1963 to develop systems capable of destroying bomb carrying satellites.[12]

In addition to announcing the existence of the two ASAT systems,

Johnson also divulged the Over The Horizon (OTH) radar programme. The next day McNamara elaborated on both systems to the press.

The impact of the disclosures on the domestic and international scene was minimal. The Republicans seemed to divide their criticisms of Johnson by berating him for disclosing hitherto classified material for political ends and for trumpeting a capability that they considered neither operational nor effective. The crisis over America's growing involvement in Vietnam also acted to eclipse the significance of the event. After this brief appearance in the 1964 election campaign, the development of antisatellite systems ceased to be an important subject for discussion until it was again brought up in the context of the Soviet FOBS "threat". In the meantime, research was authorized to improve the credibility of the two existing systems.

It had been accepted almost from the beginning that the use of nuclear antisatellite warheads in space would not only contravene existing treaty regulations but would also be counterproductive to US space operations. In short, a nuclear explosion in space was just as likely to damage or disable US satellites in its vicinity as was the Soviet target. Although the deployment of the US systems provided important political and psychological insurance against possible Soviet "space threats", their nuclear warheads were as much a deterrent to US use as they were to Soviet treaty violations. Thus the basic prerequisite for the follow-on systems was that they should have a nonnuclear warhead and therefore be more "usable" in the event of a crisis. All three services carried out research to these ends: the Air Force with a modified warhead and guidance system (Project 922), the Army with their Nike-X ABM system, and the Navy with a revised Polaris Early Spring proposal.

Despite reports that the Defense Department was on the verge of authorizing a second-generation antisatellite weapons programme, this never materialized, although modifications were made to the missiles in place. The declining space threat and the competition for funds with other higher priority systems, particularly following the escalation of the Vietnam War in the late sixties, meant that funding was kept to a minimum. By 1967 it was decided that Program 505 (Nike Zeus) was no longer required and it was decommissioned. Although the Defense Department was not prepared to fund a follow-on system, they were also not prepared to dismantle *all* the existing systems. The Thors of Program 437 continued to be tested up to 1970. The Army also had reason to believe that its Nike-X missile system would be eventually modified for the antisatellite role. Chapter 6 has further details of these programmes.

It was in the area of manned military space activities that the new administration showed signs of diverging from its predecessor. On 10 December 1963 McNamara cancelled the Dynasoar (X-20) project, but at the same time announced that the United States would proceed with feasibility studies of a near-earth Manned Orbiting Laboratory (MOL). The cancellation of Dynasoar was due to a combination of factors, principally technical and economic in nature. As McNamara's statement at the time of the cancellation implied, the project had been technically overambitious in trying to accomplish the goals of manned space experimentation, precise re-entry, and landing with one space vehicle. Furthermore, Dynasoar was only designed to carry one man into space for a limited duration. In short, too many problems needed to be solved for a limited return in capability. Having already cost $400 million, the project was likely to reach a prohibitive $1 billion if allowed to continue.[13] In its place, the MOL would consist of a modified Gemini capsule fitted to a pressurised cylinder the size of a "small house trailer" that would be launched by a Titan III booster. Two astronauts would be able to move freely without space suits and conduct observations and experiments.

Despite the decision to begin feasibility studies, Defense Department officials were still not convinced of the utility of manned military space operations. As McNamara emphasized at the press conference announcing the MOL decision:

> This is an experimental program, not related to a specific military mission. I have said many times in the past that the potential requirements for manned operations in space are not clear. But that, despite the fact that they are not clear, we will undertake a carefully controlled program of developing the techniques which would be required were we to ever suddenly be confronted with military missions in space.[14]

Also, during the FY 1965 budget hearings, Harold Brown stated:

> The problems of manned military space flights are, and generally will continue to be, more complex and more difficult and expensive to solve. I want strongly to emphasize that as of this time even the requirement for manned military operations is still in question.[15]

Both McNamara's and Brown's statements suggest that the decision to begin studies of MOL had as much to do with placating Air Force feelings over the cancellation of Dynasoar as with the belief that MOL offered additional advantages to unmanned systems.

Despite DoD scepticism, funding of MOL continued and expectations grew within the Air Force that they had at last achieved their goal of a manned military space programme. Finally, after numerous studies of the potential uses of the MOL during 1964, presidential approval was given to proceed with its development on 25 August 1965. Strategic reconnaissance by means of a huge 90-inch telescope was designated as its primary mission, but other possible uses such as satellite inspection and destruction were also canvassed by the Air Force.[16]

Johnson was wary of how the Soviets would perceive the MOL decision since they had earlier protested against the military use of the Gemini V flight. In a speech soon after his decision to allow the development of MOL, Johnson reiterated the US interest in extending the rule of law into space. In an implicit reference to the 1963 UN resolution he stated:

> We intend to live up to our agreement not to orbit weapons of mass destruction and we will continue to hold out to all nations, including the Soviet Union, the hand of cooperation in the exciting years of space exploration which lie ahead for all of us.[17]

Johnson also announced that top-ranking Soviet scientists were being invited to witness the next Gemini launch in October 1965 as a confidence-building measure. However, Johnson's assurances did not assuage the Soviet critics, who continued their condemnatory remarks about the hostile and aggressive intentions of the US MOL programme.[18]

Although some members of the US intelligence community apparently felt that a manned reconnaissance system would appear too aggressive to the Soviet Union and possibly invoke direct countermeasures, Soviet opposition to the MOL eventually ceased.[19] In September 1965 the Soviet representatives to the meeting of the Legal Subcommittee of the UN Committee on the Peaceful Uses of Outer Space made no attempt to denounce the MOL. Indeed, the Soviet Union announced soon afterwards that it too would develop a manned space station.[20] The *modus vivendi* seemed to be holding up.

Support continued for the MOL programme for the remainder of the Johnson administration. By 1968, however, there were increasing signs that the programme would be cancelled. The level of funding was not what the Air Force had either anticipated or requested as being necessary to keep the programme on schedule. As a result, the development programme had been progressively "stretched" to accommodate the fiscal constraints, by which time whatever additional benefits the MOL was supposed to offer had ceased to be attractive. It was left to the Nixon administration to deliver the *coup de grâce.*

The Soviet FOBS Threat

Speculation that the Soviet Union was developing a "space weapon" received further encouragement after a new Soviet three-stage ICBM was described as an "orbital" rocket in the May Day parade of 1965. Designated the SS-10 "Scrag" by Western observers, it was paraded again in November with the same description. As this would have technically contravened the 1963 UN resolution, Soviet officials were privately questioned on their definition of "orbital" during November 1965. The United States was satisfied with the Soviet reply that "orbital" in this context meant nothing more than "intercontinental".[21]

However, suspicions were again raised by two Soviet space launches on 17 September and 2 November 1966, which went totally unacknowledged by *Tass*. They were both launched from Tyuratam at a new inclination—49.6° to the equator—and both subsequently broke up into pieces. By November 1967 a distinctive pattern had developed with a further nine launches of this type.[22] All flew at 49.6° to 50° inclination from Tyuratam and, although they were now publicly acknowledged as part of the Kosmos series, no orbital period was ever given in the Soviet announcements. This signalled quite clearly that the payload portion had been commanded down to earth *before* one complete orbit, although debris from the test continued to circle the earth for some additional hours.[23]

The purpose of these mysterious Soviet tests was finally clarified on 3 November 1967, when McNamara announced that the Soviet Union had developed what he termed as a "Fractional Orbital Bombardment System", or FOBS. This used an SS-9 "Scarp" booster to extend the normal ballistic flight of an ICBM warhead so that its trajectory through space constituted a partial orbit of the earth. The purpose was to approach the United States from the south—the least expected direction and therefore the least defended area. Because the FOBS tests did not amount to full revolutions of the earth and were unlikely to have contained nuclear warheads, McNamara stated that they did not violate the 1963 UN resolution barring weapons of mass destruction in orbit. Furthermore, he was at pains to state that the FOBS did not constitute a great threat to the United States, although he admitted that it would shorten the warning time of an attack against the Strategic Air Command (SAC) bomber bases.

The use of the term "Fractional Orbital Bombardment System" was unfortunate as it implied a space weapon system and therefore a space threat. Concern over the implications of FOBS was still reverberating by the time the FY 1969 hearings were held, as McNamara was again questioned on whether there was a Soviet space threat. He replied that:

> I do not believe that they will develop or introduce offensive weapons in space that will provide them with capabilities greater than they would have had had they utilized the same funds for expansion of their earth-based systems. The FOBS system is an illustration. They probably developed that in order to increase their capability against our soft land-based bomber system. But it is my personal opinion that if they had put those same funds into ICBM systems of high accuracy, and with certain other characteristics, they would have created a greater threat to our bombers and missiles than the FOBS system will prove to be.
>
> I think the same conclusion will apply to potential Soviet space-based offensive weapons systems within the next five to ten years.[24]

Despite these reassuring words, the administration still felt it necessary to respond to the existence of FOBS to avoid any future criticism. Particular emphasis was put on improving early warning facilities by Over the Horizon radar and increasing the radar coverage in the southern part of the United States. There was also speculation that the Nike-X (later Sentinel) ABM system, with its combination of long-range Spartan and short-range Sprint missiles, would also be configured for an anti-FOBS role. *Aviation Week* reported before the 3 November announcement that the: "Capability of intercepting a nuclear armed de-orbited satellite following a ballistic trajectory to its target is being cranked into the Army Nike-X anti-ballistic missile system on an urgent basis".[25] Later, in the FY 1969 budget hearings, the new Director of Defense Research and Engineering, John Foster, Jr, stated with respect to FOBS:

> We have not designed the Sentinel system with any particular emphasis on this possible Soviet weapon. Rather, its capability to intercept fractional or multiple orbit bombardment systems is a side product of the system's capabilities.[26]

NEGOTIATIONS LEADING TO THE 1967 OUTER SPACE TREATY

Within weeks of Johnson entering office, much of the uncertainty over the legal status of space activities was formally wrapped up on 13 December 1963 with UNGA Resolution 1962 (XVIII), "Declaration of Legal Principles Governing Activities of States in the Exploration and Use of Outer Space". This agreement and the earlier Resolution 1884 (XVIII) in turn provided the basis for the 1967 Outer Space Treaty or, as it is formally known, the "Treaty on Principles Governing the Activities of States in the Exploration and Use of Outer Space, Including the Moon and Other Celestial Bodies".

However, the original discussions leading to this treaty had a more limited objective. In June 1965 the State Department circulated a proposal for a "Treaty on the Exploration of Celestial Bodies" to interested agencies for their comments and approval.[27] As a result, on 23 September 1965 Ambassador Goldberg suggested in a speech during the 20th Session of the General Assembly that the United Nations begin work on a comprehensive treaty on the exploration of celestial bodies.[28] Preparations were already under way in the administration when Goldberg informed the UN First Committee that the United States planned to present a proposal for such a treaty.[29] This was circulated in draft form to the relevant departments to obtain their own views and suggestions. Predictably, the Joint Chiefs of Staff pleaded for caution in the negotiation of any treaty relating to outer space. A memorandum to the Secretary of Defense from the JCS stated:

> The Joint Chiefs of Staff believe that a serious disadvantage could arise in the future if the treaty resulted in an adverse influence on the conduct of the U.S. military space effort. The idea that the potential USSR threat in space had been reduced by this treaty could lead to a diminution of the U.S. military exploitation of space. Accordingly, the provisions of this treaty should not preclude the conduct of intelligence activities deemed essential to U.S. security.[30]

NASA and the State Department also added their own recommendations.[31]

By April 1966 the differences of opinion had been settled and a request to proceed with the completed proposal had been submitted to the President on behalf of the Departmental Secretaries. On 5 April Walt Rostow, the Special Assistant for National Security Affairs, urged Johnson in a memorandum to propose a "Celestial Body Treaty" to the Soviet Union as:

> The Secretary's recommendation has become urgent because there are signs that the Soviet Union may be planning to introduce its own treaty at an early date in order to preempt this subject. It would be to our advantage to act before they do.[32]

On 7 May President Johnson announced the essential elements of the US proposal of a treaty governing activities on the moon and celestial bodies. After the President's proposal had been circulated within the UN committees and submitted to the Soviet Union, Foreign Minister Gromyko informed the UN Secretary General U Thant of their own proposal for a space treaty. As a result, on 16 June both the Soviet Union and the United States submitted draft proposals. The Soviet draft was considerably wider in its coverage, being drawn mainly from the two 1963 UN resolutions outlined above. As both these resolutions had already been approved by the United States, the Soviet draft was perceived within the State Department as being "... designed as a serious basis for negotiation with the United States..."[33] Although the Soviet Union had proposed that negotiations take place within the UN General Assembly, the United States requested that the forum of debate be moved to the Legal Subcomittee of the UN Outer Space Committee. As a *quid pro quo*, the United States accepted the Soviet proposal as the basis for discussion.

Between 12 July and 4 August the Legal Committee met and agreed on nine articles of the draft treaty. They met again in September but failed to make any further progress. The principal points of contention were the following: access to facilities on celestial bodies, the reporting of space activities, equal conditions with respect to the provision of tracking facilities, and the use of military equipment and personnel in space exploration.[34] These issues were discussed at a meeting of the National Security Council on 15 September 1966.

Between 17 September and 7 December 1966, as a result of informal negotiations among members of the Legal Subcommittee, the contentious issues were finally settled. This enabled President Johnson to announce the successful conclusion of the negotiations on 8 December, which he described as being the "most important arms control development since the Limited Test Ban Treaty of 1963".[35] The General Assembly's First Committee endorsed the agreement on 17 December, followed by the whole General Assembly on the 19th. By 27 January 1967 the treaty was open for signature.

While Article III of the treaty states that "Parties to the Treaty shall carry on activities in the exploration and use of outer space ... in the interest of maintaining international peace and security", the most specific arms control clause is outlined in Article IV:

> States Parties to the Treaty undertake not to place in orbit around the earth any objects carrying nuclear weapons or any other kinds of weapons of mass destruction, install such weapons on celestial bodies, or station such weapons in outer space in any other manner.
>
> The Moon and other celestial bodies shall be used by all States Parties to the Treaty exclusively for peaceful purposes. The establishment of military bases, installations and fortifications, the testing of any type of weapons and the conduct of military maneuvers on celestial bodies shall be forbidden. . .[36]

Other more marginally relevant clauses are contained in Articles VIII and IX. Like Article III, Article VI outlines the spirit in which space activities are to be conducted and their position within international law. It states: "Ownership of objects launched into Outer Space . . . is not affected by their presence in outer space." Thus, technically any deliberate interference contravenes the UN Charter. Article IX concerns space activities that may be harmful to those of another state:

> If a State Party to the Treaty has reasons to believe that an activity or experiment planned by it or its nationals in outer space . . . would cause potentially harmful interference with the activities of other State Parties in the peaceful exploration and use of outer space, it shall undertake appropriate international consultations before proceeding with any such activities or experiment. A State Party may also "request consultation" if it believes that such an activity or experiment is about to happen.[37]

Although the treaty contains these important clauses, it was felt by many that the treaty represented nothing more than an agreement to desist from activities which neither side had any intention of doing anyway. Certainly the United States had no intention of procuring the types of weapons that the treaty banned, and equally it had no intention of widening the coverage of the agreement. One participant in the interdepartmental discussions prior to the completion of the US proposal recalls "great concern, over at Defense that we would not get anything into it that would inhibit them".[38] One example of this was the desire not to include "Fractional Orbital Bombardment Systems" in a treaty because, as another participant recalls, the "Defense Department did not wish to foreclose this option", even though they had decided in 1965, after several years of research, to discontinue their own studies.[39] There was also no attempt to broaden the treaty to cover antisatellite systems, for fear of reopening the whole debate over the legitimacy of military satellites.[40] The dust was only just settling from the last contest over the right to conduct "peaceful" military activities in space and nobody had any intention of renewing it.

The United States Senate approved the treaty by a vote of 88 to 0 on 25 April 1967, but this unanimous support did not reflect the extensive questioning that government officials experienced during the hearings. Despite assurances that the treaty could be adequately verified unilaterally, General Wheeler stated that the "Joint Chiefs of Staff remain concerned about the assured verification capability with regard to weapons in orbit".[41] However, he later explained that the JCS supported the treaty, and that if the Soviets did develop even a limited capability its significance would be minimal.

For someone who had taken such personal interest in the US space programme, President Johnson was particularly pleased with the achievement of the treaty. While he was probably aware that in terms of arms control it was not particularly meaningful, he did predict that: "Its significance will grow as our mastery of space grows and our children will remark the wisdom to a greater degree than the present state of our knowledge quite permits today".[42] In the light of the recent trend in the militarization of space, Johnson's comments may yet turn out to be remarkably prophetic.

CONCLUSION

The years spanned by the Johnson administration represented a period of consolidation for the US military space programme. After some initial "teething" troubles, the United States was now conducting regular and generally trouble-free operations in space. Furthermore, the opportunities provided by the introduction of new launch vehicles with greater lift capacity allowed new generations of more superior military satellites to enter service. In short, the unmanned space programme developed into a routine operation that provided a wide range of services to a wide range of users. As a consequence, the United States gradually became more dependent on its military space assets.

Although there were strong indications that the Soviet Union had accepted a *modus vivendi* with the United States in the military exploitation of space, the Johnson administration still felt it necessary to continue the development and deployment of the two antisatellite projects authorised by Kennedy. As both these systems remained limited in capability and without improved replacements, US decision-makers doubtless believed that, through self-restraint if not self-denial, the arms race could still be prevented from spreading to outer space.

Although this view was given further encouragement with the passage of the Outer Space Treaty in 1967, the United States deliberately avoided widening the provisions of the treaty beyond the 1963 United Nations resolution to retain its freedom of action and for fear of reopening the debate about the legitimacy of the military use of space. This was to prove particularly ironical, as in 1968 the Soviet Union began testing an antisatellite system which would later pose a threat more real than had ever been the case with orbital bombs.

6

US Antisatellite Research and Development, 1957–1970

We have now developed and tested, two systems with the ability to intercept and destroy armed satellites circling the earth in space. I can tell you today that these systems are operationally ready, that these systems are on the alert to protect this nation and to protect the free world.

President Johnson,
Speech on the Steps of the State Capitol,
Sacramento, California, 17 September 1964

EARLY US RESEARCH ON SPACE WEAPONRY

Men began thinking about disabling satellites well before there were any spacecraft to disable. Some of the early classified satellite reconnaissance studies, for example, apparently included assessments of possible antisatellite methods.[1] While this can hardly be called extensive research, it was carried out with the understanding that the United States might one day want to prevent an enemy from using satellites. Conversely, the United States also needed to know about possible countermeasures in order to deal with them if the need arose.

Despite the underlying policy goal of encouraging the "peaceful" exploitation of space, the amount of US space weapon-related research proliferated dramatically after Sputnik. In part sanctioned by directives such as NSC 5802/1 which called for this kind of research as an insurance policy against possible hostile activities in space, the services also carried out their own independent studies in the belief that they could both influence policy and also be in a favourable position *vis-à-vis* the other services if and when policy changed.

Although ARPA was given the formal responsibility to co-ordinate and control this research, the services soon found ways of circumventing the regulations. As Kistiakowsky notes in his diary, projects funded over $500,000 (in current dollars) had to receive clearance from the Director of Defense Research and Engineering (DDR&E). Thus the services "tended to chop up dubious larger projects into $499,000 pieces which did not have to be cleared by the DDR&E".[2] A considerable number of projects were funded in this manner.[3] ARPA, however, did run their own programme—Project Defender—which investigated both ballistic missile defence and satellite interception. In June 1958 ARPA requested the Air Force's Air Research and Development Command (ARDC) to let out two contracts for the "study of weapon systems to combat hostile satellites".[4] In FY 1959 total ARPA funding for "Missile and Satellite Identification and Kill" amounted to $10 million with a further $17 million requested for FY 1960.[5] ARPA was also supported in later research on BMD and ASAT weapons techniques by NASA.[6]

Although the vast majority of this research were "paper studies", it is worth highlighting those antisatellite and ASAT-related projects that went beyond this level of development.

The High Altitude Nuclear Test Programme

The effect of a nuclear explosion in space was of obvious interest to those wanting to utilize this method for disabling satellites and also to those concerned with defending against it. The first such test programme was Project Argus.[7] This began primarily as a scientific experiment initiated by PSAC to test a theory proposed by Nicholas Constantine Christofilos at the University of California's Radiation Laboratory. The theory was that electrons produced by a nuclear explosion within the earth's magnetic field would become trapped in that field. However, when members of PSAC proposed conducting a nuclear explosion in outer space to the NSC on 6 March 1958, it was also suggested that the results of this test could be militarily significant. For example, it was considered relevant to the Nike Zeus ABM programme because of the possible "blackout" effect of high-altitude explosions on radars and communication links.[8] The President gave his approval to the project on the understanding that a US satellite—Explorer IV—would be in the vicinity of the detonation to record the results.[9]

Three tests were carried out in September 1958 from the deck of the USS *Norden Sound,* at 45° South latitude in the Atlantic. Solid fuelled rockets were used to carry the nuclear devices with yields from one to two kilotons to an altitude ranging from 125 to 300 miles.[10] The results of these tests were reported to the President on 3 November 1958 including the observation that:

> A nuclear explosion in space produces three kinds of effects of military importance. The high energy radiation including particles from the explosion produces effects on space; the whirling high energy electrons generate radio noise; and the delayed radiation from the fission products can affect radio transmission.
>
> All these effects are matters of degree, depending on yield, location and geometrical considerations.

And later in the report:

> The effect in space itself is of importance to apparatus such as satellites and ballistic missiles exposed to this effect. The high energy electrons generate X-rays when they strike any material objects; these X-rays are very penetrating and can damage electronic equipment.[11]

Although no mention was made of the implications of this experiment for antisatellite research, the Argus results were used by ARPA in later studies on satellite interception.[12]

A further series of high-altitude tests—the Fishbowl series—was conducted above Johnston Island in the Pacific in the summer and autumn of 1962. All but one of the five tests in this series were in the kiloton range and in the upper atmosphere. However, on 9 July 1962 the first shot, known as Starfish Prime, was detonated at an altitude of 248 miles with a yield of 1.4 megatons.[13] Afterwards the public discussion of the results of this test concentrated almost solely on the earthbound effects of the explosion, namely the general blackout of communications in the Pacific area and even the failure of lighting and burglar alarms 800 kilometers away on Hawaii.[14] But an equally significant, though generally ignored, result was the unforeseen damage to at least three satellites by the trapped radiation from the explosion. "Permanent effects" to the solar cells of the British Ariel, US TRAAC, and Transit IV-B satellites were later reported in congressional testimony.[15] Thus even before the first US antisatellite missiles with nuclear warheads were deployed in 1963, US officials knew that such weapons would have a limited operational utility due to the inevitable collateral effects on other US satellites.[16]

Project Bold Orion

Project Bold Orion was primarily designed to demonstrate the feasibility of an air-launched ballistic missile, but in its final demonstration stage it was tested as a possible antisatellite weapon. The project, which was called USAF 7795, was initiated in March 1958 by the Martin Company of Baltimore and supported by ARDC at Wright Field. After a number of tests using a single-stage missile, permission was received from the Air Force to add a second stage to extend the missile's range. It was the last of the two-stage missile test vehicles (No. 12 in the total series) that was used for a satellite intercept demonstration.[17] The test was an attempt to intercept the Explorer VI satellite as it passed through its apogee above the Eastern Test Range, Cape Canaveral. The missile was launched from a B-47 aircraft on 13 October 1959 and, although poor tracking facilities meant that the satellite and the second-stage vehicle could not be simultaneously tracked, an examination of the flight data indicated that all the test objectives had been met. It was later reported in *Aviation Week* that the missile had come within four miles of the Explorer satellite.[18] Encouraged by the interest of the Air Force and the results of the test, Martin subsequently presented a number of antisatellite proposals to ARDC. But by then Air Force thinking had changed to favour ground-launched over air-launched ASAT systems and, moreover, the need for satellite inspection as a prelude to destruction. The responsibility for the Air Force's research in this area was also transferred to the Ballistic Missile Division (BMD) in Los Angeles. Martin Marietta continued its studies in response to the Air Force tender for a satellite interceptor but RCA was chosen as the successful bidder in what would be known as the SAINT programme (see below). Although the single Bold Orion satellite intercept test was in itself a relatively minor event, it is significant for being the very first antisatellite test conducted by any country.[19] Table 1 in Appendix II has a complete list of all known US antisatellite tests.

US Navy Antisatellite Projects

In addition to the Air Force, the Navy also carried out some of the earliest antisatellite research. For example, at congressional hearings held in March 1961, the Navy unveiled its Early Spring ASAT proposal. This was described as a "minimum energy missile" that would be:

> . . .launched vertically with just enough power to arrive at the altitude of the satellite at zero gravity. At that point it can hover and wait for the satellite to come and then by terminal guidance, seek out the satellite and kill it with some mechanism.[20]

Early Spring became an umbrella name for a variety of ASAT proposals that were put forward by the Navy between 1960 and 1964. Most of these proposals involved the use of the Polaris submarine-launched ballistic missile (SLBM) in the belief that the direct ascent intercept method was more economical than the Air Force's co-orbital SAINT project and that mobile submarine launch platforms provided greater flexibility when targeting satellites at differing orbital inclinations.

Although the Navy provided some of the funds, most of the research was company financed, principally by Raytheon and Ling-Temco-Vought (LTV). At Raytheon, research concentrated on matching a modified Sparrow missile with a Polaris booster, to the extent that the Sparrow missile was later successfully tested in a "space chamber" for the ASAT role. Further research was carried out to replace the Sparrow missile's radar guidance system with an infra-red device that would home in on the reflected sunlight from the satellite.[21] Raytheon apparently also requested permission to demonstrate the use of a camera in the missile's nose cone but this was refused.

The Navy's Polaris missile was reviewed on a number of occasions as a possible ASAT launcher, but without approval for further development. This was almost certainly due to the fact that Polaris missiles had only just entered service and the Department of Defense did not want to procure more or reduce the number of SLBMs in the US strategic inventory. Operational factors of command and control could also have militated against submarine launch platforms.

A related if not overlapping programme was Project Skipper. This proposal involved the use of modified Scout launchers from surface ships or submarines.[22] Its kill mechanism relied on the use of hyper velocity metal pellets or rods, but this too never went beyond the drawing board.

However, another Navy ASAT-related research programme, namely Project HiHo, succeeded where Skipper had failed. In 1962 the Navy began test launches of a Caleb rocket from a Phantom F4D fighter bomber to demonstrate the feasibility of launching satellites and space probes vertically from high flying aircraft. As with the earlier Air Force Bold Orion tests, a secondary objective was to see whether similar platforms could be used for launching ASAT weapons.[23] Two tests were carried out at the Pacific Missile Range in April and July 1962, with the second missile reaching an altitude of 1,000 miles.[24]

The Navy continued with some low-level exploratory research during the late 1960s and early 1970s but with little enthusiasm. The Navy's interest in antisatellite devices would resume in the mid-1970s with the need to counter the threat from Soviet ocean reconnaissance satellites. This, however, concentrated primarily on nondestructive countermeasures.

Laser Weapons

Some of the earliest work on space weapons-related LASER (Light Amplification by Stimulated Emission of Radiation) and MASER (Microwave Amplification by Stimulated Emission of Radiation) research began in the early 1960s.[25] Among this research was at least one study related to disabling satellites—Air Force Project Blackeye, supported by the Space Systems Division. This involved studying the use of high intensity optical maser beams to disable satellite sensors such as infra-red detectors.[26] Furthermore, in 1965 Arthur Kantrowitz, Director of the Avco Corporation, made probably the first proposal for the use of particle beam weapons for satellite defence.[27]

Funding for this Directed Energy Weapon (DEW) research, as it is generally known, remained at a low level throughout this period due to technical scepticism and the conservatism of the services. However, one Air Force officer, General Curtis LeMay, predicted as early as 28 March 1962 that "beam-directed energy weapons would be able to transmit energy across space with the speed of light and bring about the technological disarmament of nuclear weapons. . ." He warned:

> Whatever we do, the Soviets already have recognized the importance of these new developments and they are moving at full speed for a decisive capability in space. If they are successful, they can deny space to us.[28]

It was a speech that was echoed 15 years later by the almost identical statements of General George Keegan (see Chapter 9).

The projects described above did not involve large amounts of effort or money. From 1957 to 1968, however, the United States committed itself to a number of ASAT development projects that amounted to a considerable level of investment and which, as noted in the previous chapters, led to a deployed antisatellite capability.

MAJOR ANTISATELLITE PROJECTS

Project SAINT

The Satellite Inspector or SAINT project is generally recognized as being the first full-scale US effort to develop an antisatellite system. Although it was often called a "satellite interceptor" and the Air Force always harboured plans to convert it to a true antisatellite weapon, this accreditation is unfortunate as SAINT remained first and foremost an in-orbit *inspection* system.

The SAINT project originated with a study by ARDC of possible defence measures to combat hostile satellites which was carried out in 1956.[29] With the reorganization of the space programme after Sputnk, ARPA assumed responsibility for this study but kept ARDC as the project supervisor. Further interest in countering the hostile use of space by an adversary led ARPA on 11 June 1959 to award the Radio Corporation of America (RCA) a $600,000 study contract to examine satellite interception techniques. A DoD press release at the time stated that the RCA contract award followed a "detailed technical evaluation by ARPA of general interceptor studies performed both independently and under government sponsorship".[30] The satellite intercept study was carried out under the directorship of Dr Robert Seamans (later Secretary of the Air Force) at the RCA's Burlington Division, and was completed within six months.[31]

While RCA was still involved with this study, the Air Force Ballistic Missile Division proposed in August 1959 a preliminary development plan for a satellite interceptor and inspector under the programme acronym of SAINT.[32] This called for a programme to demonstrate the feasibility of a variety of different satellite interceptor configurations: unmanned and ground-launched, unmanned and air-launched, and manned. Since the Air Staff did not believe, for financial reasons, it could get OSD approval of all three approaches, it recommended that the proposal be reduced to demonstrating the feasibility of just a co-orbital vehicle possessing rendezvous and inspection capabilities. The other options, however, were not discarded entirely and continued to be studied by the Air Force. After reviewing the revised SAINT proposal, the Air Staff passed it on to the Secretary of the Air Force for approval at the beginning of 1960.

The apparent catalyst for the decision to proceed with SAINT was the "crisis" caused by the unidentified satellite that was detected in December 1959. According to a former official of ARPA, it created a series of embarrassing questions directed to the Department of

Defense from the White House on the nature and purpose of the presumed Soviet satellite.[33] They also wanted to know why, after being detected in December by US tracking facilities, the unknown satellite only became a cause for concern at the beginning of February. Kistiakowsky, as described in Chapter 3, was worried that this incident might be used by the services to support their case for an antisatellite weapon. As he states in his *Diary*, he was "afraid that unwise decisions would be taken", and requested that a special NSC meeting to discuss US policy on reconnaissance satellites originally scheduled for March be brought forward to 5 February.[34] Kistiakowsky's concern proved justified as Joseph Charyk, the Assistant Secretary of the Air Force for Research and Development, presented the Air Force's plans for a satellite interceptor at this meeting, albeit emphasising its *inspection* rather than its destructive potential.[35] The unknown satellite was discussed immediately afterwards and the possibility that it was a piece of rocket debris was raised for the first time. The President initially insisted that there be no public statement on the matter, but with the subsequent leaks to the press, his decision was rescinded in a later meeting on 10 February.[36] The uncertainty over this mysterious space object continued for some time until, finally, it was confirmed to be part of the second stage of the Discoverer V satellite launched in August 1959.[37] Although Kistiakowsky maintains that support was not forthcoming for a satellite inspector/interceptor at the 5 February NSC meeting, the incident over the unknown satellite almost certainly influenced the decision to proceed with SAINT.

On 5 April the Air Force applied through ARPA to the DDR&E for "an engineering feasibility demonstration program for a co-orbital satellite Inspector".[38] After consultations between Herbert York, the DDR&E, and Jack Ruina, his Assistant Director for Air Defense, the programme was given the go ahead under the following conditions.[39] In a memorandum to Charyk dated 16 June, York stated:

> We believe that this program should proceed at an orderly pace on a strictly research and development basis. It is also premature at this time to program for more than the co-orbiting experiments with simply payloads and the development (but not necessarily flight testing) of critical components of a prototype system.
>
> After examining the ARPA-Air Force relationship in the program proposed in the referenced memorandum, we believe that it is more appropriate for the Air Force to administer and finance the program on its own rather than jointly with ARPA. Funds for the program should come from the Air Force projects of lesser priority.

> The Air Force should submit a development plan for DDR&E approval which follows the guidelines in this memorandum.[40]

The fact that funds had to come out of their own budget was the first indication that the SAINT programme would run into serious budgetary difficulties.

The Air Force submitted a "Space System Abbreviated Development Plan" (dated 1 July 1960) in a memorandum for the DDR&E on 21 July 1960. The plan was reviewed and discussed formally with representatives from the Air Force, RCA and the Space Technology Laboratories (STL) of the Ramo-Woolridge Company.[41] Finally, on 25 August 1960, the Deputy Directory of DDR&E, John Rubel (Herbert York was absent due to illness) gave the final approval for Project SAINT to proceed. However, Rubel's memorandum expressed reservations about the Air Force's intentions to launch just four vehicles to meet the demonstration objectives and about the overall costing of the project, which he considered to be too low.[42] As a result, the Air Force was informed that it should be "prepared to program additional launches and support substantial increases in funding as the program progresses. . ."[43] But this was with "the understanding that any substantial change in the scope of the program will be presented to this office [DDR&E] for approval and that the funding required to implement the program can be made available within the resources presently available to the Air Force".[44] In other words, the Air Force could proceed with SAINT but DoD was not prepared to fund it.

The Air Force Ballistic Missile Divison (BMD) was given executive responsibility for the programme and later briefed defence contractors in early September on the requirements of the SAINT vehicle and payload. RCA was eventually chosen as the primary contractor for the inspection system, while Convair and Lockheed were selected to provide the boosters.[45] The newly created non-profit Aerospace Corporation was given the task of providing system engineering for the project.[46]

The basic configuration of the SAINT demonstration system was a first-stage Convair Atlas booster, topped by a Lockheed-Bell Agena B with the SAINT vehicle providing the third stage. The planned demonstrations called for the Agena B to place the SAINT vehicle in an orbital path slightly ahead of the target satellite (reported to be inflatable spheres), after which it would use its own propulsion unit to manoeuvre within 50 feet of the target for maximum inspection. The early inspection payloads would have included a TV camera and radar sensors, although more ambitious techniques using infra-red, X-ray and radiation equipment were considered for later models.[47] If it had become operational, the data obtained from an inspection would have been transmitted to ground readout stations, processed, and then displayed at NORAD. It was also hoped, at least within the Air Force, that later SAINT models would include a "kill mechanism".[48] As a result the Air Force studied a variety of ways to disable satellites from the SAINT vehicle, including the use of small spin-stabilized high-explosive rockets. Although Charyk issued a memorandum on 15 July 1960 directing the Air Force to delete all references to a "kill" capability, the possible use of the SAINT vehicle as an ASAT continued to be discussed in public.[49]

According to the project's milestone chart, SAINT would have become fully operational by the summer of 1967 at a total cost of $1,268.5 million. Nineteen vehicles were planned for the development programme, with the first "flight" demonstration cycle of four test interceptions due to begin in March 1963.[50] The fate of the SAINT project, however, was virtually sealed from its inception with the Defense Department's insistence that Air Force funds be used to support the development programme. As a result only $6.1 million was allocated for FY 1961. While sections of the Air Force were enthusiastic about SAINT (though never as great as the support for manned space projects), it was with the expectation that extra funding would be forthcoming from the Kennedy administration and, moreover, that it would become a true antisatellite system. Unfortunately for the supporters of SAINT, the original "minimum essential" request for $31 million in FY 1962 was pared down to $26 million, and it remained a satellite *inspection* system.[51]

Even with this objective, SAINT was always vulnerable to the criticism that it contradicted the administration's emphasis on the peaceful nature of the US military space programme. It was no doubt with these criticisms in mind and possibly in an effort to assuage White House and OSD concern at the possible international repercussions of satellite inspection that General Schriever stated with respect to SAINT's projected operations:

> I have in mind not only the fact that our use of them will be solely to deter aggression but also the fact that they are entirely passive in character. They carry no armament and represent no threat to any other nation.[52]

Criticism of SAINT also came from some unsuspected quarters, namely certain religious groups who protested against the irreverent title of the programme. In response to this, the Air Force changed the project's name briefly to Hawkeye but decided that this connoted "snooping" and "espionage", which was also unacceptable. With the clampdown on the release of information, SAINT became known as Program 621A (later changed to 720).

Throughout 1961 and early 1962 contracts were awarded to develop the basic components on the inspection vehicle. The Air Force also announced that it was studying a manned version of the satellite inspector which, as mentioned earlier, apparently angered Kennedy when details were released to the press.[53] However, by the autumn of 1962 it became clear that the SAINT programme's days were numbered.

The basic problems with the SAINT programme were primarily technical, conceptual and financial in nature. As Charles Sheldon, the late Congressional Research Service space expert recalled from a day-long briefing on the SAINT project given by RCA in late 1962 at the White House (probably to the Space Council):

> We saw an enormous array of technical problems that had not been solved—with brave talk on the part of the contractors but a long, long list of difficulties.

Furthermore, there were

> ... severe conceptual problems—for example: could you get away from inspecting another satellite without creating horrendous international problems?

Moreover,

> What could you learn?; you could take photos and learn the lengths of antennas; the emission of neutrons and whether the satelite was hot but would this mean anything? Our feeling was that you could learn more from the orbital characteristics of a satellite.[54]

The Air Force finally cancelled the SAINT project on 3 December 1962. Although its supporters tried to accelerate the programme by getting further DoD rather than Air Force funds injected into the project, the combination of the problems mentioned above overcame the Air Force desire to support the programme any further. The OSD accepted this decision without any hesitation, although this may have had more to do with a desire to avoid upsetting the peaceful image of US military activities in space than any technical or financial reason.

In the ensuing post-mortem further reasons were offered for SAINT's failure. Apart from the limited financial support, it was reported at the time that "failure of the USAF to clearly define program objectives, coupled with a lack of firm management both in the USAF's Space Systems Division and in RCA had contributed to its demise".[55] A later official history of the programme indicated that, despite SAINT's primary goal of satellite inspection, the Air Force continued to place emphasis on the expectation that it would become a true ASAT system, with the result that development work got "out of balance".[56] Furthermore, conflicting views between the Aerospace Corporation and the Air Force were reported to have complicated matters. The same report also noted that NORAD (North American Air Defense Command) was critical of the SAINT concept as they believed it could easily be overwhelmed by a large number of inexpensive decoys.[57]

A more cynical explanation of the willingness of the Air Force to cancel SAINT is that a successful (or even unsuccessful) test programme might have prejudiced its ultimate goal of a manned military space project.[58] This view is given some credence by the fact that the Air Force announced at the time of SAINT's cancellation that the satellite inspection mission would be transferred to the manned Blue Gemini programme.[59]

While this project was later cancelled, the Air Force did continue some satellite inspection studies, notably in co-operation with NASA's Gemini programme.[60] It is highly likely that some of the classified DoD experiments on board the subsequent Gemini flights were related to this research.[61] It is also possible that some of the sensor technology developed in the SAINT studies were later used in Program 437 (AP), which will be discussed below.

By the mid-1960s the requirement for orbital satellite inspection had been reduced by advances in electro-optical *ground*-based systems. These systems could get a great deal of information without the added expense and possible international repercussions of in-space inspection.

Program 505: Nike Zeus (MUDFLAP)

The US Army's proposal to convert the Nike Zeus missile to the ASAT role in November 1957 and later in January 1960 marked the beginning of an almost symbiotic relationship between ABM and ASAT research and development.[62] This was inevitable given the similar requirements and methods to detect, track and intercept both missiles and satellites. Moreover, the possession of exoatmospheric ABM missiles by definition provided a limited ASAT capability or certainly a system that could be transformed into one with relative ease. With this in mind, the Army resumed its efforts to get support for the Nike Zeus ASAT variant at the beginning of the Kennedy administration. At the March 1961 hearings following McNamara's edict on space research and development responsibilities, General Trudeau requested that the Army should be given the authority for the development of a "follow on Zeus that would produce an antisatellite weapon, unless something better comes up".[63] Later in the same month General Lewis, Director Special Weapons for the Army, repeated the idea that the Nike Zeus could have an ASAT capability with "little modification".[64] Cynics, however, viewed these requests as a disingenuous way to gain full funding for the operational development of the Nike Zeus ABM programme, which was then under threat of cancellation.

By May 1962, however, the Army had received permission to develop the Nike Zeus missile for the ASAT role, although as noted above it is uncertain whether this was a directive to go into advanced engineering or operational deployment.[65] One observer of the May 1962 decision recalls a "highly secret" delegation coming from Washington, DC to Bell Laboratories (the main contractor for Nike Zeus) to set up the Nike Zeus ASAT variant under the code name of MUDFLAP, later changed to Program 505.[66]

The Nike Zeus missile was a three-stage, solid fuelled rocket, 43 feet long, and five feet in diameter. Its range was approximately 100 miles, although this was extended for the ASAT role to over 150 miles.[67] Its warhead was nuclear—probably around 1 megaton in yield. Once detonated in close proximity to the target the radiation from the resultant fireball would have been the principal method of disablement, although the X- and *gamma*-rays from the explosion would also have produced electromagnetic pulses (EMP) that, depending on the distance and the amount of protective precautions, would also have disrupted or totally disabled the target satellite. Program 505's operational base was at Kwajalein Atoll—the western extension of the Pacific Missile Range (PMR). While this was the main site for the Army's continuing ABM test programme, its geographical position relatively close to the equator was also suited for the antisatellite role.

The first Nike Zeus missile modified for the antisatellite tests was fired at the White Sands Missile Range (WSMR) in New Mexico on 17 December 1962. This successfully intercepted a designated point in space at an altitude of 100 nautical miles.[68] The missile's altitude was extended to 151 nautical miles in a second test at WSMR on 15 February 1963. Testing then switched to Kwajalein Atoll, where a further two intercept attempts at simulated satellite targets were carried out on 21 March and 19 April 1963. Both, however, were unsuccessful due to equipment failure.

The first successful satellite intercept test from Kwajalein occurred on 24 May 1963, this time against a real satellite target.[69] An Agena-D upper-stage vehicle was specially equipped with radio and radar miss distance sensors which reported that a "close intercept" occurred. From May 1964 Program 505 officially became operational with one missile "always checked out and in a state of readiness".[70] The degree of readiness, however, was the cause of some debate. At a briefing for the Secretary of Defense on "Satellite Detection, Inspection and Negation" held on 27 June 1963, McNamara "stressed that the first priority for efforts in this area was to provide an immediate capability to destroy any postulated Soviet satellite threat". He emphasized his desires in this matter by stressing that he wanted it clearly understood that he wanted the "capability to initiate destruction of the satellite by a phone call".[71] Although it is uncertain what level of readiness was reached by Program 505, it was declared operational on 1 August 1963.

Details of the subsequent tests in this programme are fragmentary and include the following:

(a) A Nike Zeus fired 6 January 1964, successfully intercepted a simulated (tape) satellite with a hit proximity of 216 metres at a 93 n mile range and a 79 n mile altitude.[72]

(b) During 1 April-1 May 1965, one Nike Zeus missile was successfully fired from Kwajalein. Objectives were to test communication components aboard the missile, to demonstrate exoatmospheric beam intercept, and to evaluate exit altitude control.[73]

(c) During 1 June-31 July 1965, 4 Nike Zeus missiles were fired at Kwajalein, 3 of which were successful. Objectives were to evaluate function of missile warhead and adaption kit components in satellite intercept role.[74]

(d) On 13 January 1966, a Nike Zeus missile, modified for the antisatellite role, was fired from Kwajalein to demonstrate capability of the operational crew to successfully prepare and fire against a taped satellite target. All test objectives were achieved.[75]

It appears as if personnel from Bell Laboratories and McDonnell-Douglas conducted most of these tests, although the Army would have been responsible for its ultimate use. According to the Bell Lab's official history of the Nike Zeus project, the "ready" requirement for Program 505 was dropped after 1964. On 23 May 1966 the Army was informed by the JCS that McNamara had decided "to phase out Project 505 expeditiously".[76] By 1967 this had been completed.

The decision to downgrade and finally phase out the Nike Zeus system was clearly due to the existing capability of Program 437. At the same briefing in which McNamara expressed his system readiness requirements he also stated that "the Zeus follow-on [Nike X-Sentinel] program as proposed would be in direct competition and duplication [deleted] THOR capability and that he felt that there should be only one such system for follow-on operational use. He felt that the Air Force was, and rightly so, the prime agency in this area."[77] However, with the resurgence of interest in ABM systems in the late 1960s and the reluctance to procure a dedicated follow-on ASAT weapon, the possibility of using the long-range Spartan missiles in the Safeguard system as an ASAT interceptor was admitted by the then DDR&E John Foster. At the FY 1970 NASA budget hearings Foster stated: "The Safeguard ABM system when deployed beginning in early FY 1974, will have an anti-satellite capability against satellites passing within the field of fire of the deployed system."[78] In the following year, Foster again stated that Safeguard offered an antisatellite system "for far less money".[79] Had Safeguard been deployed with this dual capability it would have been somewhat ironic given that it was Program 437 that had replaced its predecessor—Nike Zeus. But with the cancellation of the Safeguard ABM project in 1972 the Army's involvement in antisatellite activities virtually ceased.

Program 437: Thor

By the time President Kennedy directed the Defense Department to develop an antisatellite system "at the earliest practicable time" on 8 May 1963, the Air Force was also well advanced with its plans to test some modified Thor missiles at Johnston Atoll for this purpose. Known as Program 437, this had originated from Advanced Development Objective 40 (ADO-40) that was issued by the Air Force on 9 February 1962. This called for "the demonstration of the technical feasibility of developing a nonorbital collision-course satellite interceptor system capable of destroying satellites in an early time period".[80] This would be ground-based or possibly air-launched but prime consideration would be given to using boosters already under development or in production. The timing of ADO-40 was probably the result of a combination of factors: first, an appreciation of the high-level interest in a "quick fix" ASAT system following the "scare" of December 1961; and, second, frustration over OSD's insistence that SAINT remain a satellite inspector despite the Air Force's campaign to add a satellite negation capability.[81]

Air Force Systems Command (AFSC) was directed to review the possible choices and prepare a development plan by 30 June 1962. In the meantime the Air Defense Command (ADC) also studied this question and on 1 May 1962 released its own operational requirement for an ASAT system to be ready no later than 1964.[82] By October 1962, AFSC had completed its studies and was ready to submit its recommendation to the DoD. This consisted of four demonstration tests of a modified Thor (LV-2D) booster to be conducted at Johnston Island in the spring of 1964. The choice of the Thor was almost certainly the result of the availability of surplus boosters after their use as IRBMs in Europe had ceased in 1962.[83] Following a briefing to McNamara on 20 November 1962, Harold Brown gave "tentative approval" to the Air Force proposal in a memorandum on 13 December. This contained the statement: "Some version of your proposal appears to be the fastest way of obtaining an increased capability in range and altitude beyond that which will be available from the NIKE ZEUS installation at Kwajalein in the spring of 1963."[84] While Brown voiced some doubts about the development programme, $6 million was programmed for it in the FY 1963 budget and $11 million for FY 1964. Further confirmation of the decision to proceed with the demonstration tests came in a memorandum from Harold Brown on 7 February 1963. This laid out in detail the antisatellite research and development responsibilities for the respective services and ARPA.[85]

On the following day the demonstration test series was officially designated Program 437 by Secretary of the Air Force Zuckert. Just over a week later, on 15 February, the Air Force was informed by DoD to prepare for an "operational standby capability following the intercept demonstration".[86] As a result, a new operational concept plan was drawn up and ready by 15 March 1963. This was reviewed by Zuckert on 20 March and gained his enthusiastic support. In a memorandum for the Air Force Chief of Staff he stated:

> I have just reviewed Program 437 and wish to reemphasize the fact that development of an operational capability to negate satellites has top priority among defense programs. Action must be taken, therefore, to insure that the

> necessary resources are allocated to AFSC to expedite conduct of the research and development demonstrations and the establishment of an emergency operational capability.[87]

The new operational plan called for a twin-pad launch complex at Johnston Island with the missiles and support facilities located at Vandenberg AFB, California. Three missiles would be maintained in a state of readiness for airlifting to Johnston Island if the need arose. Once there, two missiles would be counted down simultaneously, in case the primary missile failed after the order to launch had been given. The expected deployment time from alert to launch was not to exceed two weeks. The plan had been drawn up under strict guidelines to keep costs as low as possible—hence the idea of staging from Vandenberg. However, even before Kennedy issued the 8 May directive, there were already high-level indications that Program 437's operational effectiveness would not be sacrificed for financial reasons.[88] The Air Force hoped to get the reaction time down to two or three days, but even this dramatic reduction was insufficient for McNamara. At the 27 June briefing on the US ASAT programme mentioned above, he criticized the Air Force reaction time as being too slow. When McNamara was informed that this was based on the staging time from Vandenberg he said that "this was unacceptable and did not meet his stated requirement for operational capability".[89]

In response to this criticism, the Air Force Space Systems Division issued a revised operational plan in August 1963. This now envisaged the deployment of two-launch ready missiles with all the necessary support facilities *on* Johnston Island. The launch procedure, however, remained the same, with the dual countdown of both a primary and back-up missile. This proved to be the final configuration for the operational system.[90] As a result the reaction time was reduced to between 24 and 36 hours.[91] Any further reduction was constrained by the time taken for the target to be located, tracked, and the intercept calculations to be transmitted from NORAD, as well as the need to fuel up the Thor missiles prior to launch. A later report to the President stated, however, that the system was deployed on "a fully alert basis" and could intercept satellites twice a day.[92]

After some bureaucratic infighting within the Air Force, responsibility for the programme was shifted from AFSC to Air Defense Command in November 1963. Operational control was exercised by the Commander in Chief Continental Air Defense Command (CINCONAD), with launch responsibility assigned to the 10th Aerospace Defense Squadron. Although this had its headquarters at

Vandenberg AFB, Detachment 1 comprising two officers and 110 men was located at Johnston Island. The squadron had six missiles in its inventory, two in operational readiness at Johnston and four back-up and training boosters at Vandenberg.[93]

Before the tests could begin in 1964, considerable work had to be undertaken to prepare Johnston Island. This was both very small and, despite the earlier high-altitude test programme, the facilities were very basic. As a result, the FY 1963 and 1964 defence budgets brought a massive injection of funds to lengthen the existing runway and enlarge the overall island by 270 acres—the latter completed by dredging the nearby coral.[94] In addition to the two launch pads, a ground guidance station, a payload detonation system, launch control and targeting systems, payload and liquid oxygen (LOX) storage facilities, as well as a communication centre and living quarters, all had to be constructed.

The first of the initial four-test demonstration series was conducted on 14 February 1964 with the successful interception of a Transit 2A rocket body.[95] This was carried out by contractor crews, as were the next two tests on 2 March and 21 April, both of which were also successful. For the fourth launch, carried out on 28 May, the contractor crews supervised personnel from the 10th Air Defense Squadron. Although this proved to be a failure due to a technical malfunction, General Thatcher of ADC and General Funk from SSD who watched the test were satisfied and gave the unit an initial operational capability (IOC) of one ready missile on the following day. Full operational capability came on 10 June 1964, when both missiles were placed on alert.[96] These Thor missiles could intercept targets up to an altitude of 200 nautical miles and a horizontal reach of 1,500 nautical miles. Like th Army's ASAT system, the disablement of the target would have been performed by a nuclear warhead, the MK49, with a 1.5 megaton charge.[97] With the prevailing Test Ban Treaty this was never tested, and therefore estimates of the performance of the interception were provided by instrumentation in the nose cone of the missile and from the launch site. From 1964 to 1970 16 live firings were carried out at Johnston Island. Of these, however, only six were "combat test launches". The others were either variants of Program 437 or entirely different programmes that utilized the facilities at Johnston Island. These are listed and described overleaf.

Table 6.1. Program 437 and Related Launches from Johnston Island

No.	Date	Comments
1.	14 February 1964	Successful test intercept of a Transit 2-A rocket body
2.	2 March 1964	Success
3.	21 April 1964	Success
4.	28 May 1964	Test failed but system declared operational
5.	16 November 1964	Success. First Combat Test Launch (CTL)
6.	5 April 1965	Successful CTL. Came within 0.89 n. miles of inactive Transit 2-A satellite
7.	7 December 1965	Contractor Crew tests on Advanced Program 437, also referred to as 437 (AP)
8.	18 January 1966	
9.	12 March 1966	
10.	2 July 1966	
11.	30 March 1967	Successful Combat Evaluation Launch (CEL). Part of Exercise SQUANTO-TERROR
12.	15 May 1968	Successful CEL
13.	21 November 1968	Successful CEL
14.	28 March 1970	Success. Last CEL
15.	25 April 1970	Failure. Launch of THOR missile for Special Defense Program
16.	24 September 1970	Success. Launch of THOR missile for High Altitude Program

When ADC assumed responsibility for Program 437 it was with the understanding that three Combat Test Launches (CTLs) would be completed each year to maintain the readiness of the unit. However, even before the first official CTL launch of 16 November 1964, ADC learned that the Defense Department had cut the original procurement of missiles from sixteen to eight to last through FY 1967. As four missiles were required to maintain the system in operational readiness (two at Johnston and two at Vandenberg) only four week left for training. With a further test on 5 April 1965, only two were left. This understandably caused great concern within ADC, epitomised by an appeal from the commander of the 10th ADS, Colonel Minihan, that "437 must improve or die".[98] The problem was partially resolved in September 1965 when DoD authorized the procurement of sixteen additional test vehicles for CTL use from FY 1966 to FY 1971.[99] The

5 April CTL, however, proved to be the last until March 1967. In the meantime, one of the two launch pads at Johnston Island was handed over to contractor crews for work on "an advanced 437 test program".[100]

This was known initially as Program 437 X but the suffix was later changed to AP for Alternate Payload. The object of this programme was to develop a photographic satellite inspection capability. The idea of using Program 437 for this role was originally proposed by AFSC in conjunction with the General Electric Company.[101] This reached the attention of Brockway McMillan, now Under Secretary of the Air Force, who on 9 December 1963 requested the Chief of Staff to submit a development plan based on the concept.[102] McMillan reportedly stated in this request that "while the Program 437 antisatellite capability was important to have in hand, there was in fact a low probability of it being used. On the other hand, an inspector version filling a complementary role held promise of frequent use."[103] As McMillan was also Director of the National Reconnaissance Office (NRO), he was especially receptive to the idea of in-space inspection. As a former Air Force official recalled, Program 437 X was "a heritage of SAINT" as there was still:

> ... considerable interest within the "black community" to know what some of the space objects (Soviet satellites) were. The resolution on the camera would have been good enough to determine the size of antennas on the Russian "birds" (satellites) and possibly even the size of the optical aperture which would have been of interest to us.[104]

The AFSC plan was completed by 23 December 1963 which, on McMillan's suggestion to keep its cost to a minimum (and possibly as cover to the programme), proposed that Program 437 X should share the basic 437 facilities and launch vehicles. The DDR&E was briefed on the programme in March 1964 and again after the Air Force submitted a revised plan for four demonstration launches in April 1964. Permission was subsequently given for tests to begin in 1965. The new inspection payload vehicle was a modification of the General Electric MK2 re-entry vehicle with a Corona/Discoverer-type camera and recovery system. Stored programming data within the vehicle would activate the cameras and the subsequent ejection of the film. The exposed film would then be fed into the recovery canister and dropped by parachute to be recovered by C-130s operating from Hickman AFB, Hawaii. The system was expected to give five to seven photos of the target in daylight at altitudes between 70 and 420 miles.[105]

Details of the four tests that were carried out between 1965 and 1966 remain classified. Apparently these simulated "first time pass intercepts". The operational concept or "trick" was to survey Soviet launch sites for signs of activity and then to fuel up and "hold" the Thors on a 30-minute alert in readiness for the intercept.[106] It is unclear whether photos of Soviet satellites were ever taken, as one participant recalled that McNamara gave explicit instructions denying the use of the system in this way.[107] It is therefore uncertain whether Program 437 X (AP) ever became an operational alternative.

In the interim between CTL launches, ADC set about to improve the standard 437 programme. After considerable delays and technical problems, the ground radar and guidance system was improved.[108] In order to test the newly installed systems, CTL launches were resumed on 30 March 1967 as part of Exercise SQUANTO-TERROR, conducted by Continental Air Defense Command. A piece of space debris simulating an orbital bomb was intercepted with a miss distance of just over two nautical miles, "well within the lethal range of the weapon system".[109] With tests of a follow-on system known as Program 922 (see below for details) due to begin in 1969, ADC requested three more CTL launches before that date, and an additional five before the planned end of Program 437 in FY 1974.[110] However, ADC was not granted its wishes; Program 922 was cancelled and only three more CTL launches were held, two in 1968 and the final one on 28 March 1970.[111]

The last two test launches at Johnston Island were not strictly part of the ASAT programme but none the less provided useful experience to the launch crews. One was part of the Air Force's ballistic missile defence programme, and the other was designed to test US readiness to resume high-altitude nuclear tests. With the renewed interest in ballistic missile defence in the later 1960s, the Air Force set about to demonstrate its competence in this area in the hope that the Army's jurisdictional claim to the mission would be weakened. Known as the Special Defense Program (SDP), this was designed to carry out midcourse interception of re-entry vehicles by means of infra-red homing devices. Although SAMSO (the successor to Space Systems Division) had proposed a demonstration in the spring of 1968, the test did not occur until 25 April 1970. This used the launch facililities at Johnston Island and a modified Thor missile from Program 437. The test, however, proved to be a failure after the Thor booster collided with the payload after separation. This appears to have been the only test in the SDP. On 24 September 1970, the Air Force used another Thor missile as a test vehicle for its High Altitude

Program (HAP). This was part of the National Nuclear Test Readiness Plan designed to maintain preparedness for atmospheric nuclear testing if the Soviet Union abrogated the Limited Test Ban Treaty.[112]

While Program 437 remained operational during these two tests, a decision to phase down the system had already been taken. On 4 May 1970 Deputy Secretary of Defense David Packard informed the Secretary of the Air Force that: "In view of the unlikelihood that the U.S. would even use the *deleted* with its nuclear kill mechanism, the Air Force should phase down the system by the end of FY '70 or as soon thereafter as possible."[113] Accordingly, on 13 July 1970 the Secretary of the Air Force informed Melvin Laird, the new Secretary of Defense, that this would proceed, although ADC was not told of the decision until 30 July.[114] On 1 October, the system's operational reaction time was reduced to 30 days following the withdrawal of missiles and most of the launch personnel to Vandenberg. Practice redeployments were to be made semiannually in order to test the system short of firing the missiles.[115] Although Program 437 remained nominally operational until 1975, this was effectively the end of the system.

Originally conceived as a cheap "quick fix" ASAT weapon, Program 437 was not quite cheap enough when high-level interest in the antisatellite mission declined in the late 1960s. The system needed regular modifications and testing if its credibility as a weapon system was to be maintained. In comparison to other defence projects, this did not amount to great sums of money, but in the context of the perceived threat from space and the priorities of the war in Southeast Asia they were too much. While a semblance of operational readiness was maintained by irregular combat training launches, the system as a whole suffered from inherent structural weaknesses that undermined its credibility both as a deterrent and as a weapon. Once the "bombs in orbit" threat had died down, especially after the Outer Space Treaty had been signed in 1967, the utility of Program 437 in situations other than extreme emergencies was always in doubt. Its fixed site and the limited number of available missiles meant that it would have only been able to engage a small number of targets. Moreover, while its nuclear warhead was commensurate with the threat from nuclear orbital bombs, this was clearly "overkill" with any other target. More importantly, the use of warheads to disable Soviet satellites would have jeopardized the functioning of more valuable US satellites exposed to its effects and, as exhibited in the Starfish Prime test, produced widespread disruption of communications on earth. It was with these operational shortcomings in mind that the Air Force began from the outset of Program 437 to develop a more credible follow-on system.

"Follow-on" Projects to Program 437

Both ARPA and the Air Force had in fact investigated nonnuclear satellite interception methods well before President Kennedy authorized "nuclear and nonnuclear" ASAT development in his 8 May directive. In particular, due to the similarity of techniques there was a considerable overlap and cross-fertilization with the research on air and ballistic missile defence. As nonnuclear kill mechanisms such as steel pellets or rods, impact or proximity charges required more sophisticated guidance systems, much of this research was also aimed at developing "homing" or "sensor" technologies such as radar and infra-red.

On 7 February 1963, Harold Brown informed the Air Force that $10 million had been set aside in the FY 1964 budget for nonnuclear kill satellite interception research. This was to commence after ARPA had completed its own studies worth $12 million in FY 1963.[116] By the summer of 1963, Brown was urging that the Air Force consider using the Thor system on Johnston Island as the basis of its future research on advanced guidance and nonnuclear kill technologies.[117] Despite this, the Air Force's Program 437 follow-on study—designated Program 893—favoured the use of ICBMs, with the Minuteman II emerging as their eventual choice.[118] Before Program 893 could be submitted to OSD for approval, McMillan sent a memorandum to the Chief of Staff on 9 December, stating that for financial reasons "initiation of a *separate* follow-on anti-satellite program at this time is not likely" and therefore the Air Force should "orient" its research to utilizing Program 437 facilities.[119]

With this in mind, AFSC submitted a second development plan for a follow-on ASAT weapon—now known as Program 437 Y—on 11 May 1964.[120] While this emphasized, as the name suggests, use of the Thor booster, the plan called for a nonnuclear warhead that could be fitted to other missiles. By November 1964 the Air Force Space Systems Division was requesting study proposals for "an advanced direct-ascent intercept system employing terminal homing" and informal approval had been given by Brown for this project at the end of the year.[121] Later, in a remarkably candid statement the United States acknowledged that it was pursuing advanced ASAT research in the 1964 Presidential Report to Congress on US Aeronautics and Space Activities:

> The Air Force is currently conducting exploratory, advanced and engineering development effort, plus conceptual studies, to improve the present antisatellite system capability and to provide the design concept and technology base for follow-on systems.[122]

By the following March, the programme's title was changed to Program 922 and three contracts were let out to Hughes, Ling Temco Vought (LTV), and Northrop.[123] LTV was eventually chosen as the primary contractor in June 1967 with test launches scheduled to begin in 1969.[124] In total, $20 million was allocated for Program 922 in FY 1968, but even before testing could begin half of the money was diverted for the war in Southeast Asia. Finally OSD cancelled the entire project and recommended that the technology and remaining funds be transferred to the Army's ABM programme.[125] It seems likely, however, that some of this research was used in the Air Force's SDP programme mentioned earlier, as LTV was also its prime contractor.

Undeterred by the cancellation of Program 922, Air Defense Command drafted on 24 September 1968 a new, more general ASAT requirement document (ROC-18-68) and submitted it to HQ USAF.[126] At about the same time, SAMSO also prepared an Advanced Development Plan (ADP) for an "Improved 437".[127] By 1970, neither had been given permission to start development work. Conceptual studies did continue, however, and will be discussed in Chapter 10.

Project Dynasoar

The Dynasoar project (a compound of Dynamic-Soaring) consisted of the development of an experimental, manned hypersonic glide vehicle that would be boosted into space to "skip" or "bounce" off the upper atmosphere and then be directed back to earth to land at preselected sites. Various applications were proposed for an operational Dynasoar vehicle, including strategic reconnaissance, satellite inspection/interception and intercontinental bombardment.

The origins of the Dynasoar project extend back to the theoretical work conducted by Dr Eugene Sanger in wartime Germany on boost-glide vehicles with greater lift than drag properties. This research was taken up by his colleague, Dr Walter Dornberger, who later joined the Bell Corporation and worked on the development of a manned-bomber-missile known as BOMI between 1954-7. It was from work completed by the Bell Corporation on BOMI and similar projects—

Brass Bell and ROBO as well as the NACA work on project Hywards—that led directly to the Air Force Development Directive No. 94 and System Development Directive 464 in November 1957 to proceed with the Dynasoar programme.[128] Between 1957 and 1959 Boeing and Martin carried out weapon system definition studies with contracts worth $800,000 to each company. Their conclusions were that Dynasoar should in the first instance be an orbiting hypersonic test vehicle which, if developed further, would provide an operational platform for reconnaissance and bombardment.

After a three-month review of the programme, the Defense Department gave its approval in April 1960 to the first stage of a three-phase development scheme.[129] This consisted of: Step I—suborbital research; Step IIa—orbital military testing; Step IIb—interim operational capability; and Step III—full operational capability by 1966. Herbert York, then DDR&E, restricted the programme to Step I, as he believed there was as yet no specific requirement for Dynasoar and therefore it should remain a "contingency program" until its military utility could be demonstrated.[130] Contracts were subsequently let out for the glider, its launch vehicle and engines. By 1961 the design details of the glider vehicle were finalized, culminating in the presentation of a full-scale mock up in September.[131]

Dynasoar, however, increasingly fell foul of the cost-conscious scepticism of McNamara and the rest of OSD. After the House Appropriations Committee had voted an extra, though unrequested, $85.8 million for Dynasoar in the FY 1962 budget, McNamara resisted the attempt to accelerate the programme now known as X-20 and began to emphasize that it was just experimental research to test advanced unmanned re-entry techniques.[132]

On 18 January 1963 McNamara asked the Air Force to consider the possibility of cutting back on the X-20 project so as to become more involved in the Gemini programme. During the DoD external review of the feasibility and desirability of the X-20 carried out in the summer and autumn of 1963, the Air Force argued in vain that the operational characteristics of the X-20, particularly its flexibility in orbital manoeuvres, re-entry and landing warranted the continuation of the project. McNamara had already made his mind up and, as noted previously, cancelled the project on 10 December 1963.

In justifying his decision to Congress, McNamara stated afterwards that Dynasoar (X-20) had been a "narrowly defined program, limited primarily to developing the techniques of controlled reentry at a time when the broader question of 'Do we need to operate in earth orbit?' has not yet been answered".[133] It was partially to find an answer to this question that McNamara authorized the MOL programme based on the use of hardware developed for NASA's Gemini project.

THE US SPACE DETECTION AND TRACKING FACILITIES

A vital but often overlooked part of the US antisatellite programme was the creation of a space detection and tracking network. The ability to locate and identify objects in space to determine their hostile intent and then provide accurate and timely targeting information for their interception is an essential prerequisite to any antisatellite capability.[134] The importance of this facility was emphasized by McNamara when he stated to Congress in 1963:

> Although, as I have said earlier, attacks from enemy satellites is not a very likely threat for the immediate future, it is a possibility and we must develop the necessary techniques and equipment now so that we can quickly provide a defense if the need should ever arise. The first task is to be able to detect and track all objects in orbit.[135]

The space detection and tracking system that developed during the late 1950s and early 1960s evolved in a somewhat haphazard manner, primarily the result of inter-service rivalry and the fact that many of the facilities attached to the overall system served other functions, such as ballistic missile early warning. Preparations for tracking satellites began in 1955, in anticipation of the International Geophysical Year. Three global systems were initiated: Minitrack, a system of low-cost antennas working on the principle of interferometric measurement of satellite radio emissions, which was eventually handed over to NASA; the Baker Nunn Network, a more sophisticated tracking system relying on extremely accurate cameras to photograph the reflected sunlight on satellites, which was operated by the Smithsonian Astrophysical Observatory; and Moonwatch, a network of teams of volunteers with homemade telescopes and stopwatches.

Following Sputnik, all three services began developing tracking facilities and it was left to ARPA to co-ordinate them into an effective system. This was known as Project Shepherd, which had the objective of detecting "dark" or passive satellites as they passed over the United States.[136] Three systems were involved initially: the Navy's SPASUR (Space Surveillance) System; the Air Force's Interim National Space Surveillance Control Center (INSSCC), otherwise known as SPACE

TRACK, located in Bedford, Massachusetts; and the Army's DOPLOC radars. Although the Army's DOPLOC programme was eventually terminated, SPASUR and SPACE TRACK were expanded considerably and became the backbone of the Space Detection and Tracking System (SPADATS), controlled by NORAD from its headquarters within Cheyenne Mountain, Colorado.

By 1969, the SPADATS network included the following facilities:

(i) *SPASUR* Nine field stations strung out across the United States consisting of: transmitter sites at Gila River, Arizona; Jordan Lake, Alabama (both 50kw); and one megawatt station at Kickapool, Texas; two high-altitude receivers at Elephant Butte, New Mexico; and Hawkinsville, Georgia; and four low-altitude receivers at San Diego, California; Red River, Arkansas; Silver Lake, Missouri; and Ft. Stewart, Georgia.

(ii) *SPACE TRACK (496L)*

Radars

(a) AN/FPS-85 Phased Array Radar, Eglin AFB, Florida

(b) AN/FPS-17 Detection Radar "Fans" at Diyarbakir, Turkey; Shemya, Alaska.

(c) AN/FPS-50 BMEWS Detection Radar "Fans", at Thule, Greenland; Clear, Alaska; and Fylingdales, England.

(d) AN/FPS-49 Tracking Radar at Fylingdales, England; Thule, Greenland.

(e) AN/FPS-79 Radar Tracker, Diyarbakir, Turkey.

(f) AN/GPS-10 Radar Tracker, Ko Kha, Thailand.

(g) AN/FPS-92 and AN/FPS-99 Tracking Radar at Clear, Alaska.

(h) AN/FPS-80 Shemya.

Baker-Nunn Cameras

(a) Operated by the US Air Force: San Vito, Italy; Mt John, New Zealand; Edwards AFB; Sand Island, Kwajalein Atoll; Pulmosan, South Korea.

(b) Operated by Canadian Air Force: Cold Lake, Alberta; Margarets, New Brunswick.

(iii) *Co-operating Sensors*

(a) Eastern Test Range Tracking Radars at Ascension Island, Merritt Island, Patrick AFB, Grand Turk Island, Antigua, Grand Bahamas.

(b) Smithsonian Astrophysics Observatory (SAO) Baker Nunn Cameras at Olifants, South Africa; Dionysos, Greece; Dodaira, Japan; Arequipa, Peru; Island Lagoon, Australia; Debr

Zeit, Ethiopia; Maui, Hawaii; Commodoro, Argentina.

(c) Space and Missile Test Center Tracking Radars at Canton Island, Kaena Point, Hawaii.

(d) Pacific Missile Range Tracking Radars at Point Magu, California; San Nicholas Island.

(iv) *Other Sensors*

(a) Tracking Radars at Kwajalein Atoll; Malvern, England; Millstone Hill, Massachusetts; White Sand, New Mexico; Bermuda; Woomera, Australia; Wallops Island, Virginia.

(b) 440L Over the Horizon (OTH) Tracking Radars.

(c) AN/FPS-2 Electro-optical sensor sites at Cloudcroft, New Mexico; and Maui, Hawaii (see below).

(d) USAF Satellite Control Facility, Sunnyvale.

(e) NASA Tracking Networks and the Worldwide Satellite Observation Network (amateur groups co-ordinated through the SAO).

Sources:

(i) NORAD *SPADATS Sensor Manual*, vol. 1 no. 55-13 (15 December 1971 and revised 5 May 1975).

(ii) DMS Market Intelligence Report *SPACETRACK* no. 496L (January 1970).

(iii) DMS Market Intelligence Report *SPASUR* (April 1969).

(iv) Wilkes, "Space Tracking and Space Warfare".

Space Object Identification

Apart from the detection and tracking of space objects, another key element to antisatellite missions is target identification. As noted above, it was largely advances in this area that undermined the role of orbital satellite inspection systems like SAINT. A considerable amount of information on the purpose or intent of a space object can be determined from its orbital elements, as particular categories of military satellites have distinctive orbital characteristics. However, other systems were developed to "interrogate" satellites in an "active" or "passive" manner. These comprise the following:[137]

Electro-optical Sensors. In the mid-1960s the Air Force developed a Prototype Optical Surveillance Station (POSS) using the RCA AN/FPS-2 "optical radar". This "uses electronic detectors rather than photographic film to record the light collected by telescope to yield an 'optical signature' rather than an image of a particular satellite".[138] It is based at Cloudcroft, New Mexico, with a further improved system at the summit of Halesakala Mountain, Maui, Hawaii. Both these sites also possess the means to photograph satellites with the AN/FSR-2 electro-optical trackers.[139] A further system using lasers to illuminate the optics on satellites known as LARIAT for Laser Radar Intelligence Technology was also developed at Cloudcroft and Maui.

Interception of Telemetry. Another method of determining the mission of a satellite is through the interception of its telemetry signals. This is used with remarkable success by "amateur" groups such as the Kettering Boys School in England, who have provided consistently accurate information on the Soviet space programme. All three of the US armed services, as well as the CIA and National Security Agency, intercept the telemetry from Soviet launchings and satellites in orbit.

Radar Analysis. Apart from space tracking, sensitive radar systems such as the AN/FPS-85 phased-array radar at Eglin AFB can also determine with considerable accuracy the size, shape and general configuration of objects in space. A purpose-built system, however, was developed by the Air Force at Hollman AFB, New Mexico. Known as RATSCAT (Radar Target Scatter), it was completed in July 1964 to obtain radar "signatures" of a variety of objects, including satellites.[140] This provided further information to the catalogue of identified objects that are stored and monitored by NORAD.

7

The New Soviet Space Challenge, 1968–1977

The rapid development of spacecraft and specifically of artificial earth satellites, which can be launched for the most diverse purposes, even as vehicles for nuclear weapons, has put a new problem on the agenda, that of defense against space devices—PKO. It is still early to predict what line will be taken in the solution of this problem, but as surely as an offensive one is created, a defensive one will be too.

Marshal V. D. Sokolovskiy, 1963

Introduction

In 1968 the Soviet Union began testing a satellite interceptor, or "killer satellite" as it is more commonly known, against targets in space. Although Western observers were initially unsure of the purpose of these tests, by 1970 there was little doubt that the Soviet Union had developed an antisatellite system. The tests did not appear to cause undue alarm in the United States most probably because the Soviet Union had recently signed the Outer Space Treaty and attention was anyway diverted by the war in Vietnam. Moreover, the Soviet ASAT tests abruptly ceased in 1971, apparently in response to the new climate of détente between the superpowers.

The satellite interceptor programme was just one of a number of new Soviet military space programmes that emerged in the late 1960s. Apart from the ASAT tests, the two most important aspects of the expanding Soviet space programme that received the most Western attention were the use made of photoreconnaissance satellites during the major international conflicts of this period and the surveillance of Western naval exercises by a new class of reconnaissance satellites.

Throughout the 1970s, the expanding military use of space by the Soviet Union generated fears of a different kind to those experienced by US officials in the previous decade. Whereas before the fear had been of a *direct* attack on US satellites or on the US homeland, Soviet satellites provided an *indirect* threat to US military forces on earth by enhancing the Soviet Union's overall war-fighting potential. This concern over Soviet space activities was compounded by the sudden resumption of ASAT testing in 1976 after a four-year hiatus. Furthermore, allegations of Soviet advances in, and even use of, direct energy weapons such as lasers, added another dimension to the US debate about the Soviet space "threat" during the latter part of the 1970s.

This chapter on the Soviet military space programme acts as a backcloth to the subsequent chapters on US military space policy during the 1970s. It is divided into four sections: the origins and development of the Soviet ASAT programme to 1971; the expansion of the Soviet military space programme during the 1970s; the resumption of satellite interceptor tests in 1976 and the debate over Soviet DEW activities; and lastly, an assessment of the possible motives behind the Soviet ASAT effort.

The First Phase of Satellite Interceptor Tests, 1968–1971

Although the first *full* and unambiguous test of a Soviet satellite interceptor against a target in space did not take place until October 1968, the origins of this programme can be traced back to well before that date. One can firstly identify the necessary "building block" exercises for satellite interception, namely satellite manoeuvres and rendezvous, in earlier Soviet space activities.[1] Secondly, component parts of the interceptor system were also experimented with prior to the first full test.

The basic launch vehicle of the satellite interceptor is known to be the SS-9 "Scarp" (using the NATO nomenclature), or F-1 in the designation introduced by Charles Sheldon of the US Congressional Research Service. The F-1 vehicle had been first used for the FOBS tests beginning in 1966 with a fourth stage retrorocket (hence the F-1-r designation) to keep the flight suborbital. The first use of the manoueuvring, multiple-burn fourth stage system (F-1-m) came in October 1967 with the launch of Kosmos 185. This was launched on October 27th into a slightly eccentric low orbit and then manoeuvred upward to a higher orbit. No purpose was announced for the flight. A more puzzling test of the F-1-m vehicle came with the launch of Kosmos 217 on April 27th of the following year. *Tass* announced its orbital parameters as being similar to those of Kosmos 185, but Western sensors tracked only debris at a much lower orbit. This suggested either a failure to attain a higher orbit or the possible test of the FOBS system, which had exhibited a similar flight pattern and resultant debris. The probable intention of these tests only became apparent with hindsight after the first full Soviet interceptor test six months later.

On 19 October 1968 Kosmos 248 was put into a similar orbit as that announced for, but not achieved by, Kosmos 217. On the following day, Kosmos 249 was launched into a highly eccentric orbit but one that matched both the orbital plane and apogee of Kosmos 248. In fact, within four hours of the launch a close high-speed "flyby" occurred. What was even more significant, however, was that Kosmos 249 was detonated and destroyed after the flypast. The *Tass* announcement stated:

> In addition to scientific instruments, the satellite carries a radio system for the exact measurement of the orbital elements and a radiotelemetric system for transmitting instrument readings back to earth. The scientific investigations under the programme have been carried out.[2]

Although this was not the first time that Soviet satellites had been exploded in orbit—presumably to prevent their eventual orbital decay and possible retrieval by the United States—its coincidence with the interception of another satellite was enough for Geoffrey Perry of the Kettering Group to write on October 22nd to the London Bureau Chief of the journal *Aviation Week & Space Technology*, with a note stating that the "latest launches in the Cosmos programme suggest the initiation of a new type of activity".[3]

Perry's suspicions were confirmed following the launch of Kosmos 252 on November 1st. It executed almost identical manoeuvres to that performed by Kosmos 249, and exploded after passing close to Kosmos 248. No damage was done to Kosmos 248 in either of the two experiments. Again, *Tass* announced that the aims of the launch had been fulfilled. As one congressional report stated: "Because the orbits of Kosmos 249 and 252 were so placed that they would have lasted many years, the prompt announcement of program completion was a pretty good indication that the explosions coming as they did in a pair were planned."[4]

A two-year hiatus followed between the use of Kosmos 252 and the next clearly identifiable ASAT test, although Kosmos 291, launched on 6 August 1969, exhibited similar characteristics to a target satellite but no interceptor appeared. In the meantime, speculation ripened as to the purpose of the new Soviet space activities. This almost invariably emphasized either satellite inspection or satellite interception and destruction.[5] In the first interpretation, the subsequent detonation was explained as a precaution against US retrieval. However, on 16 February 1970 *Newsweek* reported that "Pentagon sources" had confirmed that the "Russians have successfully tested a hunter-killer satellite that can seek out and destroy other orbiting spacecraft".[6]

On 20 October 1970 Kosmos 373 was launched from Tyuratam into an eccentric orbit with a high apogee. This was later modified on subsequent revolutions to circularize its orbital path in a similar fashion to the first target satellite—Kosmos 248. Three days later Kosmos 374 was launched from Tyuratam. Again, this interceptor satellite was manoeuvred to match the average orbital altitude of Kosmos 373 at its perigee and a high-speed fly pass was completed. This was followed by yet another detonation of the interceptor. Seven days later the whole exercise was repeated using the same target satellite, but with a new interceptor (Kosmos 375).

While the nature of the tests no longer left any room to doubt the basic objective of the programme, there was still some speculation over the function of the target satellite. In particular, were the targets—Kosmos 248 and Kosmos 373—the source of the destruction of the four interceptor satellites, thus suggesting a defensive mechanism, or did the interceptors carry the "kill mechanism", which would suggest an offensive system? As Perry observed at the time, the target presumably had some "active role", as it would not have been necessary to launch a second almost identical target satellite since Kosmos 248 is still circling the earth today.[7] However, it is now officially acknowledged by the US that the satellite interceptor inflicts the damage. It is likely, however, that the target satellite contains instrumentation either to measure the "miss distance" or calibrate the effects of the explosion, or both.

Just as a pattern to these satellite interception tests was becoming clear, the subsequent experiments in 1971 differed in a number of distinct ways. The target satellites began to be launched from Plesetsk by an SS-5 "Skean" or C-1 booster and at a brand-new inclination—65.8 degrees. The first of the "new" series began with the launch of the target satellite—Kosmos 394—on 9th February to be followed 16 days later by launch of Kosmos 397—the satellite interceptor—from Tyuratam. After starting in low orbit, the interceptor manoeuvred to a higher altitude to complete a similar perigee-matching exercise with

Kosmos 394. It then exploded after the intercept. However, unlike previous tests, the target satellite was not used again for a second interception. In fact, all subsequent target satellites that have been involved in a test judged to be successful have not been used again.

Three weeks later on 19 March 1971, a new target vehicle—Kosmos 400—was launched into an approximately 1,000 km circular orbit, again using a C-1 vehicle from Plesetsk. Likewise 16 days after this launch an interceptor satellite—Kosmos 404—was launched from Tyuratam atop an F-1-m vehicle. This manoeuvred into a circular orbit similar to Kosmos 400. As Perry notes, at the start of its second revolution Kosmos 404 was less than three minutes ahead of its target, and by the end of its third, only one minute behind. With their orbital elements and therefore their orbital velocities so similar, a much slower flyby would have been achieved, suggesting to Perry that the mission may have been to test inspection equipment.[8] Furthermore, no explosion accompanied this test and Kosmos 404 was deorbited back to earth.

A further variation in the test programme was exhibited later in the year. On 29 November 1971 Kosmos 459 was launched from Plesetsk, but into the lowest orbit ever flown by the C-1 launcher—a roughly circular 250 km orbit at a 65.8° inclination. Later, on 3 December, Kosmos 462 was launched from Tyuratam into an eccentric orbit and completed the now-familiar high-speed interception at their respective perigees. This time the interceptor craft exploded into fragments after the flyby. The interception occurred in darkness within direct line of sight of Plesetsk. In a quite remarkable observation, the explosion was seen in Sweden as resembling a large flare that lasted for about 20 seconds.[9] This particular interception was also more demanding as it occurred at a lower altitude (similar to photoreconnaissance satellites), which due to the higher drag of the earth complicates the prediction of the speed and likely position of the target vehicle.

This was the last apparent test until they were resumed in 1976. However, two satellites on C-1 vehicles were later launched from Plesetsk with similar orbital parameters to the earlier target satellites but, like Kosmos 291, no interceptor launches from Tyuratam were observed. These were Kosmos 521 on 29 September 1972 and Kosmos 752 on 24 July 1975.[10]

During this first phase of testing, the Soviet Union had demonstrated a rudimentary but nevertheless significant antisatellite capability that threatens an important category of US satellites. As Perry states:

"Within a period of eleven months the Russians had demonstrated their ability to place hunter spacecraft in the vicinity of targets with orbits characteristic of [US] electronic ferrets, meteorological and navigation satellites and photo-reconnaissance payloads."[11] Moreover, of the seven clearly identifiable tests during this phase of testing, five were judged to have been successful.[12]

THE EXPANDING SOVIET MILITARY SPACE PROGRAMME

Between the cessation of Soviet satellite interceptor tests in 1971 and their resumption in 1976, US attention became increasingly focused on the growing use of Soviet photoreconnaissance satellites to monitor international conflicts, and in particular on the emergence of a Soviet ocean surveillance system to track US and NATO warships. Both indicated the probable active role of Soviet space systems in the event of hostilities involving the Soviet Union.

Although Soviet photoreconnaissance satellites had most probably been used to observe certain international events immediately after the inception of the programme in 1962, the first unambiguous case where they were launched to cover a specific crisis was exhibited in the invasion of Czechoslovakia in 1968 and especially during the border hostilities with China in 1969.[13] In the latter case, the Soviets clearly demonstrated a "surge" capability to launch and retrieve an extra number of satellites in response to the border clashes in March, June and August 1969.[14]

The third generation reconnaissance satellites, though tested in 1968, do not appear to have been used in either crisis. This new class of satellite provided greater flexibility in being manoeuvrable, having a longer duration in orbit (12-13 days), and most probably a better camera system. The first demonstration of these extra capabilities to monitor an international crisis came during the Indo-Pakistan War of 1971. The use of Kosmos 463 and Kosmos 464 (both third-generation reconnaissance satellites) was significant in that they were manoeuvred (by lowering the apogee or perigee) to reduce the westward drift of the satellites' ground track. This in turn increased the number of daylight passes over the conflict zone. Both were also recovered before the end of their normal operational lifetime.[15] This particular type of satellite was again used to cover the NATO naval exercise *Strong Express* in September 1972.[16]

The most intensive use of Soviet reconnaissance satellites occurred during the Middle East War in October 1973. In all, seven Kosmos reconnaissance satellites were launched between 3 October and 27 October, specifically to cover the hostilities. As was evident during the Indo-Pakistan conflict, some of these were manoeuvred to stabilize their ground track over the Sinai and Golan areas at critical moments during the war.[17]

In addition to this further demonstration of the Soviet Union's increasing capability to launch and manoeuvre satellites at short notice to meet specific requirements, the information gained from these satellites was also reported to have played a direct role in the progress of the war.[18] According to one report, it was only after President Sadat had been shown photos, taken by the Kosmos satellites, of the full extent of the Israeli counteroffensive and imminent encirclement of the Egyptian Third Army that he appreciated the gravity of the situation and requested Soviet military aid. However, President Sadat's own account of the war tells a different story.

> While the U.S. satellite hourly transmitted information to Israel, we received nothing at all from the Soviet satellite which followed up the fighting . . . Soviet satellites did keep a watch over the battle from the start, for Syria had informed the Soviet Union of Zero Hour. The recordings made were played to the CPSU Central Committee. I asked for a copy of that videotape [*sic*] but have received no reply to this day and won't receive any.[19]

A further example of this ability to manoeuvre reconnaissance satellites for specific purposes came in the following year when Kosmos 667 was directed to observe the conflict in Cyprus in July and August 1974. During North Vietnam's offensive against the South in March 1975, Soviet satellites were also launched in quick succession and specially manoeuvred to cover the fighting.[20] In short, by the mid-1970s the tailored use of reconnaissance satellites during the major international events of this period indicated the importance of space systems for Soviet crisis monitoring and most probably Soviet war-fighting as well.

At the same time, the Soviet Union also demonstrated the capability to detect and track surface ships from space. The origins of the Soviet ocean reconnaissance programme also go back to the late 1960s. The first Soviet test of a dedicated ocean reconnaissance satellite is accredited to Kosmos 198, launched on 27 December 1967. Kosmos 209 launched on 22 March 1968, is believed to have been the second. These satellites were identifiable by the launcher—the same used by the satellite interceptor—and the orbital manoeuvres at the end of its mission. The reason behind the shift from a relatively low circular orbit (approximately 270 km) to a much higher one (approximately 950 km) only became public knowledge following the Kosmos 954 incident in 1978 which is discussed below.

After further tests in 1971, the system appears to have become truly operational in 1972. The first official US indication of its existence came in 1973, in a heavily censored statement by Dr Peter Waterman, Acting Assistant Secretary of the Navy for Research and Development, to the Senate Armed Services Committee:

> The Soviets already have in operation an extensive (deleted) used primarily for ocean surveillance. They also maintain a fleet of long-range Bear aircraft for the means of location and attack guidance against U.S. naval forces. It is evident that they have a high priority development program for (deleted). A series of heavy manoeuvrable satellites has been launched culminating in (deleted).[21]

In September 1974 a more open statement was issued in the US Navy publication "Understanding Soviet Naval Development", prepared by the Director of Naval Intelligence. This stated:

> Increasingly, the Soviet Navy is employing advanced satellite reconnaissance systems.
>
> Recent naval-associated reconnaissance satellites have improved collection rates and processing capabilities, and include electronic intelligence satellites that can lock on to intercepted signals to provide target location information. Large area radar surveillance satellites have also been identified.[22]

The latter report acknowledged for the first time that these satellites used active radar systems (hence the subsequent use of the acronym RORSAT for Radar Ocean Reconnaissance Satellite), and also the use of ELINT, or more properly, EORSAT satellites for the same purpose. Beginning in 1974 these two space systems began to track in pairs, the RORSAT at the lower altitude of 250 km and the EORSAT much higher at about 440 km. The RORSAT's shift to higher orbit was subsequently explained by the fact that the radar system is nuclear-powered, which necessitates disposing of the generator in an orbit where it will remain while its radioactivity slowly decays.[23] The consequences of failing to execute this safety manoeuvre were dramatically illustrated when Kosmos 954 scattered radioactive debris over northern Canada in January 1978, and more recently with

Kosmos 1402 in January 1983. Although the coverage by the RORSATs and EORSATs has never been comprehensive in any one year, it is still considered sufficiently capable to concern US officials.[24]

In addition to these two uses of reconnaissance satellites, the early 1970s witnessed the introduction of new classes of Soviet military space systems and an increased launch rate. Navigation satellites began to be used for the first time in 1970, while the number of electronic reconnaissance, communication, and meteorological satellites launched during this period more than doubled (see Table 4 in Appendix II). Communication satellites also began to be used for tactical purposes, while geodetic satellites started to provide additional targeting data for Soviet ICBMs.[25]

The Resumption of ASAT Testing and the Debate over the Soviet Beam Weapon Programme

To the dismay and alarm of many US observers, the Soviet Union resumed testing of their satellite interceptor system in 1976. Although, as noted above, there may have been some aborted antisatellite tests in 1972 and 1975, the interception of Kosmos 803 by Kosmos 804 in February 1976 was the first unambiguous test since December 1971.

On 12 February 1976 a target satellite (Kosmos 803) was launched from Plesetsk on a C-1 vehicle into a slightly eccentric orbit at 66° inclination. Four days later Kosmos 804 was launched from Tyuratam into a more eccentric orbit at 65.1° inclination. After several manoeuvres, its orbit was circularized and its inclination changed to match that of Kosmos 803, with the eventual interception occurring over the Soviet Union.[26] Although the *Tass* announcement stated that all objectives had been reached, this test was seen by some US observers to have been a failure as the "miss distance" between the two satellites was put at 80 n. miles, with no apparent explosion. Furthermore, the chase vehicle was brought back to earth immediately after the "interception".

Kosmos 803 was used again as a target satellite on 13 April 1976 when Kosmos 814 was launched within four minutes of Kosmos 803 passing over the Tyuratam launch site. As a result of this and some post-launch manoeuvres, the interception took place within one orbit, but again no explosion followed the interception.[27]

Warning of another test came with the launch of Kosmos 839 on 8 July, but this time it went into a much higher orbit than the previous target satellites. Although, as expected, an interceptor satellite (Kosmos 843) followed on 21 July, it appeared to have failed to reach the requisite height and re-entered the atmosphere soon afterwards. It was therefore judged to have been another failure by US observers.[28] However, it is possible, as Perry suggests, that an interception could have taken place within one revolution, and without it being recorded by Western observers.[29]

Four months later, the Soviets reverted to the rapid flypast interception technique followed by the deliberate destruction of the chase vehicle. This was completed by Kosmos 880 (launched on 9 December 1976) and Kosmos 886 (launched on 27 December). At least four intercepts had been attempted in 1976 and, although some of these may have been failures, it none the less represented the highest number of tests in any one year up to that time. This latest test proved to be the last before the Carter administration took office in January 1977. A further four tests, beginning in May, were conducted in 1977. Table 2 in Appendix II has the complete list of known Soviet ASAT tests.

Further details of the Soviet satellite interceptor emerged during this period. Its dimensions are put at between 15 to 20 feet in length, 5 feet in diameter, with a weight of about 2.5 tons. It is also reported to have five main boosters for manoeuvring in space. Later reports also noted that the Soviets were experimenting with a new guidance system. Whereas earlier interceptors utilized a radar homing system, it was reported that a new optical infra-red sensor was used for the Kosmos 880/886 intercept on 27 December 1976, possibly in anticipation of US countermeasures.[30]

Following the April 1976 test, the journal *Aviation Week & Space Technology* gave additional details of the system's readiness. It reported that between 1972 and 1975 observations had been made of ground exercises in which:

> Soviet SS-9 boosters carrying killer satellite payloads have been wheeled from their hangers at Tyuratam, erected at a launch pad, fuelled and prepared for launch on a schedule that on actual missions would result in a launch in less than 90 minutes from the start of the booster transport to the pad.[31]

Later, a classified CIA report on the Soviet ASAT system, which had been leaked to the press, was quoted as stating:

> If the booster and interceptor are stored properly, the system can probably be fired within an hour of the decision to launch. If the booster and the interceptor are maintained on the launch pad, the firing would occur within 10 to 30 minutes of the decision to launch. We believe as many as 10 boosters and interceptors can be stored in the launch area.[32]

The same report also stated that "two unique [*sic*] tracking sites in Western USSR' at Tals and Kirzhach, in combination with the space tracking network controlled by the Strategic Rocket Forces, monitor the antisatellite tests. The interception by Kosmos 804 on 16 February 1976 is also linked in the CIA report to Soviet military exercises that took place between 29 January and 15 February. The day after Kosmos 804 was launched, these exercises practised strikes by naval and long-range aircraft culminating in the simulated "launch" of strategic ballistic missiles on 19 February. The report concluded:

> If the anti-satellite tests were part of these exercises, we can infer from the exercise scenarios that the Soviets were practising attacks on U.S. satellites after a period of localised warfare in Europe in which the Soviet or Warsaw Pact forces had gained the advantage, but before the war had reached the point of a strategic nuclear exchange.[33]

The impact of these latest tests on US perceptions of Soviet military intentions in space was reinforced further by the increasing reports of Soviet progress in the development of directed energy weapons such as lasers and particle beams. The United States had known for some time that the Soviet Union was conducting a large-scale research effort into the development of these technologies. In fact, Soviet scientists had been prominent in the early development of lasers and continued to play a significant role at international scientific symposia where beam technologies were discussed. However, it was uncertain to what degree this research was militarily oriented and, moreover, what stage it had reached.

Early Soviet writings, such as Sokolovskiy's book *Military Strategy*, had actually referred to the potential use of "death ray lasers" as "anti-space" weapons, but this was in the context of likely US activities. The journal *Aviation Week & Space Technology*, which was at the forefront of later allegations of Soviet DEW developments, reported in August 1973 that Pentagon officials were becoming suspicious of the apparent disappearance from the open Soviet literature of any reference to the scientists known to be working on high-energy lasers.[34] There were other reports from US scientists who had visited the Soviet Union and been impressed with the scope and extent of Soviet laser research.[35]

The most significant event that heightened US suspicions was the reported "blinding" of three US satellites in 1975. The first incident came on 18 October, when the infra-red sensors on an early warning satellite situated in geostationary orbit over the Indian Ocean became "illuminated" by an intense beam of radiation from a source in the western part of the Soviet Union. Five such incidents were reported, one of which lasted for more than four hours. Also, on 17 and 18 November two SDS data relay satellites apparently experienced similar interference. After this information had been leaked to the press, speculation grew as to the cause of these incidents. These ranged from the use of ground-based high-energy lasers and laser tracking or interrogation devices to natural phenomena such as forest fires and volcanoes. After a Department of Defense investigation into these incidents, the official explanation was that a large rupture and resultant fire along the trans-Siberian gas pipeline had affected the sensors aboard the US satellites. Any connection with the possible use of antisatellite laser weapons was denied. However, many US observers remained unconvinced by this explanation. In particular, they were sceptical of the ability of a gas pipeline fire to produce the kind of interference that was reported. Furthermore, the number of incidents over a period of weeks was also not adequately explained.[36]

Although public attention to Soviet DEW developments diminished after the incidents had been officially explained, this issue would resurface during the Carter administration with further allegations of Soviet beam weapon advances.

MOTIVES

Assessing the motives or determinants of Soviet weapons acquisition is a notoriously difficult exercise, due principally to the paucity of reliable evidence. While information on the *performance* of Soviet weapons systems is comparatively plentiful, particularly after tests have been monitored or equipment has been tried by third parties in battle, the amount of hard data on the *origins* and *development* of these weapons is minimal.[37] As a result, scholars of the weapons acquisition process have often concentrated on the most "rational" explanation of Soviet behaviour or just descriptions of the institutional basis to decision-making. In recent years, however, analysts have increasingly eschewed these often simplistic or unicausal explanations of Soviet political behaviour and instead have broadened their perspective to utilize more complex and pluralistic models of decision-making.[38] In many respects the study of Soviet decision-making has both reflected and been influenced by a similar trend in the analysis of Western political processes. However, while Western-oriented "tools" of political analysis may prove useful in understanding Soviet decision-making, it would be wrong to ignore the distinctly Soviet (or Russian) characteristics of their politico-military process and planning. It was partly with this in mind that Arthur Alexander listed six possible "explanations" of decision-making in Soviet weapons procurement in an attempt to accommodate the complex nature of the Soviet system. These are:

(1) Historically-based culture and values;
(2) Military-political doctrine;
(3) The "objective" situation (the "threat" and the capabilities available to meet the threat) and "rational" response;
(4) Organizational relationships, decision-making practices, and bureaucratic routines;
(5) Internal political power and accommodations;
(6) Personalities.[39]

The Soviet ASAT programme is a case *in extremis* of the problems that face the "interpreters" of Soviet military behaviour. As Charles Sheldon observed: "No series of flights offers greater variety, complexity or mystery."[40] While a sizable amount of information exists on the performance of the Soviet satellite interceptor system in the unclassified domain, very little hard information is publicly available on the motives underlying the programme. Thus one can only speculate from what is available and Alexander's lists provide a useful framework to organize the information.

Historically-Based Culture and Values

One of the most pervasive attributes of Soviet society is state secrecy. This, when combined with an intense concern for national security that at times borders on paranoia (which is largely explained by its historical experience with foreign invasion), has had a particularly potent effect on the choice of Soviet weaponry. Thus, when the United States began to reconnoitre Soviet territory with virtual impunity using a variety of aircraft and then with satellites, it is understandable that this was viewed as a serious challenge to state security. As noted earlier, it was no surprise when the Soviet Union developed the capacity to intercept U-2 reconnaissance aircraft. Neither was it surprising that they should later attempt to have "observation" satellites banned by international law. However, the decision to develop an antisatellite, or perhaps more correctly an anti-reconnaissance satellite system, is often viewed as inconsistent with the Soviet Union's later, albeit tacit, acceptance of reconnaissance from space. This, as noted in Chapter 4, was first signalled in 1963 by the withdrawal of the opposing clauses in the Soviet draft resolution to the UN Space Committee and later, though again implicitly, with the provisions respecting national technical means of verification in the SALT I agreement of 1972. Moreover, the Soviet Union began developing its own reconnaissance satellite programme, which likewise indicated that this was now internationally acceptable even if it was not to be publicly admitted.

Although the official Soviet attitude towards satellite reconnaissance had changed by the mid 1960s, this does not necessarily make the subsequent development of an antisatellite system either inconsistent or surprising. First, the Soviet Union probably draws a distinction between aerial overflight and reconnaissance from space—the former being an unacceptable intrusion during peacetime while the latter is permissible in practice but not necessarily inviolate under wartime conditions. Second, the Soviet Union may also perceive a difference between satellite reconnaissance for arms control verification and its use for military intelligence gathering. Also, the SALT I agreement only refers to noninterference with *US* satellites. Third, the Soviet Union is only too aware of the *range* of benefits that military satellites give to the United States' overall war-fighting ability. Therefore, reconnaissance satellites would be only one of the potential targets during wartime.

Military-Political Doctrine

With its public stand on the peaceful uses of space and the corollary of avoiding any reference to its military space programme, the Soviet Union has not provided a clear exposition of its military doctrine relating to operations in space. As Thomas Wolfe has observed: "The development of a coherent doctrine of space warfare seems to have been inhibited by the necessity to preserve a propaganda image of the

Soviet Union as a country interested solely in the exploration of space for peaceful purposes."[41] However, statements of the significance of military space activities in Soviet planning have emerged on a number of occasions, which are sufficiently coherent to suggest a doctrinal basis for the development and employment of antisatellite weapons. Whether doctrine was a response or stimulus to technological developments is almost impossible to judge from available information.

In a series of two articles in March 1967 in the Army newspaper *Red Star*, Lt. Col. V. Larionov stated that:

> ... the creation and employment of various space systems and apparatus can lead immediately to major strategic results. The working out of efficient means of striking from space and of combat with space weapons in combination with nuclear weapons places in the hands of the strategic leadership a new powerful means of affecting the military-economic potential and military might of the enemy.[42]

In a later article Larionov contrasted Soviet space activities with the apparent development of US military space systems and stated that the Soviet Union:

> ... cannot ignore all the preparations of the American imperialists and is forced to adopt corresponding measures in order to safeguard its security against an attack through outer space.[43]

The single most important source on the evolution of Soviet military thought—the three editions of Marshal Sokolovskiy's book *Military Strategy* (*Voyennaya Strategiya*)—also contains references to the military uses of space. After following the familiar pattern of detailing the US space threat and the latter's intention "to use space to accomplish their aggressive projects", Sokolovskiy states:

> In this regard Soviet military strategy takes into account the need for studying questions on the use of outer space and aerospace vehicles to strengthen the defences of the socialist countries. This must be done to ensure the safety of our country in the interest of all socialist co-operation for the preservation of peace in the world. It would be a mistake to allow the imperialist camp to achieve superiority in this field. We must oppose the imperialists with more effective means and methods for the use of space for defence purposes. Only in this way can we force them to renounce the use of space for a destructive and devastating war.[44]

One of the first indications that an operational doctrine or mission had been adopted by the Soviet armed forces to meet this requirement came in the 1965 edition of the *Dictionary of Basic Military Terms* with the entry of *Protivokosmicheskaya Oborona* and its translation as "anti-space defense". The entry states that:

> The main purpose of anti-space defense is to destroy space systems used by the enemy for military purposes in their orbits. The principal means of anti-space defense are special spacecraft and vehicles (e.g. satellite interceptors), which may be controlled either from the ground or by special crews.[45]

Western observers have subsequently identified a separate operational unit referred to as the PKO (which is the acronym of the Russian term anti-space defence) under the overall command of the National Air Defence Forces (Protivovozdushnaya Oborona (PVO)). This, however, is by no means clear from the Soviet sources they cite.[46] Furthermore, it is not clear whether this unit is still operational, if it did ever exist.

The "Objective" Threat and the "Rational" Response

Western commentators have looked to key events or specific US space programmes to understand the real motives for the development of the Soviet ASAT system. To assess whether the Soviet ASAT programme was tailored to meet a specific threat or requirement, the analyst must turn to the characteristics of the Soviet antisatellite testing programme. In particular, the following must be studied: the identification of the programme initiation date from estimates of the development lead time; the system's capabilities against objects in certain orbits; the characteristics of the intercept; and, moreover, the attention given to potential threats in the Soviet literature. However, given this system's lengthy research and development life cycle, from inception (probably in 1959-60) to the end of the first testing phase in 1971, followed by a second phase in 1976, "rational" analysis of this kind should also accept the possibility that the Soviet ASAT's initial *raison d'être* has evolved over time.

Soviet perceptions of the value of military satellites to the US war-fighting capability or, conversely, the threat of US space systems to Soviet military operations clearly provide adequate reasons to build an antisatellite system. Here, the threat and resultant military requirement comes in two forms: the threat posed by US military satellites and the threat of US offensive operations in or from space. Whether Soviet planners see the distinction between the two is open to conjecture.

US Military Satellites. The most commonly cited category of US military satellites deemed to be threatening by the Soviet Union are those used for reconnaissance purposes. Besides their generally subversive character, these satellites provide important military benefits to the US armed forces in the event of hostilities. The Soviets were aware of the US satellite reconnaissance programme well before 1960, and with the capture of the camera from the U-2 in May 1960 they also had a good idea of the potential resolution of US photoreconnaissance satellites.

In addition to the diplomatic offensive designed to outlaw "espionage" satellites, Soviet officials also indicated that they would counter these satellites militarily. In June 1960, Khrushchev told an audience at Bucharest, in an apparent reference to US space reconnaissance, that "these efforts, too, will be paralyzed and a rebuff administered".[47] Later, in February 1963, Defense Minister Marshal Malinovskiy stated that Soviet defence forces were "assigned the extremely important role of combating an aggressor's modern means of nuclear attack and his attempt to reconnoitre our country from the air and from space".[48]

While these threats ceased after the Soviet Union dropped its opposition in September 1963, Soviet military writers continued to refer to the threat posed by US reconnaissance systems such as Discoverer, SAMOS, and later the MOL. For example, Col. A. Krasnov states in the April 1966 edition of *Military Thought* (*Voyennaya Mysl'*):

> At the present stage of development of space vehicles, combat with space reconnaissance of the enemy assumes the highest importance. The presence on satellites of various kinds of reconnaissance equipment enables the discovery of various objects, including those which have been concealed with artificial camouflage . . . Therefore combat with reconnaissance space vehicles or neutralization of their operation already is not a problem for the distant future, but assumes pressing importance at the present time.[49]

Therefore, while the Soviet Union was not going to take action against US satellites in peacetime, they still posed a sufficient threat to warrant countermeasures in wartime. An interim response to this requirement before the satellite interceptor system was deployed may have been to assign some of the Galosh SA-5 ABM missiles (possibly at their Sary Shagan facility) to the anti-space defence mission.

In addition to reconnaissance satellites, there are other possible satellite targets which may have stimulated the Soviet ASAT programme. A notable candidate is the Transit navigation system, which provides accurate position fixes for the Polaris SSBNs and ultimately their ballistic missiles. As these submarines were seen as a particular threat to the Soviet Union, any countermeasures against their overall effectiveness would have been considered valuable. While the Transit programme, which began in 1958, does not seem to have received a great deal of attention in the Soviet literature, a 1968 East German radio broadcast did state:

> Under combat conditions it would undoubtedly be possible to stop the extensive radio communication required for navigational purposes; moreover, it would not be too difficult to put the satellites out of operation with the help of weapons systems which, for example, the Soviet Army has at its disposal. And thus the entire system, would of course, be ineffective.[50]

US "Offensive" Operations in Space. US satellite inspection and interception plans have also received wide attention in the Soviet literature. The 1963 edition of *Soviet Military Strategy* noted that:

> Plans are also being made for the development and utilization of space systems to destroy ballistic missiles in the powered phase of flight, to identify and destroy enemy military satellites, etc.[51]

The *Dictionary of Basic Military Terms* also includes a reference to the "Space (Aerospace) Doctrine" (Kosmicheskaya [Vozdushno-Kosmicheskaya] Doktrina) with the notation that this is a foreign concept. It is defined as "doctrine encouraging active hostilities in space, and regarding mastery of space as an important prerequisite for achieving victory in war".[52]

US projects such as SAINT, Dynasoar, BAMBI, Early Spring and the two ASAT systems deployed in the Pacific were mentioned regularly in the Soviet literature to support these claims.[53] Although Soviet commentators later noted that Project SAINT had been cancelled or postponed, subsequent programmes, such as Blue Gemini, Gemini, and particularly the MOL, all became regular targets for criticism in the Soviet press.[54]

It is difficult to assess the extent to which the threat from these projects rather than US military satellites was the dominant stimulus to the Soviet ASAT programme. On balance, the Soviets were probably more concerned with countering US military satellites than in emulating its ASAT capability. However, the two could be linked now inasmuch as the Soviets may believe that they can deter the use of US antisatellite weapons by the threat of reciprocal action. This may also explain the resumption of testing in 1976 after the first indications that the United States was about to embark on a new ASAT programme (see Chapter 8). Alternatively, once the Space Shuttle project reached an advanced stage of development—unlike many of its predecessors—this too may have been sufficiently worrisome to the Soviets to cause them to reactivate their interceptor programme.

Other Possibilities. Some commentators have looked to the Chinese space programme as providing the *raison d'être* for the Soviet ASAT effort.[55] In particular, they point to the similar orbital parameters of the Soviet ASAT tests and the Chinese satellites. However, this argument is almost certainly spurious. Apart from the fact that the Chinese programme had not even started when the Soviet ASAT project began, the similarity of inclinations is totally unrelated. The shift from 62°–63° in 1968-70 to 65.8° for all subsequent tests was dictated by the fact that the target satellites were now being launched from Plesetsk on a C-1 booster. Due to Soviet range safety constraints and other considerations, each Soviet launch vehicle can fly only at certain azimuths. Therefore, to perform co-planar interceptions the Soviets had to find an inclination mutually suitable for the F-1-m interceptor from Tyuratam and the C-1 target vehicle from Plesetsk.[56] However, it is possible that the resumption of Soviet ASAT tests in 1976 may have been stimulated by the first tests of the Chinese satellite reconnaissance system in 1975.

The Soviet ASAT programme has also been linked with Soviet fears of direct broadcasting satellites "beaming" propaganda from the West into their territory. This became apparent on 12 October 1972 when Foreign Minister Andrei Gromyko presented a draft resolution to the United Nations on direct broadcasting satellites (DBS), which would have allowed states to take action against illegal broadcasts.[57] This concern was also reported to have been reiterated during the ASAT arms control talks in 1978. While this is an understandable worry to the Soviets, their ASAT system was certainly not designed to respond to it. First, the ASAT programme began well before this fear surfaced and, second, direct broadcasting satellites are positioned in geostationary orbit well out of the range of the Soviet interceptor.

Organizational Relationships, Internal Political Accommodations and Personalities

Due to the paucity of available information on the organizational relationships, internal political accommodations and the role of specific individuals within the Soviet military space programme, the last three of Alexander's possible explanations have been combined here.[58] Unfortunately, it is precisely these factors that could help explain some of the idiosyncrasies of the programme—particularly the apparent low priority assigned to its testing and subsequent modification. For example, it may only have begun as a demonstration on the behest of a particular individual or design bureau and only later achieved the status of a major military requirement.

The system's history may also reflect the relative power of certain groups, such as the PVO-Strany within the military requirements/procurement process. For example, in his doctoral thesis on the "Soviet Space Controversy" Sawyer identifies the existence of a "military space lobby" that campaigned for a more active Soviet military space programme between 1960-3. While this may correspond with the initiation date for the Soviet ASAT programme, Sawyer admits that evidence of the influence of this lobby is inconclusive.[59] Similarly, Clemens has also speculated on the existence of a pressure group for space weaponry. He notes that "the Soviet political and military hierarchy has tended to divide into proponents of 'traditional' and 'modernist' views, the former stressing the importance of balanced conventional and nuclear forces, the latter the role of rockets and surprise attack". Moreover, "the Soviet modernists are probably divided into a group that argues for reliance on land and sea based missiles and a more radical faction militating for weapons in space". The "radical group" most likely includes "military, scientific and economic leaders with a specific stake in space weaponry (as opposed to space technology in general), and to some hard-line politicians who hope for a superiority over the United States through outer space".[60] While the "radical" modernists appear to have lost out in their campaign for strategic weapons in space, they may have been "accommodated" by the FOBS and satellite interceptor programmes, both of which share common characteristics.

Stephen Meyer offers a similar interpretation for the latter part of the ASAT programme. He compares the frequency of ASAT testing to the development of other Soviet weapons programmes and concludes:

> The ASAT programme seems to have been pursued at a leisurely pace, one that is not characteristic of the Soviet military's priority programmes. One hypothesis is that, during the 1968-71 period, it was unable to compete for booster allocation against the Strategic Rocket Forces (SRF) SS-9 ICBM programme and FOBS programme, or for system development funds against the *PVO-Strany*'s (the Air Defence Force) ABM project. During the second period, the ASAT interceptor programme may have competed for funds with the *PVO-Strany* laser and particle beam programmes, or the new investment in anti-cruise missile systems, *and lost.*[61]

CONCLUSION

In conclusion, the late 1960s and early 1970s marked a period of considerable expansion for the Soviet military space programme. While the increase in the use and variety of Soviet military satellites reflects a clear appreciation of the military benefits of outer space, Soviet attitudes towards the development of antisatellite or "anti-space defence" weapons appear to have been ambivalent. The lengthy period of the Soviet ASAT test programme—including a four-year hiatus in testing—has produced only a relatively crude weapon system with a limited operational capability. Although this system is none the less effective against some low-altitude targets, its capability will become progressively eroded by US satellite survivability measures.

The various factors outlined above have all probably contributed in some way to shaping the pace and direction of the Soviet ASAT programme. Its inception was in part an extension of the need to deny aerial reconnaissance of the Soviet Union and counter whatever military benefits space gave to its primary adversary, the United States. During the early 1960s, the Soviets appear to have been uncertain as to whether the United States would deploy weapons in space. Thus, in many respects, the need to take out some technological insurance mirrors the US rationale for its own ASAT programme. While this particular threat declined, the utility of a satellite interceptor against reconnaissance and navigation targets in wartime remained. In either case, it was entirely compatible with Soviet doctrine at the time. The hiatus in testing between 1971 and 1976 was most probably the result of a combination of budgetary, political and technical factors. Having demonstrated a rudimentary capability, the demands of other military programmes became paramount. Moreover, the Soviets probably felt that overt testing could jeopardize, or at least complicate, the détente process. The mixed success of the testing programme may also have convinced the leadership that more basic research was required before testing could resume.

8

Nixon and Ford: Continuity and Change

The USSR is seizing a new initiative, and creating the prospect of a new dimension of military conflict—war in space. Our lead in space technology is a strong one, but we have deliberately restrained the development of an antisatellite capability. If the Soviet Union chooses to continue along the path they appear to be taking, they will find it a dangerous one.

Dr Malcolm Currie,
Director of Defense Research and Engineering, 1977

Introduction

The first manned landing on the moon at the beginning of the Nixon administration marked the culmination of the US effort to recover its self-esteem in space activities after the shock of Sputnik. Paradoxically, it also marked the beginning of a period of retrenchment and disillusionment in the US civil space programme. After the rapid expansion of the early 1960s, NASA's budget had been slowly declining during the latter part of the decade. Further financial cutbacks during the Nixon administration severely limited NASA's plans for the new decade; prestige alone was no longer a sufficient reason to fund space projects. This downward budgetary trend was even more pronounced in the military space programme. After a peak in 1969, funding plummeted during the early 1970s and only recovered to an equivalent level (in real terms) during the Carter administration (see Table 3 in Appendix I for details). While the initial drop in funding can be attributed to the cancellation of the Manned Orbiting Laboratory, other factors were influential over the long term. The maturation of the military space programme as a whole brought more capable, longer lasting satellites and a concomitant drop in the required number of space launches. The initial surge of interest in military space activities among the services had also levelled out. Furthermore the perceptions of a Soviet threat in space had declined not least due to the period of détente.

After the Outer Space Treaty, further arms control agreements were signed between the United States and the Soviet Union with important implications for the military exploitation of space, the most notable being the 1972 SALT (Strategic Arms Limitation Talks) agreements. Moreover, US apprehensions about the embryonic Soviet antisatellite (and FOBS) programme were reduced somewhat by the cessation of testing in 1971.

It was only in the latter stages of the Ford administration that concern over the qualitative and quantitative transformation in Soviet space activities really grew. The increasing use of satellites to enhance the Soviet war-fighting potential coupled with the resumption of satellite interceptor tests in 1976 precipitated a reappraisal of US satellite vulnerability and the need for a new US antisatellite weapon system. Reaction to these developments was by no means uniform within the administration. In contrast to the concern of the White House, the Air Force and the Defense Department remained ambivalent about the "threat". In part this was due to bureaucratic inertia but there was also considerable scepticism of the wisdom and utility of a US ASAT weapons programme. Thus, while funding for "space defense" research did increase it was not at the pace that the White House considered necessary. As a result, it took two explicit National Security Decision Memoranda (NSDM) from the White House to galvanize the Defense Department and Air Force into taking action and for US policy to change. This chapter traces the evolution of US policy under Nixon and Ford and in particular the events leading up to the second major decision to develop an antisatellite weapon in January 1977.

Nixon and the Military Space Programme

On 13 February 1969, less than a month after the inauguration of Richard Nixon, a presidential Space Task Group (STG) was established to review the future space programme of the United

States. The STG, which was chaired by Vice President Spiro Agnew, was organised on two levels: a senior group consisting of Agnew, Thomas Paine (NASA Administrator), Robert Seamans (Secretary of the Air Force—representing Secretary of Defense Melvin Laird) and Dr Lee Du Bridge (the President's Science Advisor and head of the Office of Science and Technology); and below this a working group of assistants from the respective Departments. Three observers—Glen Seaborg (Chairman of the Atomic Energy Commission), Robert P. May (Director of the Office of Management and Budget [OMB]), and U. Alexis Johnson (Under Secretary of State for Political Affairs)—were also active in the discussions.[1]

The recommendations of their report, which was ready by September 1969, were clearly tailored to fit the pragmatic emphasis of the new administration.[2] While exploration of the moon and planets was included, the main recommendations were: an overall cost-reduction of the national space programme, extending man's capability to live and work in space, practical space applications, and international co-operation.[3] Similarly, the recommendations for the military space programme reflected the administration's cost-conscious mood. As a later Aeronautics and Space Report of the President stated:

> Decisions to initiate new space programs will be carefully structured in the context of the threat, economic constraints, and the national priorities placed on defense. In particularly the DoD will embark on new military space programs only when they can clearly show that particular mission functions can be achieved in a more cost-effective way than by using more conventional methods.[4]

The Task Group's recommendations largely confirmed what was already being carried out in practice by the administration. The initial optimism in some quarters that Nixon would place greater emphasis on military space systems—reminiscent of the early expectations for Kennedy's space programme—was brought to an abrupt halt with the cancellation of the Manned Orbiting Laboratory in June 1969. The project had been struggling from insufficient funding for a number of years and with the new emphasis on budget limitation it became, in the words of a senior Air Force officer, "an ideal target for OMB".[5] The cancellation of MOL had also been recommended by the new DDR&E John Foster, with the concurrence of Defense Secretary Melvin Laird. When the decision was announced, Laird's Deputy, David Packard, gave the DoD's reasons: "In order to reduce the defense research and development budget significantly, it was necessary to cut back drastically on numerous small programs or to terminate one of the larger, more costly research and development undertakings". Furthermore, "since the MOL program was initiated, the Department of Defense has accumulated much experience in unmanned satellite systems for purposes of research, communications, navigation, meteorology".[6] What Packard could not state was that the planned "Big Bird" reconnaissance satellite, that would use the same Titan III launcher, would provide just as much intelligence information as the MOL but without the exorbitant cost. Big Bird or Project 612 (later KH-9) was originally planned as a back-up to the MOL but, with the latter's cancellation, it naturally came to the forefront. Apart from its enormous high-resolution camera developed by Perkin-Elmer, the Big Bird satellite could carry other payloads, such as electronic intelligence equipment. In short, it completely undermined the *raison d'être* of the MOL programme. By 15 June 1971 the Big Bird (KH-9) satellite programme had become operational.[7]

The Defense Department's arguments were reiterated during the FY 1971 budget hearings, where DDR&E John Foster argued that:

> . . . despite the overall decrease in the Defense Space Budget as a result of the deletion of MOL, the remainder of the military space program has actually enjoyed a slight increase.

Furthermore:

> We have found in the last decade that by increasing the capability of the satellite, we can substantially reduce the price of equipment on the ground. So we see a growth in research and development of space activities in the next few years, simply on the basis of achieving economy in the operation of our forces.[8]

The operation of the military space programme had indeed become more cost-efficient since the early 1960s. The growth in lift capacity through the development of larger boosters and the increasing miniaturization and sophistication of equipment meant that the United Sates could deploy ever more capable and multifunctional payloads in space. This in turn influenced the number of launches required each year; the choice of orbits that could be utilized; and the operational lifetime of the satellites. For example, in 1966 the United States successfully launched 66 military satellites; by 1971 this figure had fallen to 22 and to 16 three years later.[9] Higher altitudes could also be exploited or, alternatively, if the function of the satellite benefited from being in low-earth orbit, its lifetime could be extended by extra on-board propellant to reduce the rate of orbital decay. Thus by the 1970s most US communication, navigation, early warning and some ELINT satellites had moved to higher orbits. Moreover, the original life expectancy of the photoreconnaissance satellites (four to five days and two to three weeks for the Corona and SAMOS satellites, respectively) had risen to 250 days for the Big Bird and two years for the CIA's KH-11 satellites. The drawback for space defence purposes was that it produced higher levels of dependency on fewer, more capable satellites.

Further confirmation of the administration's policy came with the report of the Space Science and Technology Panel of the President's Science Advisory Committee. Entitled "The Next Decade in Space" and published in March 1970, it did not review the military space programme in depth, but did make some general recommendations. For example:

> ... we are convinced that a gradual growth in military uses of space systems will prove to be justified on a case-by-case basis within the normal budget process. A projection of a 50 percent increase in budget for uses of space in the next ten years seems consistent with the execution of the tasks ...[10]

Later, with respect to manned military systems, the Report noted:

> We do not believe that military requirements have been identified for which man in space is either uniquely capable or clearly most cost-effective in comparison with alternative technologies.[11]

The Nixon administration, however, was prepared to fund the manned Space Transportation System (STS), otherwise known as the Space Shuttle, which had been recommended by the Space Task Group. In 1970 NASA and the Defense Department jointly carried out extensive feasibility studies and final approval was given on 5 January 1972 by President Nixon.[12] Although the Shuttle was a manned system with a sizable Defense Department interest, the expected cost savings, from being a reusable launch vehicle, made the project compatible with the administration's overall space policy guidelines.

Some changes in the organization and management of the US space programme also occurred during the Nixon administration. The NSAM 156 Committee no longer functioned after 1969, and in 1973 the National Aeronautics and Space Council (NASC) was also disbanded along with the Office of Science and Technology.[13]

McNamara's 1961 directive on the management of the military space programme (5160.32) was also amended in September 1970, though only marginally. While the Air Force increased its responsibility for military space operations, it was still not the sole operator.[14]

Apart from the cancellation of MOL and the initiation of the Space Shuttle project, the military space programme was not a major issue for internal debate during the Nixon administration. Even the Soviet ASAT tests failed to excite much interest, which would have been unthinkable had they started during the more sensitive period of the early 1960s. Various factors account for this. In the first instance, the initial tests of the Soviet interceptor satellite coincided with the run-up to the 1968 presidential election and were naturally eclipsed by that event. Attention was also focused on the growing US involvement in the Vietnam war. The US response can be characterised as a "wait and see" policy; first to see whether the tests were in fact ASAT-related, and second to see whether they represented a serious Soviet commitment to ASAT deployment. Despite speculation in the press in December 1968 that the mysterious Soviet satellite manoeuvres were part of an antisatellite research and development programme, this was neither known for a fact nor likely to be of major concern to the outgoing administration. Furthermore, nearly two years went by between the second test on 1 November 1968 and the third on 23 October 1970. While speculation increased in the press, there was still no official US confirmation that the Soviet Union was conducting antisatellite tests. In fact the Nixon administration appears to have paid very little attention to these earlier tests.

This began to change slowly. In what appears to have been a reference to the Soviet ASAT tests, DDR&E John Foster stated in early 1970 that he did not consider them to be a threat.[15] Later at the NASA Authorization hearings for FY 1971 he stated that there had been no interceptions of US satellites, though his comments on the nature of the targets were deleted from the record.[16] By the following year, however, the Soviet Union was accredited with an "Orbital Antisatellite System" in a table of comparative weapons systems included in the FY 1972 congressional hearings. This was possibly the first official public recognition that such a system was being tested.[17]

By the time Soviet ASAT testing resumed in October 1970, the administration had become sufficiently concerned to review US policy on the issue. As a result, the President's National Security Advisor, Henry Kissinger, called—probably in the form of a National Security Study Memorandum (NSSM)—for a "quick study" of what the United States should do in response to the tests. The NSC request was passed to the Office of the Secretary of Defense (OSD), who in turn gave the DDR&E responsibility for organizing the study. Two groups were set up: a Steering Group, consisting of high-level officials from various agencies; and a Working Group, chaired by Manfred Eimer (Assistant Director, Intelligence, at DDR&E), which consisted of their more junior representatives.[18] ACDA, curiously, was not represented in either of these groups. Although Kissinger had called for a quick study, progress was often delayed by the need for the Steering Group to review the findings and recommendations of the Working Group. Often months would elapse before the Working Group could once again return to its deliberations. Both groups concerned themselves with a number of key and often overlapping questions, in particular:

(a) What had motivated the Soviet ASAT programme?
(b) Why had the Soviets chosen this method?
(c) Which US satellite targets were the Soviets most concerned with?
(d) What should the United States do in response?

By the time the Working Group met there was a general consensus that the Soviet Union was testing an antisatellite system. While the purpose of the tests was well understood, the Group could not reach a common position on why the Soviets had embarked on this programme or what the circumstances of its use would be. Some participants believed that the Soviet programme was more the result of internal bureaucratic "politicking" than any coherent strategy. It was concluded, however, that the Soviet effort did not represent a "crash" programme.

There was some controversy within the Group arising from the perceived advantages and disadvantages of the co-orbital method for intercepting satellite targets. For example, why had the Soviets not chosen a direct ascent system similar to the Nike Zeus or Thor systems? This led to extensive discussions over whether a potential antisatellite capability resided in other Soviet systems, particularly their ABM and air defence missiles. Similarly, the advantages and disadvantages of the apparent kill mechanism were also debated. The discussion of the chosen ASAT method inevitably overlapped with the question of which US targets most interested the Soviets, and what countermeasures should the United States adopt.

Apparently, those present who had a vested interest in space operations believed that their particular satellite programmes were the reason or "target" for the Soviet system. As one participant observed: "Nobody wanted to accept that their programme was not of sufficient importance to be driving the Soviet programme." This proved to be a fruitless avenue of discussion, as it was recognized that certain US satellites "had not even been born" when the Soviet tests had begun. Furthermore, it was also acknowledged that the Soviet tests could be just part of a general research programme in case further development was required later. As a result, they concluded that they could not isolate a "specific target" or "identifiable reason" why the Soviets had begun their programme.

Apart from the general recommendation that the United States should take the necessary steps to reduce the vulnerability of its satellites to the Soviet ASAT system, the Eimer group also considered whether the US should respond with its own ASAT programme. In particular, would this stimulate further Soviet ASAT development and would a US ASAT deter its use? After some debate, it was agreed that this was "not an area where deterrence works very well because of dissimilarities in value between US and Soviet space systems". As the United States depended on its satellites to a greater extent than the Soviet Union, it was believed that the threat of retaliation in kind for an attack on US satellites was unlikely to deter the Soviet Union. Thus deterrence of Soviet ASAT use was not considered at this stage to provide a viable rationale for the development of a new US ASAT system. The Working Group did recognize, however, that the asymmetry of dependency might change, especially with the latest indications of the growing Soviet military use of space. John Foster, Jr acknowledged this "trend" in an apparent reference to the Eimer group during congressional testimony in 1972:

> We also have a trend in the use of *deleted* satellites for tactical purposes that has become much more important than in previous years. So our dependence in the future on satellites may be considerably different than we had thought. We are now in the process of thoroughly reworking this whole basic question.[19]

As a result, the Working Group discussed future Soviet space activities and also possible "conflict scenarios" where a US ASAT system could come in use. They eventually drew up a list or "matrix" of current and potential Soviet space threats and attached specific recommendations to counter them. The "bottom line" was to propose that the United States maintain a minimum ASAT development lead time, which would be proportional to the warning that the United

States could expect to have before a particular threat materialized. Thus ASAT research and development could continue but along these prescribed guidelines. The Group's report also made some recommendations on the technology to be pursued—presumably nonnuclear kill and terminal homing warheads. A Defense Science Board Panel convened at the time provided more detailed guidance for this research.

The report was finally submitted in 1973 to the NSC staff but with the upheaval created by the Watergate scandal combined with the fact that Soviet ASAT testing had apparently ceased in 1971, the problem was no longer considered to be of high priority. There is no *available* evidence to suggest that a National Security Decision Memorandum (NSDM) or any other similar directive was issued by the Nixon administration on the basis of the Group's recommendations. However, there is evidence to indicate that some of the participating agencies did begin to take the problem of satellite vulnerability more seriously. Satellite survivability studies began to be funded albeit by small amounts.

In addition to the halt in Soviet ASAT testing, the general climate of détente epitomised by the SALT talks undoubtedly contributed to the limited concern and reaction to the Soviet ASAT threat. Moreover, the two most important treaties of this period—the Treaty on the Limitation of Antiballistic Missiles and the Interim Agreement on the Limitation of Strategic Offensive Arms, both signed in May 1972 provided some constraint on the use of antisatellite systems. As Articles XII and V, respectively, state:

> Each Party undertakes not to interfere with the national technical means of verification of the other Party . . .
>
> Each Party undertakes not to use deliberate measures which impede verification by national means of compliance . . .[20]

Although the term "national technical means of verification" (often shortened to NTMs) only *implies* the use of reconnaissance satellites (among a range of intelligence-gathering methods), the approval of these Articles by the Soviet Union was greeted by many Americans as not only a great leap forward for verification but also the final acceptance of the principle and practice of satellite reconnaissance. Certainly, as Cohen has noted, there have been subsequent references in the Soviet literature to suggest that they hold "national technical means of verification" to include observation satellites.[21]

Some observers have also interpreted these provisions as banning antisatellite activities against US and Soviet reconnaissance satellites. While this is a valid interpretation, none of the detailed accounts of SALT I indicates that a common definition of what constitutes "interference" with national technical means of verification was ever reached at the negotiations.[22] In fact the antisatellite issue did not arise explicitly in the discussions on interference with NTMs. Apparently Soviet negotiators did informally refer to satellite reconnaissance as a main means for carrying out verification but the US delegates were enjoined by security guidance from doing so.[23] Cohen adds another note of caution in interpreting these clauses too liberally as he further states:

> ... satellites can conduct reconnaissance that may be perceived by the Soviets as irrelevant to arms control monitoring. Hence the Soviets may consider them illicit and not subject to the "non-interference" clauses ...
>
> Furthermore, there is little evidence with which to judge Soviet propensities for non-destructive jamming of perceived illegal peacetime satellite activity conducted by otherwise legitimate NTMs.[24]

Thus, while the US administration may have comforted themselves that antisatellite activites against certain US satellites would be banned with the passage of SALT I, the meaning of the non-interference clauses was never made explicit at the time.

It is also worth mentioning some of the other pertinent provisions of the ABM Treaty in the light of later concern about directed energy weapons, such as lasers and particle beams. Article V is particularly relevant. It states that "Each Party undertakes not to develop, test or deploy systems or components which are sea-based, air-based, *space-based* [my emphasis] or mobile land-based", and thus represents a significant addition to arms control in space. Furthermore, "Agreed Statement D" also states that limitations on "ABM systems based on other physical principles..." that may develop in the future would be the subject of further discussions. Although again the Treaty does not explicitly define "other physical principles", discussions within and between the US and Soviet SALT delegations leading to this agreement did refer to such technologies as "lasers, particle beams, and electromagnetic waves".

In fact, future ABM methods, referred to as "futuristics" by Gerard Smith and "exotics" by John Newhouse in their own accounts of SALT I, became a controversial issue within the US SALT bureaucracy. Smith goes so far as to state that: "The negotiation about future

ABM systems was as significant as any part of SALT".[25] The origins of this issue stem from NSDM 117, issued by the White House on 2 July 1971. NSDM 117 outlined a new US SALT position on a whole range of issues, including a proposal to ban "exotic" or "future" ABM systems. This memorandum was cabled to the delegation in Helsinki, who greeted the new proposal with less than unanimous support: Ambassador Gerard Smith (the head of the US delegation) and Harold Brown both supported a ban; Paul Nitze (representing the DoD) also favoured one but with the caveat that sensor technologies be excluded; while at the other extreme Ambassador Parsons (representing the Department of State) and General Allison (the JCS representative) both believed that these systems should be "unconstrained until such time as the technical and strategic possibilities became clearer and one could better judge the effects of constraint".[26] However, Raymond Garthoff—also representing State—joined Smith and Brown in opposing Parsons' position. Garthoff's viewpoint was eventually supported by the State Department in Washington.[27]

Because NSDM 117 proved so unsatisfactory to the US delegation for a variety of reasons, it was repealed and replaced on 20 July by NSDM 120. While this took into account some of their criticisms, the new memorandum made no reference to the "exotics" issue. Instead, the delegation was told to leave a blank space in the draft US proposal on ABMs pending the outcome of a SALT Verification Panel study which would discuss the matter. This met on August 9th, with Kissinger chairing the discussion. Here, the President's Scientific Advisor, Dr Edward David, was called in to give his views on the matter and imparted what one observer described as a "Buck Rogers flavor to the meeting". The differing views of the agencies represented at the SALT delegation surfaced again at this meeting but this time the White House view prevailed: the resultant NSDM 127 proposed banning "everything other than research and development on fixed land-based exotics".[28]

This proposal was put to the Soviets in Helsinki, who, apparently, were not convinced that a provision covering "undefined" systems would by itself be meaningful. The Soviet military representatives were also suspicious that this was a "fishing expedition for intelligence" rather than a serious US initiative.[29] However, the US delegation persisted with these proposals and the Soviets eventually agreed in January of the following year.

Overall, while the SALT "package" augmented the Outer Space Treaty in important ways, it failed *explicitly* to address the ASAT issue at a time when both superpowers were seemingly uninterested in further development. A major opportunity was therefore lost.

It is also worth mentioning two other agreements signed in the detente era which have some relevance to ASAT activities. First, the "Agreement on Measures to Reduce the Risk of Outbreak of Nuclear War" or, the "Accident Measures" agreement, as it is more commonly known (signed on 30 September 1971), states that the Parties agree to notify one another in the event of interference with missile early warning or related communications systems (Article III). Second, the "Agreement ... on Measures to Improve the USA-USSR Direct Communications Link" (the so-called "Hot-line Modernisation Agreement", also signed on 30 September 1971), introduced the use of Molniya and Intelsat satellites to upgrade the "Hot-line". Both parties agreed "to take all possible measures to assure the continuous and reliable operation of the communications circuits and the system of terminals of the Direct Communications Line ..." (Article II). While these agreements are not generally considered to be part of the space arms control regime they do add to the NTM provisions of SALT I.

The Ford Administration: Towards a Change in US Policy

It was only during the Ford administration that the growing Soviet military use of space, and in particular its antisatellite programme, became a significant issue within the US bureaucracy. However the attitudes, and with it reactions of the three principal actors—the White House/NSC staff, the Department of Defense and the Air Force—to the Soviet space "threat", were by no means the same:

The White House/NSC Grouping

After the disruption of the Watergate scandal and the fall of President Nixon in 1974, US policy on the military use of space was again reassessed by a panel of experts organised by the NSC. This panel was one of a number of studies carried out by a pool of outside consultants and co-ordinated by the Military Technology Advisor to the Presidential Assistant for National Security Affairs. This system of *ad hoc* panels had been established as a result of growing pressure to replace the now-defunct President's Scientific Advisory Committee with another form of scientific advice to the President. The first occupant of the position of Military Technology Advisor was Gordon

Moe, who had served on the President's Science Advisory staff between 1970-3. Moe headed a group of 50 to 60 consultants that were called upon to study particular questions as the need arose. There were typically two to three panels reviewing a particular problem at any one time, with the average duration of their studies being between six to eight months. Although suggestions for the subjects to be studied would often originate from a variety of sources, final approval was given by Kissinger and his deputy, Lt. Gen. Brent Scowcroft. These men were also the recipients of the final reports.

On Gordon Moe's recommendation a panel was established to examine the US military use of space, with a focus on satellite reconnaissance and communications for tactical purposes. This subject had apparently been spurred by the increasing evidence that the Soviet Union was pursuing a more "adventurist" policy in the projection and use of its military forces. As a result, regional military commanders had become more concerned with the problem of responding to provocative localized incidents while the White House had become more sensitive to the problem of preventing unnecessary superpower confrontations in the Third World. Thus, the panel, which was chaired by Charles Slichter from the University of Chicago, studied ways of improving and safeguarding the flow of information to and from commanders in the field. This inevitably involved assessing the threat of Soviet electronic and physical disruption to US satellites and communications facilities in general.[30]

The completion of the Slichter Panel's report in 1975 coincided with the replacement of Gordon Moe by Robert Smith as Military Technology Advisor. In the words of one participant, the report warned that "the US dependence on satellites was growing and that these satellites were largely defenceless and extremely soft to countermeasures".[31] As a result of these findings, Bob Smith initiated another study panel, to be led by Solomon Buchsbaum, former Chairman of the Defense Science Board. This had the specific task of addressing the satellite vulnerability problem. While the "blinding" incidents in October and November 1975 had highlighted this problem, it was the resumption of Soviet ASAT tests in February 1976 that most concerned the panel. Their efforts were immediately accelerated and an interim report was provided to Scowcroft. He in turn briefed President Ford with a summary of the report. Ford evidently became "very concerned" about the problem and "asked the DoD for their own analysis".[32] When the requested report showed that the DoD's satellite survivability efforts had been negligible, Ford issued National Security Decision Memorandum (NSDM)-333 in the autumn of 1976 to redress the vulnerability of US military satellites. Extra funding for satellite survivability projects was immediately earmarked for the Air Force's space defence programme and a specific System Program Office (SPO) was established at the Space and Missile Systems Organization (SAMSO)—formerly the Space Systems Division. A Defense Science Board study on satellite vulnerability was also commissioned.

Following NSDM-333, the Buchsbaum Panel continued its studies, but addressed the broader question of US military space policy and particularly the need for an antisatellite system. It was briefed by the State Department, DoD (including the findings of the Eimer Study), CIA, ACDA and the Air Force. By late 1976 the Panel had finished its report. This repeated the warning that the United States was becoming dependent on military satellites for a variety of critical functions, and that very little provision had been made for their survival in wartime. In particular, the satellite ground support facilities were viewed as being dangerously vulnerable. The Panel concluded that a range of specific actions should be taken to ensure the operation of these systems including the provision of sensors aboard US satellites to detect interference.

On the question of whether the United States should develop an antisatellite system, the report stated that this would *not* enhance the survivability of US satellites. It echoed the conclusion of the Eimer study by stating that as the US relied more heavily on space, a US ASAT would not necessarily deter the use of a Soviet ASAT. The only use for a US ASAT system would be to counter the threat posed by Soviet ocean reconnaissance and other intelligence-gathering satellites. Furthermore a US ASAT programme, it was argued, could also serve as a "bargaining chip" to gain what the Panel considered the more satisfactory solution of an ASAT arms control agreement with the Soviet Union.

During the Panel' discussions of ASAT arms control it was recognized that this would require a qualitatively different approach to the normal one. As one participant observed: "ASAT arms control is not like ordinary arms control" because, while a "handful of ICBMs make little difference to the overall Soviet capability, a handful of Soviet ASATs could do immense damage out of all proportion to their number". Moreover, "it is also substantially easier to deploy an antisatellite system without being detected".[33] The last point seems to have been a reference to the resumption of Soviet ASAT tests in 1976. The hiatus from 1971 had illustrated the problems of detecting a "residual capability" and was used by those who argued that ASAT verification was impossible. As a result, the Panel recommended that prospective ASAT arms control negotiations should not aim solely to restrict deployment but should seek comprehensive limitations that prohibited future ASAT research and development as well as the dismantlement of existing systems.[34]

The Panel's final report was presented in December and condensed into a brief summary by the NSC staff. Various policy options were also attached, including one to develop an antisatellite system. Scowcroft then briefed the President on this issue; Ford in turn requested a report from the Secretary of Defense on the current level of DoD antisatellite research and development. When President Ford was eventually provided with a list of the relevant DoD space defence activities in late December 1976, he apparently became "very upset and concerned about the relaxed approach of the Defense Department", to the extent that "he thought the only thing to do was to issue a formal directive".[35] Ford's reaction appears to have been a response to the low priority of US ASAT research and also to the statements by some Defense Department officials offering further restraint in return for reciprocal Soviet behaviour. Yet another test of the Soviet satellite interceptor on 27 December was most probably crucial in convincing Ford of the bankruptcy of this approach.

As a result, the NSC staff drafted a National Security Decision Memorandum in January which directed the Defense Department to develop an operational antisatellite system. The draft directive was then sent out to the Defense Department, State and ACDA and most probably other agencies for comment. The State Department officials responsible for reacting to the draft NSDM were concerned that the arms control option had not been considered.[36] It was probably as a result of their intervention that the final directive—NSDM 345—which was signed on 18 January 1977 did call, according to the account of Donald Hafner, for "a study of arms control options, but it did not include any concrete proposal for inviting the Soviets to ASAT talks".[37] Apparently, Kissinger believed that the United States should redress the US-Soviet asymmetry in ASAT capabilities before arms control negotiations could be considered. As Hafner points out, Kissinger "may have felt it was premature to make such a proposal; or indeed, he may not have favoured negotiations at all".[38] Thus, while the United States was now committed to developing an ASAT capability, the possibility of arms control was left to the Carter administration to consider.

The Defense Department

For at least the first half of the 1970s, the attitude of the Defense Department towards the antisatellite issue can be summarized by the following statement made by John Foster, Jr to Congress in 1972:

> . . . we in the Department of Defense, are not very clear in our own minds about what we ought to do. We have looked at this matter for a number of years. It has never been clear to us that we ought to go out and develop a system that would cost hundreds of millions of dollars.[39]

The Soviet ASAT threat in space certainly did not loom large. After its inclusion in a table of a comparative military systems in the FY 1972 congressional hearings, no further mention was made of a Soviet ASAT capability until testing was resumed in 1976. Concern over the growing use of military space systems to enhance the Soviet Union's overall war-fighting capability—while acknowledged in this period—also did not surface in any significant way until 1976. Accordingly, the Defense Department continued to fund only modest amounts in support of antisatellite research and related space defence measures. Some also believed within the Defense Department that a tacit arrangement could be reached with the Soviet Union whereby space would remain a sanctuary for the unhindered operation of their respective military space systems. Continued unilateral restraint by the United States in the development of antisatellite weapons, in addition to the closure of the Johnston Island facility in 1975, was viewed as the best way of signalling the US conviction that an arms race in space was in neither side's own interest. By supporting a low-level research and development effort with emphasis on satellite survivability, the DoD also believed it could hedge against hostile Soviet activities in space without appearing aggressive.

By the beginning of 1976, the Defense Department had decided to step up its support for "space defense" research albeit by small amounts. Both satellite survivability and space surveillance received extra funding in the FY 1977 budget but more significantly $12.8 million was requested for "space defense systems" (that is, antisatellite) research which marked a threefold increase over the previous year's request.[40] Whether the Soviet decision to resume testing in February 1976 was connected in any way with the growing US interest in antisatellite research that had been reported in the aerospace press throughout 1975, is obviously impossible to ascertain.[41] While the Soviet tests undoubtedly stimulated the US ASAT programme, it should not be overlooked that research was already under way—including identifiable elements of the current US system—*before* the renewed Soviet effort.

The changing attitudes to satellite survivability appear to have been the result of the Slichter Panel's recommendations and the "blinding" incidents of the previous autumn which—if not ASAT related—at least focused attention on the growing dependency of the United States on a relatively small number of vulnerable satellites. As Donald Rumsfeld stated in his FY 1977 Posture statement:

> As space technology matures, space based systems will play an even more important role in support of US and Soviet military operations. In the future, dependence on these systems may increase to the point where their loss could materially influence the outcome of a conflict. Consequently, it is important to know of any threat to US space activities which threaten our overall military posture. Defense is continuing R&D efforts to develop technologies for detecting, tracking and identifying objects out to geostationary orbit and for enhancing the survivability of satellite systems, at the same time abiding by the provisions of the various space treaties to which the US is a signatory.[42]

At the NASA authorization hearings, the Director of Defense Research and Engineering Malcolm Currie also stated, with the resumed Soviet ASAT tests almost certainly in mind:

> . . . satellite vulnerability has to be a major issue for us, a major topic of study and of planning over the next few years. The question is, can we maintain space as a sanctuary or not? We are taking some action in this direction. I had a Defense Science Board task group study this area during the last year, and I think we understand some of the problems. But there are many question marks that still exist.[43]

As for the growing interest in antisatellite research, Currie's testimony also gives an indication of why this issue was receiving more attention in the Defense Department:

> . . . the Soviets are investing increasing resources in space technology for military purposes. Their level of activity reached an all time high in 1975. The systems they put into orbit are significantly more sophisticated than those deployed in the past. The trend signified by these activities indicates that they are reaching a point where their space systems will contribute substantially to the effectiveness of their command-and-control systems, and directly to the performance of their strategic and general purpose forces. Soviet space technology must be taken into account in the strategic equation, and in calculating the balance of forces for conventional war.[44]

Despite the heightened interest in satellite survivability, the level of effort was, as described earlier, still considered unsatisfactory to the White House. As a result of NSDM-333, $200 million was apparently earmarked for a collection of programmes administered by the new Space Defense Systems Program office at SAMSO. However, the full amount did not materialize; funding did increase, but only moderately. The reason for this appears to have been the result of a combination of factors: the Air Force was still ambivalent about the Soviet ASAT threat while its satellite design and development teams were reluctant to add survivability measures to their programmes because of the likely penalties in weight and cost. This problem will be discussed in greater detail below. Moreover with the PPBS (Planning Programming Budgetary System) the Defense budget becomes heavily committed at least two years in advance which makes it difficult to add relatively large items at short notice. Also the military space budget had in real terms plummeted by the mid-1970s, which further encouraged conservatism among decision-makers when new space projects were proposed.

As noted earlier, the White House had also become dissatisfied with the level of antisatellite weapons research (despite the previous year's increase in funding) and with the Defense Department's general approach to this issue, epitomized by a speech given by Malcolm Currie to the Air Force Association at the end of October 1976. Described as a "clear signal to the Russians" by the journal *Aviation Week & Space Technology*, Currie stated:

> The Soviets have developed and tested a potential war-fighting antisatellite capability. They have thereby seized the initiative in an area which we hoped would be left untapped. They have opened the specter of space as a new dimension for warfare, with all that this implies. I would warn them that they have started down a dangerous road. *Restraint on their part will be matched by our own restraint, but we should not permit them to develop an asymmetry in space.*[45]

Although US policy changed with Ford's decision authorizing antisatellite development in January 1977, the FY 1978 Defense Department posture statements continued to emphasize US restraint. For example, Secretary of Defense Rumsfeld stated: "Current US space defense policy is to abide by our space treaties, exercise our rights to the full and free access to space, and limit our use of space to nonaggressive purposes."[46] While Rumsfeld went on to say that in response to the Soviet ASAT tests "we have decided to increase significantly the US space defense effort" including "space surveillance" and "satellite system survivability", he made no mention of the US antisatellite programme. There was also no description of the "space defense systems" line item apart from the fact that it had leapt to $41.6 million in FY 1978.[47]

Currie's own report to Congress also reiterated his earlier "carrot and stick" speech:

> For the US viewpoint, perhaps the most portentous Soviet activity in space is the resumption of their antisatellite development program, after a hiatus of more than four years. The USSR is seizing a new initiative and creating the prospect of a new dimension of military conflict . . . war in space. Our lead in space technology is a strong one, but we have deliberately restrained the development of an antisatellite capability. If the Soviet Union chooses to continue along the path they appear to be taking, they will find it a dangerous one. We cannot let them obtain a military advantage in space through antisatellite weapons, because the consequences to the future military balance between the US and USSR could be no less than catastrophic.[48]

Again no mention was made to the US ASAT programme. This and the conspicuous absence throughout 1976 of any reference to US antisatellite research and development suggests that it was the result of deliberate policy. This was later confirmed after portions of a classified report for Congress from Currie was leaked to the press. This reportedly included the statement: "National policy is that [*sic*] the very fact that we are investigating means by which to neutralize foreign satellites will be classified."[49] He also reported that since 1974 the United States had been trying to develop the components for a nonnuclear interceptor but that the level of funding had been inadequate to support prototype interceptor development and testing before the mid-1980s. But:

> Given the increased concern over the importance of space assets to military operations and the asymmetry that has developed between the US and the Soviets in antisatellite capability, I am increasing the emphasis on this program to provide for interceptor flight tests in 1980 which will preserve the option to deploy a system with an initial capability by 1982.[50]

Whether the latter was in response to Ford's NSDM is hard to tell as both are apparently dated 18 January 1977. However, the posture statements and the reference in the classified document to preserving the "option" suggests that the Defense Department was still not convinced of the wisdom of deploying an antisatellite weapon system.

Air Force Attitudes

With the exception of a small group of officers, the Air Force attitudes towards space had changed dramatically since the 1960s. The earlier enthusiasm for military space projects had dissipated by the 1970s to a point where there was often resistance to new space proposals. Various factors account for this transformation.

Although the Air Force had gained primary responsibility for the space mission, other services and the intelligence community were still active in the development of their own space systems. This prevented space from becoming a "mission area" to rival those for which the Air Force had *sole* responsibility. Also the cancellation of the Air Force's manned space programmes in the 1960s not only eroded the basis of support for space systems among the more traditional "flying" elements of the Air Force but also ultimately prevented the creation of a specific "user" space community. Without the equivalent of a Strategic or Tactical Air Command there was neither the career incentive to specialize in space systems nor the necessary bureaucratic "clout" for programme procurement. It was a process that perpetuated itself with the declining military space budgets of the 1970s. This in turn encouraged conservatism when new projects or missions like "space defense" were proposed. With barely enough support for satellite programmes there was even less for satellite defence. Moreover, the wider Air Force belief was that as satellites benefited the entire national security community, space defence should be a DoD initiative with DoD funds otherwise the inevitable opportunity costs would jeopardize the traditional mission oriented programmes within the Air Force budget.

Some of these factors may be overly simplified but they have frequently been cited to explain the underlying Air Force attitudes towards space and space defence in particular. This led in some instances to what appeared to be surprising proposals and arguments when space defence issues were discussed. For example, supporters of antisatellite weapons argued in favour of disbanding Program 437 in the belief that its very presence was blocking progress towards a more advanced replacement. In effect, the existence of an "operational" ASAT capability was being used by others in the budget battles as evidence that a successor was not required.[51] Furthermore, those officers who were in favour of utilizing space often argued against adding satellite survivability measures out of fear that the extra penalties in cost and payload would in turn jeopardize their already vulnerable projects.

After the first series of Soviet ASAT tests had ended in 1971, the question of what the Air Force should do to meet the threat was brought up in the 1972-3 budget reviews. Ironically, the Soviet tests were seen as further grist to those who opposed developing and depending upon more space systems. Proponents of antisatellite weapons were also accused of adopting the simplistic argument that "because the Soviets have one we must have one too". The result was that, low-level research was allowed to continue but without any real enthusiasm or funding.

With the arrival of Brigadier General Henry Stelling as Director of Space in the Office of the Deputy Chief of Staff for Research and Development in September 1972, the supporters of an antisatellite programme within the Air Force had gained a highly placed ally amongst the Air Staff. Stelling, with his assistant, Lt. Colonel Heimach (a Program Element Monitor [PEM] on the Air Staff), believed that the only way to expand the space defence programme was to establish a Systems Program Office (SPO). They thought this would generate the requisite funding and momentum towards their overall objective. Usually when a SPO is formed it is in recognition that the programme has moved from the research project stage to the acquisition cycle. To facilitate this, they decided to group three existing Program Elements (PEs)—"space defense", "space surveillance", and "satellite survivability"—under the overall title of "Space Defense". This suggestion was outlined in the form of a Program Management Directive (PMD) or, in this case, the "Advanced Space Defense Program Management Directive". This was signed by Stelling and sent to Air Force Systems Command, who in turn passed it on to SAMSO. SAMSO, however, elected to keep the space defence programme as separate research projects. As the Air Staff had not issued a specific requirement for a Systems Program Office, SAMSO no doubt believed that insufficient support existed within the Air Staff to warrant such a reorganization. It was only after NSDM-333 had been issued in 1976 that a Space Defense System Program Office along the lines envisaged by Stelling was established.[52] Thus, it was ultimately a White House rather than an Air Force initiative that brought about the change.

The interest in antisatellite systems within the Air Force was not just confined to the Directorate of Space. Aerospace Defense Command (ADCOM—formally ADC), in particular, saw the "space defense" mission as a lifeline to reverse their diminishing status within the Air Force hierarchy. With the end of Program 437 and the declining mission importance of air defence of the continental United

States there was understandable concern within ADCOM about their future. Thus, it was its Commander, General James, who briefed the Buchsbaum panel in 1976 on the Air Force's requirement for an antisatellite system. Although the Air Force was eventually directed to develop the "Miniature Homing Vehicle", the programme proved to be insufficient to save Aerospace Defense Command as a separate entity and it was disbanded in April 1980.

Overall then, while certain elements of the Air Force campaigned for a US ASAT capability, this was by no means considered a high priority requirement or even a significant area of concern by the majority of the Air Force.

CONCLUSION

The period from 1969 to 1977 marked a gradual transformation in the attitudes of US policymakers towards the "space defense" question. Although the first full test of the Soviet satellite interceptor had been observed as early as 1968, its subsequent irregular test programme and limited performance meant that, by the time the first phase of testing had ended in 1971, it was causing only minor concern among a relatively small group of US defence planners. With the continued absence of testing until 1976 it became even less of an issue. In addition to these perceptions of the Soviet threat other factors help explain US policy during this period.

First, the Soviet ASAT system posed a qualitatively different security problem than the earlier orbital bomb "threat". Given the extra demands in identifying, tracking and targeting the Soviet interceptor, preemption of its use would be extremely difficult, if not impossible. Furthermore, as various study groups recognized, ASAT deterrence by the threat of retaliation against Soviet satellites was unlikely to be credible while the United States depended more heavily on its own satellites. Therefore, until a sufficiently threatening Soviet space capability arrived that the United States could *directly* negate or, alternatively, the level of satellite dependency became more balanced to make deterrence work, there was little incentive for the United States to develop an antisatellite weapon.

Second, the prevailing funding squeeze on defence outlays made "space defense" a low priority budget request, particularly during the period of US involvement in Vietnam. This acted to reinforce the bias towards the traditional missions of the Air Force when their budget was under consideration.

Third, the established US policy of preserving the sanctuary status of space for the unhindered use of "peaceful" military satellites continued to influence US decision-makers. Many believed that by exhibiting unilateral restraint in the development of antisatellite weapons the Soviet Union would in turn reciprocate and thus prevent the extension of an arms race into space. Indeed from 1971 to 1976 it seemed as if the Soviet Union had accepted the logic of this argument, a view encouraged by their acceptance of the NTM provisions of the SALT I agreements.

The sudden resumption of Soviet ASAT testing in 1976 provided a rude challenge to this wishful thinking. Although the tests were clearly the catalytic factor in the US decision to respond with its own antisatellite programme, US attitudes towards space defence had already begun to change. Concern over the vulnerability of US satellites had produced modest increases in funding for satellite survivability, while the growing Soviet use of space had rekindled interest in ASAT research. In this regard it was largely inevitable that the US would eventually reconsider the potential utility of an ASAT system. The net effect of the Soviet ASAT tests was to bring this issue to the forefront at a much earlier date than would have been the case. As we have seen the Air Force interest in an ASAT programme at this time can only be described as ambivalent, while the Navy does not appear to have been especially worried about the Soviet ocean reconnaissance satellites—supposedly the most threatening category of Soviet space systems.[53] Moreover, while the Defense Department was more concerned with Soviet space activities, it was still willing to forgo ASAT development in return for Soviet reciprocity.

However, meeting the "threat" from Soviet satellites does not appear to have been the primary motive behind Ford's ASAT authorization. Despite the expert opinion of a series of study groups which questioned the deterrent value of an ASAT system, Ford and his immediate national security advisors considered it essential for the US to match the Soviet ASAT capability regardless of its limited effectiveness and poor testing record. It was a decision that also appears to have been coloured by the imminence of the administration's departure. According to one observer in the White House, a number of hasty decisions were made in the final days of the Ford administration in order to influence the agenda for the incoming Carter administration.[54] Unfortunately, as is often the case with weapon system decisions, it is easier to authorize development than rescind it later. Ford's last minute decision still stands as the primary enabling act for the current US ASAT programme.

9

Carter and the Two-Track Policy

... The Soviets have conducted a series of tests of an anti-satellite interceptor. The United States is approaching this subject on two tracks. We have a program to develop on anti-satellite interceptor of our own; we are also pursuing discussions with the Soviets to limit such systems.

Dr James Timbie,
Chief, Strategic Affairs Division, ACDA
29 March 1979

INTRODUCTION

During the transition period, President-elect Carter expressed his desire for "real" arms control and called on his team of advisors to study ways to revive existing negotiations (particularly SALT) and also to propose new ones. With the growing concern that space would become a new arena for a superpower arms competition—heightened by the latest Soviet ASAT test just a month before the inauguration and then Ford's last minute decision—the Carter administration was immediately receptive to a State Department initiative to seek limits on antisatellite weapons. However, getting ASAT arms control on the agenda was relatively easy compared with the problems and delays that the administration experienced in reaching a common negotiating position. Part of the reason for this delay was that the question of ASAT arms control was caught up in a comprehensive review of the organization and objectives of national space policy. Thus it took over a year before the first round of negotiations were held with the Soviets on this subject.

In parallel with this arms control effort, the Carter administration continued with the ASAT research and development programme authorized by Ford. Despite the apparent contradiction, the ASAT programme was justified on the grounds that it would support the US bargaining position at the negotiations and act as a hedge against their failure. This classic "bargaining chip" rationale became known as the two-track policy. While it was hardly a novel approach or for that matter confined to the antisatellite issue, it did—at least in theory—represent a commitment to link future US ASAT testing and deployment to the progress of arms control. The first part of this chapter traces the formation of Carter's space policy.

A major reason why the US ASAT programme was not curtailed was the increasing pace of Soviet military space activities. In addition to its expanding military satellite operations, the Soviet Union continued to test its satellite interceptor system. Although it later exercised what appeared to be a test moratorium during the period of the negotiations, there were further allegations of Soviet advances in the development of directed energy weapons. These activities are described below in the section entitled "The Continuing Soviet Space Challenge" and provide a backcloth to a more detailed discussion of the progress made towards reaching an antisatellite arms control agreement with the Soviet Union.

The Formation of Carter's Space Policy

Within days of Carter taking office, a working level State Department official who had been involved in the interagency deliberations prior to NSDM-345 took advantage of the new administration to suggest a US ASAT arms control initiative. After consultations within State and with other departments, this proposal was presented to the White House where it reached a receptive audience.[1] The first official notification that the administration was seriously considering ASAT arms control and had in fact broached the matter with the Soviets came during a general press briefing given by President Carter on 9 March 1977. Here he announced:

> I have proposed both directly and indirectly to the Soviet Union, publicly and privately, that we try to identify those items on which there is relatively close agreement—not completely yet, because details are very difficult on occasion. But I have for instance, suggested we forego the opportunity to arm satellite bodies, and also to forego the opportunity to destroy observation satellites.[2]

The subject was again raised with the Soviets during the visit of Secretary of State Cyrus Vance to Moscow in March. While the centrepiece of Vance's mission was to propose "deep cuts" in strategic weapons, both ACDA and State prepared briefing papers that outlined a range of options for antisatellite arms control. Foreign Minister Gromyko rejected Vance's main proposal but—probably to salvage something from the meeting—they agreed to make arms control in other areas the subject of superpower discussions. At the press conference following the meeting on March 30th, Vance stated that both sides had agreed to set up "working groups" to discuss specific areas of arms limitation, including one for antisatellite weapons.[3]

Just prior to Secretary Vance's departure for Moscow, President Carter issued Presidential Review Memorandum PRM/NSC-23 that directed the newly created NSC Policy Review Committee (PRC) to "thoroughly review existing policy and formulate overall principles which should guide our space activities".[4] This directive had apparently come on the initiative of Major General Robert Rosenberg who had also been on the staff of the NSC during the Ford administration. Rosenberg was concerned about what was later euphemistically described as the "growing interaction" among US space activities but really referred to the poor co-ordination and friction between the four major sponsors of US space programmes: the intelligence community, the Defense Department, the federal space agencies (NASA) and the National Oceanic and Atmospheric Administration (NOAA) and private space organizations such as COMSAT.[5]

The Policy Review Committee worked on long-term issues and comprised of cabinet-rank officials from the relevant departments. In response to PRM/NSC-23 a steering committee was established which consisted of representatives from the Departments of Defense, State, Agriculture, Commerce, Interior, as well as ACDA, CIA, JCS, NSC, NASA, NOAA, and the Office of Science and Technology. Below this was a study group that worked full time on issue papers that were forwarded to the steering committee and eventually to the PRC for final approval.

After the initial meetings of the steering committee it soon became apparent that, due to its sensitivity, the antisatellite issue could not be discussed fruitfully in such a large and diverse group. As a result, an *ad hoc* Antisatellite Working Group made up of representatives from just the State Department, DoD, CIA, JCS and ACDA and chaired by Walter Slocombe (Principal Deputy Assistant Secretary for International Security Affairs) was separated from the main PRM-23 group to discuss ASAT related questions.[6]

Given the presidential interest in antisatellite arms control, the ASAT Working Group inevitably discussed US negotiating strategy and its relationship to the US ASAT programme. While none of the group wanted to curtail the programme, there was no support for a "crash" effort either despite the general acceptance that the Soviet Union possessed an "operational" capability (to be announced publicly by Secretary of Defense Harold Brown in October 1977). Rather, it was hoped that if the US programme proceeded in an "orderly way" it would facilitate the arms control process with the Soviet Union.

This argument was both logically persuasive and bureaucratically convenient. The prospect of a US ASAT capability would provide the Soviet Union with a real incentive to negotiate and give the United States bargaining leverage once the talks began. Furthermore, in the event that an acceptable arms limitation agreement proved unattainable, the United States would still have the capability to deal with threats in space. While it was questionable whether the Defense Department and the services would have forgone the US ASAT programme once it reached an advanced stage of development (as this would have been tantamount to accepting that the threat posed by Soviet satellites was less than the threat posed by Soviet ASATs), the two-track policy of pursuing ASAT arms control with ASAT R&D gave each of the departments represented at the Working Group what they wanted.

The idea of demonstrating an interim ASAT capability to bolster the US bargaining position was also considered by the Working Group. This came in the form of a proposal to test a co-orbital intercept vehicle which would disable its target by metal pellets.[7] This relied on "off the shelf" technology and could have been relatively easy to achieve within a few years. Although a "conventional" system of this kind did eventually become part of the US ASAT programme, its primary purpose was to provide a low-risk back-up in case the more complex Miniature Homing Vehicle project ran into difficulties. The Defense Department, however, apparently opposed this proposal, probably on the grounds that a successful demonstration of an interim capability might preclude the further development of the favoured MHV programme. The other members of the group—most notably the State Department and ACDA representatives—probably felt that this would unduly complicate the arms control process.

It was the question of what kind of limitations the United States should pursue that caused the most disagreement within the Working Group. The Defense Department appears to have initially favoured seeking the complete dismantlement of the Soviet ASAT system but, either as a result of growing scepticism about the verifiability of a comprehensive ban on antisatellite weapons or a desire to preserve some ASAT capability for the United States (or both), it increasingly favoured reaching a "rules of the road" agreement with the Soviet Union that would ban "hostile acts" in space. In contrast, the State Department and ACDA representatives were more optimistic of a ban on testing and deployment. This fundamental division of opinion only became important once the negotiations got under way.

By the autumn of 1977, the two-track policy was beginning to take shape. On September 3rd, after an extensive review of the Miniature Homing Vehicle programme, the Defense Department awarded the prime contract for its development to the Vought Corporation.[8] The ASAT Working Group also presented a range of arms control options to the President and in the resultant PRM/NSC-23 Decision Paper of September 23rd, Carter indicated his preference for comprehensive limits. The new DDR&E, William Perry, later gave a brief synopsis of this directive at the FY 1979 defence budget hearings:

> The PRM/NSC-23 Decision Paper dated September 23, 1977, requires that we seek a comprehensive ASAT agreement prohibiting testing in space, deployment and use of ASAT capability ... To reduce the possibility of a future space conflict, the President has directed that we seek an effective and adequately verifiable ban on anti-satellite systems with the Soviets. As a consequence of this decision an interagency group—of which DoD is a part—has been making the necessary preparations for negotiating with the Soviets.[9]

At the same set of hearings, Harold Brown indicated his own preferences when questioned about the US response to Soviet ASAT activities:

> I think that the preferable situation, even though we would be foregoing the ability to knock out some Soviet military capabilities, would be for neither country to have an ability to knock out the other's satellites. However, as you say, the Soviets have some slight capability now. How good it is is not so clear, but it certainly can threaten some of our satellites. Under those circumstances, I think we have no choice but to provide some kind of deterrent by moving ahead at least with R&D and if we cannot reach an agreement with them that satisfies us as to its verifiability, then have a deployed capability of our own.[10]

At the same time Zbigniew Brzezinski, Presidential Assistant for National Security Affairs, asked Harold Brown to bring together the various conclusions of the PRM-23 study group and draft a Presidential Directive (PD) on national space policy that could be placed before Carter for signature. A draft was duly completed and circulated to the participating agencies in November. This received strong criticism from the civilian space agencies for its apparent bias towards national security issues.[11] It was probably as a result of their opposition that the ensuing President Directive on National Space Policy (PD/NSC-37) was not signed until 11 May 1978.[12]

The public version of this directive, released on 20 June 1978, set out the "basic principles" guiding the US space programme. This included the statement:

> The United States holds that the space systems of any nation are national property and have the right of passage through and operation in space without interference. Purposeful interference with space systems shall be viewed as an infringement upon sovereign rights. The United States will pursue activities in space in support of its right of self-defense and thereby strengthen national security, the deterrence of attack and arms control agreements.

Although the details of the national security components of the Presidential Directive remain classified, the press release did give some indication of what had been decided. The relevant section noted:

> The Secretary of Defense will establish a program for identifying and integrating, as appropriate, civil and commercial resources into military operations during the national emergencies declared by the President. Survivability of space systems will be preserved commensurate with the planned need in crisis and war, and the availability of other assets to perform the mission. Identified deficiencies will be eliminated and an aggressive, long term program will be applied to provide more, assured survivability through evolutionary changes to space systems. The United States finds itself under increasing pressure to field an antisatellite capability of its own in response to Soviet activities in this area. By exercising mutual restraint, the United States and the Soviet Union have an opportunity at this early juncture to stop an unhealthy arms competition in space before the competition develops a momentum of its own. The two countries have commenced bilateral discussions on limiting certain activities directed against space objects, which we anticipate will be consistent with the overall US goal of maintaining any nation's right of passage through and operations in space without interference. While the United States seeks verifiable comprehensive limits on antisatellite capabilities, in the absence of such an agreement, the United States will vigorously pursue development of its own

> capabilities. The U.S. space defense program shall include an integrated attack warning, notification, verification and contingency reaction capability which can effectively detect and react to threats to U.S. Space Systems.[13]

Like Malcolm Currie's statement to Congress in January 1977, the White House Press Release was a clear offer of further US ASAT restraint in return for reciprocal action from the Soviet Union. Similarly, there was also the unambigous threat of a vigorous US space defence programme if the Soviet Union failed to comply.

It is interesting to note that PD-37 also reduced the constraints on the internal transfer of information gained from US intelligence satellites to those government agencies who would not have otherwise benefited from it. While the extent of the subsequent dissemination is unknown, it did lead, none the less, to the first official public acknowledgement since President Kennedy's "blackout" order in 1962 that the United States operated military reconnaissance satellites. President Carter chose the Congressional Space Medal of Honor awards ceremony at the Kennedy Space Center, Florida on 1 October 1978, to remark that:

> Photoreconnaissance satellites have become an important stabilizing factor in world affairs in the monitoring of arms control agreements. They make an immediate contribution to the security of all nations. We shall continue to develop them.[14]

Since the Kennedy administration, successive US presidents had debated whether to acknowledge the use of reconnaissance satellites. Indeed Carter's disclosure had come after a similar review in which the reaction of certain foreign governments was also reported to have been sought. While Carter's motives were primarily to improve the public's confidence in US arms control verification facilities, it is also likely that the disclosure was designed to strengthen the legitimacy of reconnaissance from space and indirectly pressure the Soviet Union to desist from further antisatellite tests.[15]

An NSC Policy Review Committee—subsequently known as PRC (Space)—was also established by PD-37 to provide "a forum to all federal agencies for their policy views to review and advise on proposed changes to national space policy; to resolve issues referred to the Committee; and to provide for the orderly and rapid referral of open issues to the President for decision as necessary". Frank Press, Director of the Office of Science and Technology Policy, became Chairman of this Committee. This was subsequently directed to assess future civil space policy.

Two Presidential Directives resulted from the Committee's deliberations. The first—PD-42 was signed on 10 October 1978 and set out in greater detail the administration's policy towards civil space applications, space science and exploration, as well as the utilization of the shuttle.[16] The second—PD-54, dated 16 November 1979—was more specific in that it concerned policy towards civil remote sensing.[17]

The Continuing Soviet Space Challenge

In addition to the general expansion of the Soviet military use of space, the most worrisome aspects for US defence planners were the continuing tests of the satellite interceptor and further reports of Soviet advances in directed energy weapons.

The Soviet Satellite Interceptor Tests: 1977-80

The first test of the Soviet satellite interceptor during the Carter administration began on 19 May 1977 with the launch of Kosmos 909 from Plesetsk into a highly elliptical orbit at 66° inclination. Four days later the interceptor (Kosmos 910) was launched from Tyuratam into an orbit with the same inclination. It appeared that a fast flyby interception would be attempted but the interceptor satellite returned to earth within one revolution and the test was judged by Western experts to have been a failure. The planned sequence for this last test only became apparent on 17 June 1977, when another interceptor—Kosmos 918—was launched against the previous target satellite (Kosmos 909). Kosmos 918 was initially launched into a 197 x 124 km orbit at the same inclination but, in a rapid manoeuvre, the interceptor suddenly "popped up" to pass by the target satellite at its apogee. In the same movement, the interceptor also returned to earth. This demonstrated both a greater degree of flexibility in the use of the Soviet antisatellite system and a capability to intercept satellites at higher altitudes.

Just prior to the next test, Secretary of Defense Harold Brown chose the 20th anniversary of the launch of Sputnik (4 October 1977) to declare at a press conference that the Soviet Union had developed the "operational capability to destroy some U.S. satellites in space".[18] As if to underscore Brown's announcement, the Soviet Union launched another target satellite (Kosmos 959) on October 21st. Five days later

Kosmos 961 was launched and exhibited "a low-orbit demonstration test of the high orbit pop up technique".[19]

Just as a new pattern of satellite interceptions seemed to be emerging, the Soviets reverted back to the earlier intercept method followed by the destruction of the chase vehicle. On 13 December 1977, Kosmos 967 flew up from Plesetsk with the interceptor (Kosmos 970) following on December 21st. This time a "slow-fly past" was completed after the original orbit of the interceptor had been circularized. The chase vehicle was destroyed afterwards. The same target vehicle was used in the last test in this series. On 19 May 1978, just before the start of the first round of antisatellite arms control negotiations in Helsinki, Kosmos 1009 was launched from Tyuratam and manoeuvred for a close interception before being commanded back to earth.

Although a further two Kosmos satellites were launched in 1979 with characteristics similar to target spacecraft (Kosmos 1075 on February 8th and Kosmos 1146 on December 5th), no apparent attempt was made to intercept them. It now seemed likely that the Soviet Union was observing a testing moratorium to avoid jeopardizing the outcome of the negotiations. After it became obvious that the talks would not be resumed in 1980, the Soviets again tested their interceptor satellite. On 18 April 1980 Kosmos 1174 intercepted Kosmos 1171, which had been launched on April 3rd. This test appears to have been a failure as the interceptor did not pass closer than the 8 km which is considered to be the lethal radius of its shrapnel filled warhead. It was also the last test before President Carter left office.

Opinions vary on the potential effectiveness of the Soviet ASAT system. Since 1968, three distinct types of interception had been demonstrated: the fast flypast, the slow flypast and the "pop up" technique. While some US satellites in low earth orbit were now considered vulnerable, the Soviet system had not demonstrated an ability to attack satellites at the crucial "mother lode" of orbits—the geostationary orbit—which contains the vital early warning and communications satellites. Indeed, in his October statement, Harold Brown had described the Soviet ASAT's operational capabilities as only "somewhat troublesome". This low appraisal was echoed by the Air Force Chief of Staff, General Lew Allen who testified at the SALT II Treaty Hearings in 1979, that:

> The system that they had tested so far has the potential of being effective against our low-altitude satellites. It was tested in that kind of a mode, and it has had some successful tests. On the other hand, it is difficult to assign it a

> very high degree of credibility because it has not been a uniformly successful program and they have changed parameters with many of the different launches they have made.

With regard to the Soviet system's readiness, General Allen stated further:

> They have the systems that are more or less at the ready. It is not a very quick reacting system. The systems that are at the ready are located in the missile test areas. So, I think our general opinion is that we give it a very questionable operation capability for a few launches. In other words, it is a threat that we are worried about, but they have not had a test program that would cause us to believe it is a very credible threat.[20]

However, others were rather more concerned, believing also that the Soviet system was "evolutionary" and would develop into a more capable and reliable antisatellite weapon. During the internal debates over US arms control strategy the conflicting views of the current and potential Soviet ASAT capability became a major source of contention.

Soviet Directed Energy Weapons Developments

After the official explanation had been given for the mysterious interference with certain US satellites in 1975, concern over Soviet advances in lasers and particle beam weapons had become—certainly in the public domain—somewhat muted. This all changed in March 1977 following the sensational and alarming reports of the journal *Aviation Week & Space Technology*. These were based on information supplied by the retired former Director of USAF Intelligence, Major General George J. Keegan, to the effect that the Soviet Union was now in the advanced stages of developing an operational particle beam weapon. While ballistic missile defence was generally seen to be the principal mission for this new weapon, its possible antisatellite applications did not go unremarked. Indeed, later reports would specifically single out this as one of the driving motives behind the Soviet directed energy weapons programme.

These reports not only widened and intensified the bureaucratic battle concerning the interpretation of Soviet scientific research and development but also inevitably led to a reassessment of the adequacy of US efforts in the same field. This debate, which often spilled over into the public arena, was complicated by the highly technical nature of the subject matter which relatively few could claim to comprehend fully. As in many technical intelligence debates, the available evidence was patchy and inconclusive. As noted in Chapter 7, Soviet scientists are renowned for their laser and particle beam research. The extent of their scientific progress in this area is evident both from their contributions to scientific literature and symposia and also from regular exchange visits with US scientists. The problem lay not just in finding out whether Soviet scientists had achieved a breakthrough in an area that US scientists considered to be unpromising in the near term but also in assessing the extent to which Soviet beam research was weapons-related. Guidance on the latter could only come from classified intelligence reports. Unlike the relatively public arena of outer space, where the Soviet antisatellite tests could not go unnoticed, possible directed energy weapon developments within the Soviet Union were much more difficult to detect.

Thus both the participants and external observers of the debate operated in an atmosphere of claims and counterclaims that made a clear picture almost impossible to obtain. In short, it was a situation that was ripe for manipulation depending on who had access to the relevant information and how it was used. Unfortunately such situations lend themselves to alarmist reports and rumours which are, for the reasons mentioned, very difficult to disprove. Given the reputation of the journal *Aviation Week & Space Technology* as a conduit for "hawkish" leakers, their reports were viewed by many as being wildly exaggerated. None the less, they still created considerable public and congressional concern.

The first of these reports came in May 1977, after General Keegan had earlier declared that the Soviets were building particle beam weapons that would be able to negate the US nuclear deterrent.[21] The article noted that as early as 1975, Keegan and other USAF Intelligence officers had been concerned with Soviet beam weapon developments and had reported their findings to CIA Director William Colby. In response, Colby apparently convened the Joint Atomic Energy Intelligence Committee to consider whether their concern was justified.

The basis to the claims of Air Force Intelligence was the evidence of suspicious Soviet activities at a plant near Semipalatinsk in Soviet Central Asia. For example, a Defense Support Program early warning satellite with scanning radiation and infra-red detectors had reportedly detected large-scale emissions of gaseous hydrogen with traces of tritium in the upper atmosphere close to the Semipalatinsk area. Air Force Intelligence officers believed this to be related to the testing of a charged particle beam device. There were also reports of a similar test site at Azgir in Kazakhastan, near the Caspian Sea. After reviewing this evidence the Joint Intelligence Committee concluded that, while a sizable Soviet particle beam technology effort was under way, there was not enough evidence to support the Air Force's allegations about Semipalatinsk and, moreover, a near-term Soviet beam weapons capability. A panel of the US Air Force Scientific Advisory Board was also convened and after a three-day meeting at the Lawrence Livermore Laboratories they also rejected Gen. Keegan's hypothesis. Keegan apparently refused to accept the CIA and SAB evaluations and collected together a group of "young physicists" to analyse the available information and "probe the basic physics of the problem". The conclusions of this group, according to the article, questioned many of the early scientific objections to Keegan's claims. In the face of continuing opposition, however, Keegan resigned in January 1977 to pursue his crusade in public.

Keegan's allegations of Soviet beam weapon advances had a considerable impact on the media. A fortnight after the May *Aviation Week* article Secretary of Defense Harold Brown was asked to comment on the reported Soviet breakthroughs at a meeting of the National Press Club. He stated:

> It is, in my view, and that of all the technically qualified people whom I know who've looked at the whole thing, without foundation; the evidence does not support the view that the Soviets have made such a breakthrough or indeed that they are very far along in such a direction.

He went on to state that the "laws of physics are the same in the U.S. and in the Soviet Union. And in this particular case, I'm convinced that we and they can't expect to have such a weapons system in the foreseeable future."[22] The debate continued however.

In May 1978, a comprehensive study completed at RAND by Simon Kassel on "Pulsed Power Research and Development in the USSR" concluded, after reviewing the publicly available information, that:

> . . . the scope of Soviet pulsed power R&D appears to be much larger than that required for the explicitly stated purposes of Soviet activity in this area, namely, the conversion of energy from controlled thermonuclear reactions . . . it seems reasonable to infer the existence of other objectives of the ongoing pulsed-power R&D, including such military objectives as directed-energy weapons.[23]

Later reports in *Aviation Week* cited further evidence of Soviet advances in beam weapon technology.[24] These included *inter alia*: Soviet ground-based charged particle beam propagation tests against targets at Sarova near Gorki; detection in February 1978 of high levels of thermal radiation and nuclear debris vented from a test facility near Semipalatinsk; and the testing of a compact hydrogen fluoride high energy laser at Krasnaya Parka—a large-scale installation about 30 miles south of Moscow. In July 1980 another Soviet test site at Sary Shagan in Kazakhastan was reported to contain an early prototype of a charged particle device.[25]

Despite these reports, the Carter administration remained sceptical of a Soviet near-term operational beam weapon throughout its term in office. The undeniable existence of some Soviet beam research did, however, motivate the Pentagon to review this basic assessment at least once a year from 1977 onwards.[26] The conclusion remained the same: despite an extensive research effort into particle beam weapons, the Soviets were not even close to an operational system. This was an opinion shared by the OSD, the majority of the Air Force, and the various independent advisory groups that reviewed Soviet activities.

As regards laser developments, there was evidence that the Soviets were building prototypes for ground and ship-based applications but not for *deployment* in space. Despite these conclusions, pressure grew for the US to accelerate its own research programme if only to understand why the Soviets were expending so much effort in this area (see Chapter 10).

A brief mention should be made here of two other related sources of concern: first, the ability of the Soviet Union to interfere electronically with the command and control links of US spacecraft; and second, the use of the Salyut space station for military purposes. At the FY 1979 Senate Appropriate Hearings, Air Force General Alton Slay stated that "the Soviet Union has electronic warfare facilities which could be employed against certain U.S. satellites".[27] The Salyut space stations were seen as an indirect threat, as they could aid targeting and post-attack assessment. Both Salyut's 3 and 5, launched in June 1974 and June 1976 respectively, were used for photoreconnaissance purposes.[28]

THE PURSUIT OF ASAT ARMS CONTROL

After the President had expressed his preference for comprehensive limits on antisatellite weapons with the September PRM-NSC-23 Decision Paper, the NSC began preparing for the negotiations. An Antisatellite Negotiating Working Group was subsequently established and chaired by the NSC's Director of Policy Analysis, Victor Utgoff. The departments represented were the same as the earlier ASAT study group, with the result that its membership was virtually identical. However, it did not meet regularly until the negotiations drew near in the spring of 1978. As most of the important decisions were also taken at this time and before the later rounds, the seniority of the group generally increased as well. Each of the departments and agencies represented also had their own institutional support groups that would prepare their representatives for the Working Group's meeting. For example, the DoD's group was known as the Executive Committee for ASAT Arms Control with Walter Slocombe as its chairman.

The Working Group was directly responsible to the Special Co-ordinating Committee (SCC) of the NSC, which was composed of the principals of the departments involved. Whenever a major decision on the US negotiating position was required, particularly if the Working Group was in disagreement, the SCC would convene to discuss the matter. If the SCC remained divided, Utgoff as chairman of the Working Group would then write up a summary of the discussion, with the respective views of the principals in the form of a separate decision memorandum for the President. This took the form of a number of "boxes" next to the available options, which the President would tick according to his preference. However, there seems to have been relatively few meetings of the full SCC on the question of ASAT arms control strategy. Most of the discussions appear to have taken place within the Working Group.

Once the negotiations began, another layer to the decision-making structure was added with the US Antisatellite Arms Control Delegation. When in session, this had its own Backstopping Committee to handle the day-to-day affairs of the negotiations and provide a channel for the interagency debate. Although it provided tactical advice and acted as a clearing house for proposals from the delegation, overall strategy was directed by the NSC Working Group.

When the Working Group began to consider the US goals and strategy for the ASAT negotiations, it soon became clear that this was not such a straightforward question as many had believed. While PRM/NSC-23 had called for a "comprehensive" ban on the testing, deployment and use of ASAT weapons (subsequently reaffirmed by

PD-37), it had not specified what this actually included and how it was to be achieved.[29] This was understandable as, until the negotiations were definitely on the agenda, such details were not necessary and, for the sake of bureaucratic harmony, were actually avoided. As the negotiations drew closer and the necessity for detailed planning increased, so the inherent problems associated with the limitation of ASAT weapons surfaced and divided the Working Group.

The most important problems stemmed from the question of coverage: what activities and devices were to be banned? As one participant to the discussions observed, "it was like pulling the thread on a sweater". While a "dedicated" antisatellite weapon was relatively easy to define, an enormous grey area surrounded the systems that have an ASAT potential. Electronic jammers and dual-capable systems such as ABMs and ICBMs could all be used for this mission. The Soviet Galosh exoatmospheric ABM system around Moscow, for example, had by definition a rudimentary ASAT capability. Also innocuous space activities like satellite docking and manoeuvring could be a disguise for satellite intercept tests. In short, how could an agreement cover the whole spectrum of possible ASAT acts and devices? How could it be drafted without leaving loopholes for potential abuse and, moreover, how could it be verified? A further element to the coverage problem that became more significant once the negotiations began was which states should be party to the terms of the agreement. In other words, should the coverage be strictly bilateral or extended to include other nations?

Another concern raised in these discussions was whether ASAT negotiations would reopen the whole question of the legitimacy of military activities in space and with it upset the tacit acceptance of satellite reconnaissance. Similarly, would it affect the force or legality of the existing treaties that directly or indirectly regulate military activities in space?

The more these problems were discussed, the more it convinced some—particularly in the Defense Department—that a comprehensive agreement as originally envisaged would not be possible, or even desirable. As a result, the debate in the Working Group centred on the relative benefits and drawbacks of trying to get the best possible coverage or settling for more limited but less problematic objectives. The group became increasingly divided on this issue, the principal coalitions being the State Department and ACDA versus the DoD and JCS.[30] The former favoured a complete prohibition on the testing, deployment and use of dedicated antisatellite weapons, while the latter increasingly argued for a noninterference/no-hostile act agreement within a more general "rules of the road" package similar to those applicable to the high seas.

The respective arguments were as much about verification as they were about the usefulness of comprehensive versus limited agreements. The Pentagon felt that a comprehensive agreement would be impossible to verify for the reasons discussed above. Even verification of the dedicated Soviet ASAT programme, they argued, would be difficult, as the SS-9 booster was used for other missions, notably for launching ocean reconnaissance satellites. Given the small number of high-value US satellites, a small covert Soviet ASAT capability would be extremely dangerous. The argument that the Soviets would not have confidence in a small, covertly deployed ASAT system because it could not be tested was also countered by reference to the resumption of testing after the hiatus between 1971 and 1976. There were also ways to disguise an ASAT test under the cover of innocuous activities like spacecraft docking. Moreover, the Soviet directed energy weapon programme provided a long-term ASAT potential that would be very difficult to verify. In contrast, the State/ACDA coalition argued that the planned upgrade in US monitoring capabilities—both for space surveillance and also ground-based laser testing from a specially designed facility in Iran—would provide "adequate" verification of Soviet treaty compliance. Furthermore, the problem of covert ASAT capabilities could be alleviated by increasing the level of US satellite survivability and redundancy. More importantly, they argued, a no-hostile act/"rules of the road" agreement would do nothing to diminish the ASAT threat over the long run and be meaningless once war broke out. In short, the State and ACDA representatives believed that the benefits of reaching an agreement that at least curbed the more overt antisatellite systems and hopefully an arms race in space outweighed the risks of Soviet covert ASAT deployments or "breakout".

Although both the CIA and ACDA carried out their own ASAT verification studies, the interagency group remained divided on the issue.[31] By March 1978, President Carter appears to have become impatient with the group's progress and decided to initiate formal discussions with the Soviets. As a result, Vance reportedly called in the Soviet Ambassador, Anatoliy Dobrynin to inform him that the US was prepared to begin talks in April.[32] Whether this was a deliberate ploy to expedite the deliberations of the Working Group is open to speculation. By the end of the month, the Soviets had accepted the US initiative and talks were expected to commence in May.[33] This proved to be premature as the formal State Department notification of the negotiations was not released until May 8th. In a terse statement they announced that the talks were expected to be "short and preliminary in nature" and would begin in Helsinki on 8 June.[34]

The US delegation for the first round of talks was led by ACDA Director Paul Warnke. Ambassador Robert Buchheim, the deputy head of the delegation, later took over from Warnke during the first round in Helsinki and continued to head the US team for the following two meetings. The US delegation was also made up of Herbert York; Maj. Gen. David Bradburn, USAF (Ret.); and Michael Michaud of the State Department. In charge of the Soviet delegation was Oleg Khlestov, Head of the Treaty and Legal Affairs Division of the Soviet Foreign Ministry. He was accompanied by Boris Mayorski of the Foreign Ministry and Lt. Gen. Ivan Pisarev, representing the Soviet Ministry of Defense. Both delegations were supported by a group of "advisors". Three sets of talks were eventually held: in Helsinki from 8-16 June 1978; Bern from 16-23 January 1979; and in Vienna between 23 April and 17 June 1979.

As the first round was expected to be an exploratory rather than a substantive negotiating session, the Working Group was not required to hammer out *specific* recommendations. In fact, the US delegation left for Helsinki without formal instructions. Only when they arrived did they receive their guidelines in the form of a Presidential Directive cabled to the US Embassy. This was short—a single page—with the general recommendation to explore the extent of Soviet interest and thinking on the issue. Warnke was apparently satisfied that his hands had not been tied by Washington and began by proposing a complete prohibition of antisatellite weapons.[35] The US delegation hoped the negotiations would become a frank exchange of views; in reality they were one-sided. The Soviet delegation noted the US views and asked some questions in return. According to one US participant, it appeared as if the Soviet delegation only began to give serious thought to the question of antisatellite arms control once the talks had commenced, as they asked for more time before the start of the second session to consult with Moscow.[36] Another hindrance was that the Soviet civilian negotiators appeared not to be well-informed about the details of their own antisatellite programme, which US negotiators had noted at other talks, for instance SALT. While the Soviet delegation did not admit the existence of their ASAT system they did not deny it either. This caused some problems in the wording of the final communiqué at the end of the talks, with the result that the subject of the negotiations was deliberately "fudged".

In addition to a complete prohibition, the US delegation explored various interim agreements with the Soviet delegation, including a moratorium on the testing of antisatellite systems and a "noninterference" or hostile act agreement. These were later outlined in some questions put to the Soviet delegation at the end of the first round in Helsinki. Similarly, the Soviets submitted their own questions for the US to consider. In particular, the Soviet Union wanted to know how the United States could guarantee that the space shuttle would not be used as an antisatellite weapon.[37] The US delegation had already been ordered to keep the space shuttle a nonnegotiable subject, in the belief that the Soviets would use the negotiations to place limitations on its future operations.[38]

The bilateral talks were expected to resume in the autumn but this proved to be too soon. One apparent reason for the delay was that the Defense Department decided to conduct its own internal study of antisatellite arms control. To some State Department and ACDA officials this lasted longer than was necessary and heightened their suspicion that the DoD's attitude toward ASAT arms control stemmed less from an appraisal of the problems of verification and more from a desire to maintain the United States' freedom of action to deploy ASAT weapons. When the ASAT Interagency Working Group reconvened, the same fundamental differences existed among the key players.[39] The DoD's position had now hardened in favour of just a non-use/noninterference ban with additional "rules of the road". However, they were prepared to consider a moratorium on ASAT testing for a set period. This was arguably disingenuous on their part, as the US ASAT programme was nowhere near the testing phase and therefore a short-term moratorium would not have been an impediment. The State Department and ACDA, on the other hand, continued to press for more comprehensive limitations. The chances for this could only be assessed after the next round of negotiations at Bern, Switzerland.

At Bern, the US delegation was encouraged by the fact that the Soviets had upgraded the status of the talks from "preliminary" consultations to "negotiations seeking concrete results".[40] After each delegation had presented their replies to the questions posed at the end of the first round, the United States again probed Soviet views on the range of possible agreements. It became clear that, while the Soviets were willing to discuss an ASAT testing moratorium, they were not willing to talk about the dismantlement of their own ASAT system. This was unacceptable to the United States, as it would have left the Soviet Union with an ASAT system intact and prevented the United States from developing its own to a comparable level. The only common ground between the two sides seemed to be a "non-use" agreement. But even here there were problems. The Soviet delegation expressed its wish to restrict the coverage to just US and Soviet satellites, which would have excluded vital allied satellites. They also continued with their objections to the space shuttle. Altogether twelve sessions were held in Bern from January 23rd to February 16th. While these were generally constructive, it was only at the third, and what proved to be final, meeting in Vienna that tangible progress was made towards an agreement.[41]

In preparation for the Vienna negotiations, a compromise was reached within the Working Group. In view of the Pentagon's opposition to a comprehensive agreement and the Soviet Union's unwillingness to discuss the dismantlement of its satellite interceptor, a two-stage strategy was adopted. In the short run the United States' goal at the negotiations would be a no-use/noninterference agreement—possibly in assocation with a testing moratorium—to be followed in the long run with an effort to ban the "hardware", in addition to the "act".[42] ACDA and State were willing to go along with this, as it appeared to be the best that could be hoped for at the time. They rationalized this compromise by arguing that the limited first step would provide a confidence-building measure for the longer-term comprehensive effort.

At Vienna, it appears that both sides drew up draft "no-use" treaties of their own, which were presented for discussion. As a result, a considerable part of this round of talks was spent on arguing the relative merits of each draft. However, progress was made in bringing the two together. For instance, it was reported that the Soviet Union redefined the United States' "hostile act" language to cover activities designed "to damage, destroy or displace" satellites. Test moratoria were also discussed; the United States and the Soviet Union respectively favouring ones of short-and long-term duration, for reasons described earlier.

Further progress beyond the preliminary redrafting was stymied by the now-familiar problems. The Soviets repeated their objections to the space shuttle and also reiterated their wish to restrict a "no-use" agreement to US and Soviet satellites. While this was perceived to be related to Soviet apprehensions of the embryonic Chinese satellite reconnaissance programme, it would none the less have left NATO and other allied satellites uncovered by an agreement.[43] The Soviets also reserved the right to circumvent a "no-use" agreement if "hostile or pernicious" acts by a foreign satellite infringed their national sovereignty. Some Western observers interpreted this to be part of the long-standing Soviet desire to curb the potential use of direct broadcasting satellites for propaganda purposes, but the only example given by the Soviets at the time was the use of satellites for delivering poison gas![44]

As the ASAT negotiations coincided with the culmination of SALT II, there was some hope in the United States that an ASAT agreement—if only symbolic—could be announced at the signing ceremony, which was also due to be held in Vienna. While President Carter reportedly arrived in Vienna for the summit with President Brezhnev fully briefed on the state of the ASAT talks in case a breakthrough could be reached, it seems that the Soviets wanted to keep the SALT II ceremony a "one-issue" summit.[45]

The ASAT talks were adjourned the day before the signing ceremony with the expectation that a fourth round would be held in the autumn. However, further progress towards the next round was delayed by the SALT II debate in the United States. The push to ratify this treaty became an all-consuming activity and overriding priority for the Carter administration. Some administration officials also felt that JCS support for SALT II might be jeopardized if they were pushed too quickly towards accepting ASAT limitations. While the JCS appeared to support limited ASAT arms control, they none the less had strong reservations that a "no-use" agreement would rebound on their effort to improve the survivability of US space assets. They feared that Congress would be less inclined to fund extra survivability measures in the mistaken belief that US satellites had become fully protected by such a treaty. JCS doubts about the adequacy of verification also gained wider currency with the loss of monitoring stations in Iran following the revolution in November 1979.

However, it was the Soviet invasion of Afghanistan in December 1979 and the resultant rupture in East-West relations that brought to a halt all further progress on ASAT arms control during the Carter administration. There were, however, a number of what have been described as "feints" from within the US administration in 1980 to start up negotiations again in the belief that further delays would exacerbate the problems of reaching an agreement. These came principally from ACDA and the State Department, although they were supported by some NSC staff representatives.[46] The Soviets also made some informal comments that suggested that they would be interested in another round of talks. But, with the resumption of Soviet ASAT testing on 3 April 1980, it seemed that they had given up hope of an agreement before the election.

CONCLUSION

While the Carter administration can claim it inherited the US ASAT programme, it made no effort to rescind Ford's decision or curb its subsequent development. Rather, the programme was allowed to proceed steadily as part of a "two-track" strategy that, in theory at least, linked its eventual deployment to a parallel effort to achieve ASAT arms control. This was a policy of convenience in that it appealed to the two main factions within the administration: it gave the Defense Department and the services an ASAT research and development programme and it offered the State Department and ACDA the prospect of arms control. The inherent contradiction of this policy could be reconciled by the assertion that the US ASAT programme provided vital bargaining leverage for the negotiations and an insurance policy against their failure.

But, like all "bargaining chip" arguments, the two-track policy could be maintained so long as the question of what the US wanted to prohibit, or put differently, what it was prepared to forgo, did not need to be addressed. Once negotiations with the Soviets began in earnest, the basic incompatibility of goals within the administration made conflict inevitable. Even though the guiding Presidential Directives appeared to support the State Department/ACDA desire for a comprehensive prohibition on the testing, deployment and use of ASATs, a combination of Defense Department resistance to this idea and Soviet intransigence towards the dismantlement of its own system meant that a no-use/noninterference agreement was about the best that could be achieved at the time.

By the end of the third round of negotiations, the framework of a preliminary draft agreement of this kind appears to have been accepted by the two sides. Although the talks became a victim of the "crisis" in arms control following the Soviet invasion of Afghanistan, there was still a general expectation that they would be resumed and most probably successfully concluded if President Carter was re-elected in November 1980. The election of Ronald Reagan, however, made many think otherwise.

10

US Antisatellite Research and Development, 1971–1981

Space-based systems continue to play an increasingly important role in support of United States and Soviet military operations. As this exploitation continues, the utilization and dependence on these systems may increase such that their loss could alter the outcome of a conflict. Hence, we must continue our research and development efforts to provide the eventual technologies for preventing an adversary from gaining an advantage from the use of space.

Malcolm Currie,
DDR&E, 1976

The Fall and Rise of the US ASAT Programme

The Demise of Program 437

Although Program 437 remained nominally operational until April 1975, its credibility as a weapon system had virtually disappeared with the transfer of launch personnel to Vandenberg on 1 October 1970. This reduced its operational readiness from 24 hours to 30 days. Moreover, the first of the planned semiannual exercises to test the ability of launch crews to redeploy and reactivate the missiles was not carried out until 5 November 1971. While successful, this appears to have been the last such exercise; the next one planned for April 1972 was cancelled for lack of funds.[1]

Despite the declining military interest in Program 437, it proved remarkably resistant to closure even from natural causes. For example, on 19 August 1972 Hurricane Celeste passed sufficiently close to Johnston Island to cause considerable damage to the launch and support facilities. After a complete damage assessment, the island's facilities were repaired and Program 437 was put back into commission on 13 September 1972. This proved to be premature as the damage, particularly to the computer systems, was more extensive than had first been thought. As a result Program 437 was taken out of service again on 8 December 1972 and did not return until 29 March 1973.[2]

That it finally returned to service, albeit only for a short time, is particularly surprising given both the cost of repairing the Johnston Island facilities and the commitment to phase out Program 437 by the end of 1974.[3] Apart from a probable rear guard action from Air Defense Command, the major reason was that the Johnston Island facilities were deemed necessary for other non-ASAT programmes: the High Altitude Program (HAP) component of the National Nuclear Test Readiness Program (NNTRP), the Ballistic Missile Defense Test Program, and the planned Site Defense Prototype Demonstration. Details of these programmes were subsequently laid out in the Program Management Directive for Program 437, dated 10 August 1974.[4] Although this directive formally ended the "satellite-intercept system" it was not until 6 March 1975 that CINCONAD informed the JCS that "The Program 437 Anti-Satellite Option is terminated effective 0001Z on 1 April 1975."[5]

Project SPIKE

At the same time as Packard proposed phasing out Program 437 in his memo of 4 May 1970, he also suggested that he would "like the Army and the Air Force to consider the possibilities for a U.S. non-nuclear capability against Soviet satellites".[6] While the Army does not seem to have responded to this suggestion (the Safeguard system would have used nuclear tipped Spartan interceptors), Air Defense Command took this opportunity to renew its efforts to gain a follow-on ASAT system.

In April 1971, ADC proposed the development of an air-launched missile capable of being able to "intercept and negate satellites prior to their first overflight of the continental United States".[7] Known as Project SPIKE, its design philosophy was to use existing systems and state-of-the-art technology wherever possible. In this case, the proposed system consisted of an F-106 fighter and a modified Standard AGM-78 anti-radar missile with a terminal homing warhead. After ADC had briefed HQ Air Force on the project, SAMSO was directed to examine its technical feasibility. Both the commanders of ADC and NORAD—Generals McKee and McGehee, respectively—hoped that work could start on the development of SPIKE with FY 72 funds.[8] However, the decision to proceed was deferred pending the outcome of SAMSO's review.

By 17 September 1971 SAMSO was ready and briefed senior Air Force officials on the feasibility of the project. SAMSO was generally positive in its recommendations and called for six space intercept demonstrations at an estimated cost of $30 million.[9] John Hansen, the Assistant Secretary of the Air Force for Research and Development, could not attend the briefing but his deputy, Joe Jones, was present to raise Hansen's questions about the project. In particular, Hansen asked to know: first, why ADC wanted a satellite interceptor; second, could the envisaged missile reach "operationally meaningful altitudes"; and, third, was "the concept too sophisticated to attain without major technical advancements"?[10] Jones reported back afterwards that he considered the briefing to be "fairly shallow" and that while Project SPIKE was "a fairly interesting concept, it was by no means a simple state-of-the-art engineering development but was a rather sophisticated development whose potential risk could not be easily assessed without detailed review of the assumptions used in the study".[11] This sentiment also appears to have been felt by the senior Air Force officers present at the briefing, as Project SPIKE was not included in the FY 73 "New Initiatives" budget submission as ADC had hoped.[12] However further research on terminal homing warheads for ASAT and BMD purposes did receive some support.

The Origin of the Space Defense Programme

It was not until the end of the Ford Administration that research or antisatellite weapons, satellite survivability measures and space surveillance were grouped together to form the "space defense" programme. The origin of these individual programme elements, however, can be traced back to the early 1970s.

For example, research in "Miniature Homing Vehicle Technology", which is the basis of the current US ASAT development programme, began as early as 1970 under the "Missile and Space Defense" programme element, although clearly this was an outgrowth of the research on nonnuclear kill mechanisms carried out in the 1960s.[13] The following year also witnessed the introduction of a variety of "New Initiatives", including "survivable satellite" technology, "on board sensors" and the "Aerospace Defense Program".[14]

The survivable satellite programme was designed to reduce the vulnerability of satellite communication links, while the "on board sensors" were aimed at providing warning of physical and electronic attack to US satellites. The inclusion of both these "new initiatives" in the budget was almost certainly the result of the heightened concern over the recent series of the Soviet ASAT tests. The $10 million requested for the Aerospace Defense Program appears to have been connected primarily with Project SPIKE but with its deferment only $3.1 was eventually authorized. The Air Force repeated this request for the FY 1973 budget but under the new title of the "Space Defense Program". This included continuing work on the design of "interceptor systems".[15] However, the final authorization amounted to only $199,000. In fact, funding declined dramatically during FY 1973 and 1974 and only picked up in 1975. Table 10.1 gives the requested and actual funding totals for the Space Defense Program from 1972 to 1978.

Table 10.1: Funding for Space Defense (PE64406F), FY 1972-78 ($ millions)

FY	*Requested*	*Actual*
1972	10.0	3.1
1973	10.1	0.19
1974	0.1	0.1
1975	3.5	2.75
1976	4.0	4.0
1977	12.8	12.8
1978	41.6	41.6

Source: US Congressional Hearings.

The level of support for satellite survivability measures was also not as great as hoped for reasons described in Chapter 8. Work did continue, however, on impact sensors, laser illumination warning devices and the Lincoln Experimental Satellites (LES 8 and 9). The latter utilized a variety of novel antijamming techniques in addition to radiosotope thermoelectric (nuclear) generators rather than vulnerable solar panels.[16] Survivability measures also began to be integrated into existing satellite programmes, albeit in marginal ways at this stage.[17]

Public references to the Air Force's antisatellite research were extremely rare until March 1975 when an article in *Aviation Week & Space Technology* indirectly referred to the miniature homing vehicle

(MHV) programme. This reported that the Air Force was beginning a three-year "technology effort" into a small ground- or air-launched ASAT interceptor that would be directed by a Long-Wave Infrared (LWIR) homing system and rely on its impact velocity to disable the target.[18] It was further reported that three companies—Rockwell International (Autonetics Group), LTV and General Dynamics were competing for what had now become a four-year effort organized by SAMSO to develop antisatellite-related technology.[19] Later reports referred to competitive design work on miniature destroy-by-impact interceptors between General Dynamics (Pomona Division) and a team of contractors led by LTV and including Boeing and Hughes.[20]

Despite heavy censorship, this research effort was outlined for the first time in the FY 1977 budget submission. Of the $12.8 million requested, $9.9 was budgeted for "Miniature Systems"; $1.4 million for "Advanced Systems" that examined the use of lasers for the ASAT mission; $1.4 million for an "Instrumented Vehicle" to test the effectiveness of the MHV interceptor; and $100,000 for "Special Experiments".[21] Thus, as noted in the previous chapter, the component parts of the antisatellite programme that was later adopted by the Carter administration were present before the Soviets resumed testing in February 1976. However, it should be noted that at this stage $19.4 million was expected to be funded for FY 1978, in contrast to the eventual appropriation figure of $41.6 million.

Satellite survivability also received more attention in the FY 1977 budget, though not nearly enough for the White House, which led to Ford issuing NSDM 333. A breakdown of the various research activities within this programme included a laser receiver to detect interference and Soviet tracking of US satellites, a radar receiver to detect target acquisition tracking by ground radar, manoeuvrable systems to avoid attack, decoys to draw off an attack, laser survivability tests, ground station internetting to improve ground station redundancy, and survivable launch systems for "minimum mission replacement with small payloads".[22] Some of the earlier research efforts also began to reach fruition in 1976. On March 14th, after a year's delay, the LES 8/9 spacecraft were launched from Cape Canaveral. In addition to nuclear RTE power sources these satellites were also a test bed for Ultra High Frequency (UHF) and K band communication equipment using a variety of antijamming devices.[23]

The third element of the emerging space defence programme was space surveillance. This, as noted earlier, is critical to both satellite survivability (for warning) and antisatellite activities (for detection, tracking and targeting). Since the early 1970s there had been a number of significant additions and improvements to the SPADATS system listed earlier. These included the Cobra Dane phased array radar at Shemya in the Aleutians; the Perimeter Acquisition Radar (PAR) from the cancelled Safeguard ABM system; and the Pave Paws phased array SLBM radars at Otis AFB, Massachusetts, and Beale AFB, California. Research on more advanced systems was carried out under the "Space Surveillance Technology" programme. Among the projects funded here were the development of satellite-borne LWIR sensors and the Ground-based Electro-Optical Deep Space Surveillance (GEODSS) system to improve coverage up to the geostationary orbit.

SPACE DEFENCE RESEARCH UNDER CARTER

The Antisatellite Programme

After the Carter administration had agreed to continue ASAT research and development as part of the "two-track" strategy, it still needed to decide which programme to pursue. With work already underway on the Miniature Homing Vehicle, this seemed the most promising option. As described in the previous chapter, Carter's advisors initially discussed demonstrating a co-orbital satellite interceptor with an exploding pellet warhead to redress the perceived imbalance in ASAT capabilities and provide some bargaining leverage for arms control. For reasons discussed above, tests of this system were discounted but the programme was retained as a "low risk backup" to the Miniature Homing Vehicle.[24] Despite unforseen difficulties this remained the primary US development effort.

Both Vought and General Dynamics competed with separate designs for the MHV programme and after a review of their respective merits by the Office of Defense Research and Engineering and Air Force Space Division, Vought was finally awarded the prime contract, worth initially $58.7 million in September 1977.[25]

The Miniature Homing Vehicle is small (12" x 13") in size and basically consists of a cluster of rockets surrounding eight cryogenically cooled infra-red telescopes. After the MHV has been launched and manoeuvred into the vicinity of its target, these heat-seeking sensors lock on and track the satellite. Then by means of the small rocket thrusters, the interceptor homes in to ram the target. The velocity of the intercept is considered sufficient to destroy the target without necessitating a special explosive device.[26] Although Vought and General Dynamics had considered a variety of ways to deploy the MHV—on air-launched missiles, aboard spacecraft and on ground-based ICBMs—the final preference had not been decided by the beginning of 1978.

On 20 January 1978, Harold Brown directed the Deputy Under Secretary of Defense (Policy) to define US ASAT requirements. This prompted a series of directives that resulted in a two-phase ASAT requirements study.[27] Initially the Air Force considered using a ground-based launch vehicle such as the Minuteman III ICBM, with an aircraft-launched system as an alternative.[28] With this in mind the Air Force Flight Centre at Edwards AFB, California, had already announced that it was seeking proposals for an air-launched space booster.[29] By the end of the year, the Air Force had reversed its preference and decided to rely on an air-launched missile for the MHV. A modified Boeing Short-Range Attack Missile (SRAM) was chosen as the first stage and a Vought Altair III booster, which is normally used as the fourth stage on the Scout space launcher, was picked to be the second stage behind the MHV. The resultant missile is approximately 17 feet long, 18 inches in diameter and weighs about 2,600 lb.[30] The Air Force also evaluated a variety of possible launch platforms and the F-15 was finally chosen. Its high operational ceiling and its rapid rate of climb made it an obvious choice.

Apart from considerations of vulnerability, air-launched systems offer significant advantages in operational flexibility over fixed-site installations. The combined range of the F-15 aircraft and the reach of the missile permits attacks against targets at different orbital inclinations and, unlike the Johnston Island facility, they do not have to wait for the target to come "overhead". As General Thomas Stafford Air Force Deputy Chief of Staff for Research Development and Acquisition stated, the use of the F-15 permits "more attacks per day".[31]

The choice of launch vehicles for the MHV was also apparently heavily influenced by studies completed in May 1978 under the direction of the Joint Chiefs of Staff. The final JCS study report established a "prioritized target list" of Soviet satellites and according to General Stafford this provided the first rational basis for cost-benefit trade-offs among competing antisatellite concepts. The JCS target list referred to two "threat" time frames: the current threat (1978-85) and the future threat (1985-90).[32] The May 1978 study was only an interim report as details of the final JCS Directive were leaked to Jack Anderson of the *Washington Post* in 1981.

A maximum of 45 satellites were identified as posing some degree of threat to the United States to warrant their destruction in the event of war. They were organized into the following priority categories:

> Priority 1—Soviet Electronic Ocean Reconnaissance Satellites. These are used "for detecting and tracking of Naval Vessels in open or coastal waters" and "in land or air warfare (they) could also be used to detect airborne warning and control systems, radar sites and operating airfields". These are to be destroyed "as soon as possible".
> Priority 2—The Soviet Salyut space station which can detect missile launches and provide ICBM targeting data; photographic reconnaissance satellites which can be used in tactical situations; navigation satellites which can help direct missiles; and communication satellites.
> Priority 3—Soviet meteorological and early warning surveillance satellites.
> Both Priority 2 and 3 targets are to be destroyed within 48 hours.[33]

Given the time urgency of the JCS target list, the flexibility of the air-launched system was certainly a more attractive proposition. However, as some of these Soviet satellites occupy high altitude orbits—most notably the early warning and the *Molniya* communication satellites—the possibility of using Minuteman or even Polaris boosters with their extra reach was considered as a future option. The additional cost to achieve this extra capability was, however, not insignificant. Whereas the expected cost of the F-15/MHV system over its lifetime was put at $1.8 billion (current), a high altitude system was estimated to cost in the region of $2.5 billion.[34]

In addition to these design decisions, 1978 appears to have been a watershed year for the ASAT programme as a whole. On July 5th, the Secretary of the Air Force designated the Space Defense Systems Program for "semiannual review", which represented its transition to the status of "a major weapon system".[35] Moreover, the Under Secretary of Defense, Research and Engineering approved an Air Force proposal for "accelerating the Miniature Vehicle program". Later the Secretary of Defense approved the option to bring forward its "initial operating capability (IOC)" date.[36] Whether this move was in any way designed to pre-empt the parallel arms control effort is open to speculation. The commitment to a test programme was further illustrated with the contract award to Avco Everett on 25 June 1979, to develop the "Instrumented Test Vehicles" (ITVs). Consisting of inflatable metallic spheres, they can simulate the thermal "signatures" of a variety of potential targets.[37]

However, the technical challenges of the MHV proved to be greater than anticipated which resulted in some "slippage" in the programme's schedule. At the end of 1980 the programme was still awaiting its first DSARC (Defense Systems Acquisition Review

Committee) review. This had been originally planned for June 1980 but was postponed until 1981.[38]

As noted earlier, the Defense Department continued to support the development of a satellite interceptor with a pellet warhead as a back-up to the MHV programme in case it became impracticable. Referred to as the "conventional" programme, it relied on "low risk off the shelf technology". Three satellites weighing 700-1500 lbs. were initially scheduled to be built for the engineering development phase.[39]

In addition to the MHV and the "conventional" ASAT programmes, research also proceeded on what was titled the "Advanced System". Details of this programme are sparse but it is certain that it involved the use of lasers to disable satellites. Funding for the various elements of the Space Defense Program during the Carter administration is provided in Table 10.2 below:

Table 10.2: Funding for Space Defense by Programme, FY 1978-81 ($ millions)

	FY 1978	*FY 1979*	*FY 1980*	*FY 1981*
Miniature Homing Vehicle	20.9	44.8	63.0	82.5
Conventional Systems	3.5	5.2	1.0	—
Advanced Systems	9.8	14.5	.5	2.1
Instrumented Test Vehicle	6.7	7.9	16.0	25.3
Wargaming	.7	.6	—	—
Total	41.6	73.0	80.5	110.4

Source: Miscellaneous Congressional Documents, Arms Control Impact Statements, FY 1978—1981.

Satellite Survivability

Despite continuing concern at the vulnerability of US space assets, the level of effort to improve satellite survivability remained low relative to the value of the satellites. The financial and payload considerations identified earlier continued to hinder the incorporation of survivability measures into the design of future space systems although the situation was now markedly better than under Ford. Another major hindrance continued to be the diversity of "users" in the national security community. Without a strong and single constituency advocating change, it continued to be an uphill battle for those interested in improving the level of satellite survivability. As one exasperated Air Force officer stated: "You would think when the Defense Department needed a space asset we would make it survivable but we don't."[40] This is perhaps an overstatement but even Dr William Perry (DDR&E) admitted in a speech on election day 1980 that the United States was "nowhere close to having solved the problem of satellite survivability". The programme, he stated was moving "too slowly" and "too timidly".[41] This is reflected in the funding for the programme element "satellite systems survivability", which remained relatively constant in real terms during Carter's term in office.

Table 10.3: Funding for Satellite Systems Survivability, FY 1978-81 ($ millions)

	FY 1978	*FY 1979*	*FY 1980*	*FY 1981*
Satellite Systems Survivability (PE 3438F)	19.0	23.2	27.0	33.3

Source: US Congressional Hearings.

The satellite survivability measures funded under the programme can be divided into three areas of activity: improvements to (a) the space segment or "orbital component" that covers the satellites; (b) the "link component" comprising the communication "uplinks" and "downlinks"; and (c) the "ground component" which includes both the launch facilities as well as the satellite tracking and control systems. The first category included defensive measures (early warning sensors, the use of higher altitudes and better manoeuvrability to avoid attacks, improved hardening against nuclear and kinetic kill mechanisms, radar profile reduction, antijamming devices, decoys), and measures designed to increase the level of redundancy (the use of inert or "dark" satellite spares and cheap replacements for emergency use). The research aimed at reducing the vulnerability of the link component included improved antenna designs and the use of wider bands in the radio frequency spectrum (in particular Extremely High Frequency [EHF]) to reduce the potential for jamming; increasing the independence of satellites to minimize reliance on ground commands; manipulation of radio signals to either spread information over a variety of frequencies (frequency hopping), or compress it into very short bursts at high frequencies; and the introduction of operational codes to gain entry to satellite services so as to reduce "spoofing" and unauthorized use.[42]

Considerable concern also began to be voiced over the vulnerability of the ground component, particularly the satellite launch and control facilities. The need for greater survivability and redundancy in these facilities was one of the major recommendations of a Defense Science Board study led by Dr Michael May in the summer of 1980.[43] With respect to the former, the fear was that either the Eastern or Western Test Ranges would be destroyed in a nuclear attack thus eliminating the satellite replenishment capacity of the United States. Similar fears were also voiced about investing too great a launch dependency in the space shuttle. In response to this, studies were also carried out to see whether either the Minuteman or Trident missiles could be used to launch satellites in an emergency.[44] Some believed that proliferating the number of satellites through cheap and expendable replacements was the optimum route to ensure services and counteract the Soviet threat. Another method that gained attention was the use of civil satellites for communication and navigation purposes in the event of war. Extensive use is already made by the DoD of such satellites through leasing arrangements and considerable capacity exists for further usage if the need arises. As noted earlier, PD-37 called for preparations to expedite the use of civil satellites in an emergency.

The US satellite command and control facilities were also criticized, in particular the Satellite Control Facility (SCF) at Sunnyvale, California. Not only is this in an unhardened building and therefore vulnerable to attack, but it is also located near the San Andreas Fault, and a major highway which makes it exposed to a possible terrorist attack. As a result, the construction of a Consolidated Space Operations Center (CSOC) was authorized to provide a more modern and survivable satellite control facility. This is currently being built at Peterson AFB, Colorado Springs.[45] Action was also taken to provide mobile emergency control facilities for various satellite constellations.

Much of the design and co-ordination of the satellite survivability measures was based on "survivability cost-benefit calculations" which relate the cost of improving satellite survivability to the nature of the threat and the value of the satellite to the US military forces. This is facilitated by the annually updated Threat Workbook and Space Mission Survivability Implementation Plan (SMSIP) which contains schedules, plans and programmes for various survivability options to meet a range of specific threat scenarios.[46]

Space Surveillance Improvements

The space surveillance research and development programme under Carter was divided into two efforts: the SPACETRACK Improvement Program and Space Surveillance Technology Program. Funding for both programmes during the Carter administration was as follows (in current millions $):

Table 10.4: Funding for Space Surveillance Programs, FY 1978-81 ($ millions)

	FY 1978	*FY 1979*	*FY 1980*	*FY 1981*
Spacetrack (PE 12424F)	46.3	22.1	13.2	14.2
Space Surveillance Technology (PE 63428F)	110.0	36.1	42.1	44.3

The improvements to SPACETRACK included the following: providing a Pacific Radar Barrier (PACBAR) on Kwajalein Atoll, the Philippines and Guam; converting and integrating DARPA's existing Space Object Identification (SOI) facilities on Maui, Hawaii; providing a five site global Ground-based Electro-Optical Deep Space Surveillance (GEODSS) system to detect and track satellites out of geosynchronous orbits; and upgrading the range and general capability of the existing radar system. The Space Surveillance Technology programme covered longer-term research and development to improve satellite tracking and prediction, attack assessment and warning, and US ASAT targeting information. Another major component to this programme was the study of space-based surveillance platforms utilizing Long Wave Infra-Red technology (LWIR).[47] DARPA also continued to study more advanced methods within its SOI programme.

To co-ordinate all the US antisatellite, space surveillance and satellite survivability operations, it was decided that a Space Defense Operations Center (SPADOC)—initially referred to as the NORAD Combat Operations Center—be established. This became operational on 1 October 1979, though only in its preliminary stage of development. This is located in NORAD's Cheyenne Mountain Complex near Colorado Springs. Planning for a Prototype Mission Operations Center to support the testing and initial operations of the US ASAT system was also begun.[48]

THE US DIRECTED ENERGY WEAPONS PROGRAMME

From the late 1960s onwards, all three of the US armed services in addition to the Defense Advanced Research Projects Agency (DARPA) began to investigate the potential military uses of lasers and particle beam weapons. However, it was only towards the end of the Ford administration that significant attention was drawn to their potential for space defence purposes. The satellite blinding incidents of November 1975 not only raised the spectre of Soviet progress in this area but also increased the level of US interest. Malcolm Currie's FY 1977 posture statement reflected this when he noted the advantages of using lasers in space:

> Almost from the inception of the high energy laser, people have speculated on the possibility of deploying them in space. The technical problems are formidable, requiring major advances in chemical laser devices, precision pointing and tracking; and large, high power optics. Nevertheless, space is a favorable environment for chemical lasers. The pressure recovery problem that terrestrial and airborne applications must face does not exist in the vacuum of space. Nor are there propagation problems due to the atmosphere which can distort the beam and lessen its effectiveness.[49]

By the time the Carter administration entered office, the level of DEW research in the United States was becoming a hotly debated issue. Allegations of Soviet advances by General Keegan and alike brought sharp criticisms of the US effort and spurred frequent demands throughout Carter's term in office for a "crash" development programme.

This debate, however, was characterized as much by the arguments over the *direction* in which support should be given as it was about the level of funding. Thus the respective advantages and developmental problems associated with particle beam weapons and high energy lasers were hotly debated by proponents of DEW technology. This is turn generated further highly technical arguments over the efficacy of certain techniques for certain missions *within* these two basic areas of research. The use of DEWs in space for either ASAT or BMD purposes was after all just two of a range of potential strategic and tactical applications being canvassed.

Despite the optimistic assessments of some members of the Air Force, scientific community and Congress, the Carter administration remained sceptical of the near-and medium-term prospects for DEW technology. They believed that it had not matured sufficiently to warrant high levels of funding and that "conventional" methods were likely to remain more cost-effective in the foreseeable future. As a result, the level of funding remained relatively low and, in the case of some projects, even declined. As a result, despite calls for further development, the character of the US DEW research effort remained primarily exploratory. Its two principal components are discussed below:

US Particle Beam Research

In the autumn of 1978, partly in response to the speculation in the press concerning Soviet particle beam weapon (PBW) development, Dr Ruth M. Davis, Deputy Under Secretary for Research and Advanced Technology, sponsored a group of 53 physicists and engineers to assess the feasibility of PBWs and to outline a five-year developmental programme. The Particle Beam Technology Study group, as this was known, was directed by Edwin Chapin at the Los Alamos Laboratory. Their report, which was completed in March 1979, recommended further research in five main technology areas: power systems, accelerator technology, beam propagation, pointing and tracking and beam interaction with target materials. These recommendations were tailored to meet certain potential endo- and exoatmospheric applications.[50] An Office of Directed Energy Technology in the Defense Department was also established, which to some indicated a reversal of general attitudes to beam weapons research.[51] If this was the case, however, it was not translated into higher levels of funding or a "crash" development programme.

Two major projects were pursued: the US Navy's Chair Heritage charged particle beam programme and the Army's White Horse (formerly SIPAPU) neutral particle beam programme. Both projects were transferred to DARPA's control at the Lawrence Livermore and Los Alamos Laboratories, respectively. While the Chair Heritage programme was primarily designed for fleet air defence, the White Horse programme is tailored more for exoatmospheric applications, particularly ballistic missile defence. Neutral particle beams are believed to be particularly suited for this role because they would not be affected by the earth's magnetic field.[52]

A Defense Science Board "task force" carried out another review of US particle beam research in 1980 and apparently recommended that it be kept as a "technology programme" and therefore without a dramatic increase in funding. The funding during this period was as follows: FY 1980—$24.2m; FY 1981—$28.6m; and FY 1982—

32.5m.[53] Optimistic projections of a workable particle beam weapon system put it at least 25 years into the future—in other words, a twenty-first century phenomenon.

US Laser Beam Research

In contrast, high energy lasers are generally considered to be more practicable as a weapon system. As a result the Laser Technology Programme is much broader in its range of applications and at a more advanced stage of development. Both the Air Force and the US Navy have developed and tested high energy lasers against various targets. The Air Force's Airborne Laser Laboratory (ALL) is located in a Boeing NKC-135 test bed aircraft while the Navy Sealite programme is still land-based.[54]

In FY 1980, funding began for the development of lasers for use in space. Known as the laser "Triad" this consists of three interrelated projects: the development of a Hydrogen Fluoride Laser (Project Alpha); the demonstration of a laser optics system using 4-metre diameter mirrors (Project LODE) and the development of a laser acquisition, pointing and tracking device (Project Talon Gold).[55] In August 1980, it was reported that Harold Brown had redirected high energy laser research to emphasize space-based applications. As a result, funds for both the ALL and Sealite programmes were later reduced. This announcement had come after reports that a recent Defense Science Board High Energy Laser Task Force, led by Dr John S. Foster, had in fact recommended that *near-term* applications like shipboard defence should be emphasized. As a result, there were subsequent allegations that the Task Force's recommendations had become distorted.[56] Later, Foster presented the findings of the Defense Science Board to Congress. He reported that the DSB were unanimous in recommending that: "it is too soon to attempt to accelerate space-based laser development toward integrated space demonstration for any mission, particularly for ballistic missile defense" . . . as these tasks "can be performed more cheaply by other technologies such as miniature homing devices and ground-based lasers".[57]

While funding was increased slightly for DARPA to study the use of lasers in space, it was left to the Reagan administration to decide whether this should become a major development programme.

11

The Reagan Presidency: Towards an Arms Race in Space, 1981–1984

... no arrangements or agreements beyond those already governing military activities in outer space have been found to date that are judged to be in the overall interest of the United States and its Allies.

President Ronald Reagan,
31 March 1984

Introduction

The future development of ASAT weapons was still very much in the balance when President Reagan took office in January 1981. Although the US ASAT programme was now entering the advanced stage of development and the Soviet Union had resumed the testing of its satellite interceptor, neither represented irreversible commitments. Furthermore, while the earlier bilateral ASAT negotiations had raised some significant obstacles to arms control, these were by no means insurmountable. In short, an agreement that at least alleviated the problem was still within the superpowers' reach. By 1984, this opportunity appears to have been lost; an arms race in space now seems inevitable. The final chapter in this study on the evolution of US space policy under successive presidents examines the policy of the Reagan administration and the circumstances that lead one to this depressing prognosis.

US Military Space Policy Under Reagan

During the transition period after the election in November 1980, there was little indication of where the incoming administration stood on space issues in general, and even less on the more specific question of antisatellite weapons. While a complete reappraisal of US policy was inevitable after a changeover of administrations, there were other more pressing subjects for the Reagan administration to consider before yet another study could begin. As a result, it was not until August 1981 that Reagan directed the National Security Council to review national space policy. An interagency working group chaired by Dr Victor H. Reis (Assistant Director of the Office of Science and Technology Policy) was subsequently formed to conduct the study effort.[1] This consisted of representatives from the Departments of State, Defense, and Commerce, as well as from the Central Intelligence Agency, the Joint Chiefs of Staff, ACDA, NASA, OMB and the NSC. The group was directed to address both civil and military issues, including: "(1) launch vehicle needs; (2) adequacy of existing space policy to ensure continued satisfaction of United States civil and national security program needs; (3) shuttle organizational responsibilities and capabilities; and (4) potential legislation for space policy".[2] At the same time the Defense Department began its own internal study.[3]

However, outside observers did not have to wait for the findings of these reviews to discover the new administration's attitude to ASAT arms control. In August 1981, the Reagan administration rejected a Soviet offer to discuss a draft space weapon treaty that was submitted to the UN General Assembly. While the draft treaty contained significant flaws, the US made no effort to make a counter proposal or give any indication that it was interested in resuming bilateral talks on this subject. Just over a month later on October 5th, Caspar Weinberger, the new Secretary of Defense, unveiled the Reagan "Strategic Modernization Program" and declared before the Senate Armed Services Committee that the United States would "continue to pursue an operational antisatellite system".[4] In fact, military space systems in general received added emphasis in the modernization programme. The requirement for effective and survivable early warning, communication and attack assessment systems was considered essential if the administration's declared policy of being able to fight and "prevail" in a nuclear war was to be credible.[5]

By the summer of 1982, the space policy review had been completed. On July 4th, President Reagan took the opportunity of the homecoming of the fourth space shuttle flight at Edwards AFB to make his first major speech on space policy.[6] Apart from vague references to "establishing a more permanent presence in space" there was little of substance in Reagan's speech. However, as Reagan was speaking, the White House issued a detailed fact sheet on United States Space Policy. This was a sanitized version of National Security Decision Directive 42 that Reagan had signed a few days earlier.[7]

In outlining the "principles underlying the conduct of the United States space program" the fact sheet was no different from the declarations of previous administrations:

> The United States is committed to the exploration and use of space by all nations for peaceful purposes and for the benefit of mankind. "Peaceful purposes" allow activities in pursuit of national goals.
>
> The United States rejects any claims to sovereignty by any nation over space or over celestial bodies, or any portion thereof, and rejects any limitations on the fundamental right to acquire data from space.
>
> The United States considers the space systems of any nation to be national property with the right of passage through and operation in space without interference. Purposeful interference with space systems shall be viewed as an infringement upon sovereign rights.
>
> The United States will pursue activities in space in suport of its right of self-defense.[8]

The commitment to improve satellite survivability and develop a space attack warning system was also identical to Carter's PD-37. However, the tone and emphasis of the Reagan administration's objectives for antisatellite deployment and arms control were clearly different. While the fact sheet noted that the United States would "continue to *study* space arms control options" and "*consider* verifiable and equitable arms control measures that would ban or otherwise limit testing and deployment of specific weapon systems . . ." (my emphasis), this was a lot different from stating, as Carter had, that ASAT arms control *per se* was desirable.[9]

The primary rationale for the US ASAT programme had also changed:

> The United States will proceed with development of an antisatellite (ASAT capability), with operational deployment as a goal. The primary purposes of a United States ASAT capability are to deter threats to space systems of the United States and its allies and, within such limits imposed by international law, to deny any adversary the use of space-based systems that provide support to hostile military forces.[10]

This was also repeated in an outline of DoD Space Policy that was released a few days later by the Pentagon. Although this stated that "DoD space policy contains no new directions in space weaponry", a significant shift in official statements about the role of the US ASAT system had none the less occurred.[11] In contrast to the Carter administration, the White House and Pentagon now placed deterrence as its primary rationale. While Harold Brown and others had also mentioned ASAT deterrence, this was mainly in the context of improving satellite survivability. In fact, the possibility of deterrence working in space had been rejected outright by the then Deputy Under Secretary of Defense, Research and Engineering, Dr Seymour Zeiberg. As he stated before the Senate:

> The idea of tit for tat with satellites to my mind does not make sense, because of the great asymmetry in the value of the satellites, the deployment of the satellites, et cetera.[12]

The belief that the threat of US ASAT retaliation could deter the Soviet Union from using its own satellite interceptor would later be reaffirmed during the Reagan administration in subsequent posture statements by the Secretary of Defense and the JCS.

Another novel, though less defined, feature of the Reagan administration's military space policy was the discussion on the projection of force *from* space. This became apparent after a series of classified documents were leaked to the press. For example, the Fiscal 1984-8 Defense Guidance reportedly stated that: "The Department of Defense will vigorously pursue technology and systems both to provide responsive support and to project force in and *from* space as needed."[13] However, there were no indications of what kinds of space-based systems would be used to achieve this. The Defense Guidance for 1985-9 that was leaked in the following year said much the same: "We must achieve capabilities to ensure free access to and use of space in peace and war; deny the wartime use of space to adversaries . . . and apply military force from space if that becomes necessary."[14]

In a similar vein, an Air Force document known as the "Space Master Plan", which apparently sets that service's space policy and objectives to the year 2000, also called for a "space combat" system both to protect Air Force assets and to deny the enemy unfettered access to space.[15] Later General Robert T. Marsh, Commander of Air Force Systems Command, stated before the Investigations Subcommittee of the House Committee on Armed Services that, ". . . we should move into war-fighting capabilities—that is ground-to-space war-fighting capabilities, space-to-space, space-to-ground . . ."[16]

The Air Force's revitalized interest in space was both reflected and reinforced by organizational changes introduced at this time. On

21 June 1982 Air Force Chief of Staff General Lew Allen Jr announced the creation of an Air Force Space Command. This was a considerable achievement for those who had been campaigning for a restructured national military space organization and, more significantly, an *operational* command within the Air Force hierarchy with equal status to its other commands. By September 1982, the headquarters of Space Command had been established at Colorado Springs.[17]

The creation of Space Command marked the culmination of number of space related organizational initiatives. These included: (i) the formation by the DoD of a Space Operations Committee, to be chaired by the Secretary of the Air Force, to deal with all space operations within the Department of Defense; (ii) elevation of the Commander in Chief NORAD (CINC NORAD) to a four-star position; (iii) establishment of a Deputy Commander for Space Operations within the Space Division; (iv) continued construction of a Consolidated Space Operations Center close to NORAD; (v) formation of a Directorate for Space Operations within the Office of the Deputy Chief of Staff/Plans and Operations; (vi) establishment of a General Officer Space Operations Steering Committee (SOSC); and (vii) establishment of the Air Force Manned Space Flight Support Group at the Johnson Space Center to develop the expertise to use the space shuttle.[18]

On 15 June 1983 the United States Navy also announced that it was creating its own space command, which was later activated on 1 October 1983 at Dahlgren, Virginia.[19] While this move was intended to consolidate the Navy's existing space activities, it was also clearly designed to resist the Air Force's attempts to control all DoD space assets under a Unified Command. The Air Force, however, is apparently supported by the Defense Department, the NSC and the JCS, and will almost certainly get its way.[20] At the executive level NSDD-42 established a Senior Interagency Group (SIG) on space that is chaired by the President's National Security Advisor.[21]

ARMING FOR THE HIGH FRONTIER

Although the new administration waited until October 1981 before publicly announcing its commitment to the US ASAT programme, further contracts worth $418.8 million were awarded to Vought and Boeing within a week of Reagan taking office. This was to cover the necessary R&D through to the end of FY 1985.[22] Ground testing of both the missile and the homing vehicle also began in 1981.[23]

However, contrary to public pronouncements, the programme was not running according to schedule. In the autumn, an Air Force Science Advisory Board reviewed the ASAT programme—now known as the Prototype Miniature Air Launched System (PMALS)—and recommended further development research. They believed that the earlier laboratory demonstrations of the homing vehicle had been insufficient as a proof-of-concept and that the operational challenges of intercepting satellites by this method were more demanding than the Air Force was prepared to admit.[24] It was almost certainly as a result of this review and its recommendations that the estimated date for an ASAT initial operating capability (IOC) was changed from 1985 to 1987.

Further developmental problems meant that it was over a year before the first "captive" flight tests were carried out in December 1982. Additional exercises mating the ASAT missile to the F-15 were performed in the spring of 1983.[25] Moreover, while the Air Force continued to state that the programme met all its requirements for an ASAT system, this was singled out for criticism in a General Accounting Office (GAO) report published in January 1983.[26] As the unclassified digest of this report stated:

> When the Air Force selected the miniature vehicle technology as the primary solution to the antisatellite mission, it was announced as a relatively cheap, quick way to get an antisatellite system that would meet the mission requirements. This is no longer the case. It will be a more complex and expensive task than originally envisioned, potentially costing in the tens of billions of dollars.[27]

The GAO called for a new assessment of alternative ASAT weapons particularly ground, air and space based laser systems. Further criticisms came to light when parts of the classified report were leaked to the press. In particular, the PMALS was criticized for being unable to meet the revised JCS ASAT requirements document that had been completed in August 1981. "As a result", the report stated, "it will not be able to negate 122 of the 175 threat [*sic*] satellites—70 per cent of the projected threat—because of lack of growth potential, and the inability to attack the Soviet co-orbital ASAT."[28] Furthermore: "The air-launched miniature vehicle depends on existing surveillance systems which have limited capability, availability and survivability to accomplish the ASAT mission."[29] In response, the Defense Department defended its decision and criticized the GAO for using misleading data. They stated that the JCS target list was a "fiscally unconstrained statement" and "does not necessarily reflect what either DoD or the Air Force can realistically afford. Accordingly, the Secretary of Defense has chosen to apply available resources to only a subset of the JCS document at this time".[30] They also stated that the programme would cost in the region of $3.6 billion—nowhere near the amount the GAO had projected.

None the less, some of the GAO's criticisms seem to have had an effect, as by February 1984 the Reagan administration had decided to carry out research on an advanced high altitude ASAT system. In testimony before the House Armed Services Committee, DDR&E Richard DeLauer stated: "We have directed a comprehensive study to select a follow-on system with additional capabilities to place a wider range of Soviet satellite vehicles at risk".[31]

The planned sequence to the initial tests of the PMALS from an F-15 called for the first to be a test of the booster without the homing vehicle against a "point in space", the second would include the MHV, and the third would be a full demonstration of the complete missile against the instrumented test vehicle in space. Ten further tests of the PMALS are planned before an IOC is declared in 1987.[32] While it was expected that the first test launch would be carried out in the summer of 1983, further technical problems delayed this until 21 January 1984.[33] As a result, the first in-space intercept of a target vehicle is not likely to occur now until after the presidential election in November. Two Air Force bases—Langley in Virginia and McChord in Washington state—have been designated to receive the F-15s modified for the ASAT mission once the system becomes operational.[34] Depending on the nature of the conflict, and the targets to be intercepted, these aircraft may well be moved to other airfields capable of taking F-15s. Similarly, F-15s at other sites could be converted relatively quickly for the ASAT role.

In justifying the US ASAT programme, the Reagan administration highlighted the "threatening" nature of Soviet space activities. In fact, the new administration had hardly started when the Soviets tested their satellite interceptor for the eighteenth time. On 21 January 1981 the launch of Kosmos 1241 from Plesetsk into a slightly elliptical orbit at 65.8 degrees inclination was immediately identified as a potential ASAT target satellite. As anticipated, on February 2nd, Kosmos 1243 was launched from Tyuratam and intercepted the target at the beginning of its third revolution. The interceptor was then commanded to deorbit.[35] This was reported to be yet another failure of the new optical-thermal guidance system.[36] The same target vehicle was used again in another test on 14 March 1981, following the launch of Kosmos 1258. This time reports of the intercept stated that the chase vehicle had used the radar guidance system of earlier models and had exploded sufficiently close to the target satellite for it to have been disabled.[37] However, the latter seems unlikely as NORAD tracking data indicates that the orbit of the target satellite was not altered during the interception.[38] There were no further tests of the co-orbital interceptor in 1981, although there was speculation that Kosmos 1267 that docked with Salyut 6 in June possessed "firing ports" for miniature infra-red homing missiles.[39] The Pentagon, however, quickly denied these reports.[40]

The next clear test of the Soviet ASAT occurred on 18 June 1982 with the interception of Kosmos 1375 by Kosmos 1379.[41] Although this was apparently a failure, it was significant in that it formed part of a major exercise of the Soviet strategic forces in many ways similar to the 16 February 1976 test. During the exercise the Soviets conducted test firings of two ICBMs, two ABMs, one SLBM, and one SS-20 IRBM. Moreover, during the pre-intercept manoeuvres a navigation and photoreconnaissance satellite were launched that, according to one observer, "may have simulated the replacement of Soviet satellites negated by Allied forces during the war scenario".[42] This was the last test of the Soviet ASAT system before President Andropov announced a unilateral moratorium on testing in August 1983 (see below).

Concern over the possible use of larger boosters, such as the Proton, to extend the range of the satellite interceptor to threaten US satellites in geosynchronous orbit also grew at this time. More ominous to some observers, however, was the reported development of a new heavy booster capable of lifting 300,000 to 400,000 lbs. into low-earth orbit. While this launch vehicle could be used for a variety of missions, such as space station development, its potential ability to lift a prototype laser ASAT system into space received the most attention.[43]

In contrast to its predecessors, the Reagan administration also credited the Soviet Union with a near-term directed energy weapons capability. In March 1982 DDR&E Richard DeLauer inadvertently (though the Soviets maintained it was deliberate) read out in an open session a classified evaluation that the Soviet Union might have an operational space-based laser by as early as 1983.[44] This was immediately denounced as being totally unrealistic by scientists outside of the administration.[45]

By March 1983, official estimates of a Soviet operational laser ASAT capability had been revised in the Defense Department publication "Soviet Military Power". This identified "a very large, directed energy research program including the development of laser-beam weapon systems which could be based either in the USSR, aboard the next generation of Soviet ASATs or aboard the next generation of Soviet manned space stations".[46] Later, it stated: "The Soviets could launch the first prototype of a space-based laser antisatellite system in the later 1980s or very early 1990s. An operational system capable of attacking other satellites within a few thousand kilometers range could be established in the early 1990s."[47] The 1984 edition went still further in presenting an artist's impression of a "directed energy R&D site" at Sary Shagan and stated that a "prototype space based particle beam weapon intended only to disrupt satellite electronic equipment could be tested in the early 1990s".[48]

Despite these new estimates, there was still widespread scepticism that the Soviets could overcome the necessary technical hurdles and build a workable system. This was reflected, in part, by the conflicting recommendations for the US research programme. As noted earlier, a Defense Science Board study concluded in 1981 that "it is too soon to attempt to accelerate space-based laser development toward integrated space demonstration for any mission" due to technical uncertainties surrounding the development of high energy laser power sources and optics.[49] In February 1982, however, the GAO conducted another classified review of the DoD's laser programme and recommended that DARPA's "Triad" programme (ALPHA, TALON GOLD, and LODE) be accelerated. They also called for a restructured development plan to bring these projects to fruition.[50] Two months later, a House Armed Services Committee report recommended the reverse—that DARPA's chemical laser programme be de-emphasized, and even terminated. Instead, they argued that the United States should reorient its efforts towards the development of short wavelength lasers.[51] As this was likely to entail still further delays, the House report was strongly criticized by the proponents of beam weaponry in Congress. Later Dr Robert Cooper, Director of DARPA, testified before the House Armed Services that shorter wavelengths were indeed more efficient, but he discouraged additional funding for laser research in the FY 1983 budget.[52] A similar approach was taken for the following year's (FY 1984) budget. This was also supported by a White House Science Council study completed in early March 1984. Apparently this study did not foresee any scientific breakthroughs in directed energy technology over the next ten years and recommended that the current pace of research be maintained.[53] However, in a dramatic change of policy, President

Reagan announced in the now famous "Star Wars" speech of 23 March 1983 that he was initiating a major ballistic missile defence study effort. This was subsequently entitled the Strategic Defense Initiative (SDI), and includes both a reorganization and an acceleration of the US beam weapon research effort.

THE STRATEGIC DEFENSE INITIATIVE

Although it was clear that the Reagan administration was becoming more interested in the possibility of ballistic missile defence, very few would have predicted that Reagan would announce a major new initiative in this area. The "Star Wars" speech caught almost everybody by surprise, including those most closely associated with the existing BMD research effort.[54] Even Reagan's most senior aides, such as Secretaries Weinberger and Schultz, were only told in the final stages of its preparation. As Presidential Science Advisor George Keyworth II stated afterwards: "This was not a speech that came up; it was a top down speech . . . a speech that came from the President's heart."[55]

The televised speech in the main concerned the need for further increases in military spending, and it was only at the end that Reagan announced his intention to "embark on a program to counter the awesome Soviet missile threat with measures that are defensive". Other than stating the long-term *desideratum* of moving away from an assured destruction posture to a policy of strategic defence to deter Soviet attacks against the United States and its allies, this section of the speech was remarkably vague, both in outlining the goals of the initiative and the means to achieve it. This was a prescription for confusion and contradiction afterwards. When Secretary Weinberger was asked whether the plan called for complete population defence he replied: "The defense systems the President is talking about are not designed to be partial. What we want to try to get is a system which will develop a defense that is thoroughly reliable and total."[56] A few days later, however, Weinberger began to backpedal; the President was not looking for a "single system" that would intercept and defend flawlessly against all missiles and attacks, but a series of BMD layers "which taken together" would provide a "reliable defense".[57] The truth is that nobody had a clear idea of what the final configuration and effectiveness of a BMD system would be. While the press dubbed it the "Star Wars" initiative, neither beam weapons nor space were actually mentioned in the speech.

A clearer idea of what the President was calling for could only come after an extensive technical feasibility study, something the critics of the initiative said should have been carried out beforehand. Two days later, Reagan signed NSDD 85, entitled "Eliminating the Threat from Ballistic Missiles". This directed the bureaucracy to conduct "an intensive effort to define a long-term research and development program aimed at an ultimate goal of eliminating the threat posed by nuclear ballistic missiles".[58] Furthermore: "In order to provide the necessary basis for this effort, I further direct a study be completed on a priority basis to assess the roles that ballistic missile defense would play in future security strategy of the United States and our allies."[59]

To oversee these two study efforts, Weinberger established a Defensive Technologies Executive Committee, chaired by Deputy Secretary of Defense Paul Thayer and comprising of the Chairman of the JCS, the Under Secretaries of Defense, the Assistant Secretary of Defense (Comptroller), and the Director of Program Analysis and Evaluation.[60] Further guidance for the two studies came with National Security Study Directive 6-83, entitled "Defense Against Ballistic Missiles", issued by the White House in early April. This established the Defense Technologies Study Group, otherwise known as the Fletcher Panel after its chairman, James Fletcher, to carry out the technical evaluation; and, secondly, the Future Security Strategy Study (FS^3) or the Hoffman Panel, after its chairman Fred Hoffman, to assess the political and strategic implications.[61]

The Fletcher study was organized into a number of subgroups to examine the requirements for surveillance and acquisition; battle management and data processing; directed energy weapons and interceptors, fire control and guidance systems. These assessments were applied to specific operational concepts—boost phase interception (during the missiles' ascent), mid-course interception (during the missiles' ballistic flight through space), and terminal interception (after the missiles' warheads have re-entered the atmosphere). A "Red Team" also assessed possible Soviet countermeasures to both the technologies and defensive strategies. By October 1st, the Fletcher report—all eight volumes of it— was finally ready.

According to leaked parts of the report, it was generally optimistic of the long-term feasibility of achieving a ballistic missile defence system.[62] It recommended that, to reach a point where a more informed judgement could be made of the optimum approach, an intensive research and development effort, including demonstrations of key technologies, needed to be carried out. As the study reportedly stated: "Our goal was to provide the basis for selecting the technology paths to follow when a specific defensive strategy is chosen for implementation."[63] However, the report appears to have favoured a multilayered missile defence system with an emphasis on boost-phase interception to reduce drastically the number of warheads for subsequent interception. The report apparently took "an optimistic view of new emerging technologies and with this viewpoint concluded that a robust multitiered ballistic missile defence system can eventually be made to work".[64]

The Hoffman report was also completed at the same time. It too was generally enthusiastic about the need for more BMD research and development. In fact, it went further in advocating "intermediate options" in the belief that a comprehensive "leak proof" system "may prove to be unattainable".[65] These included: defence against tactical ballistic missiles in Europe; defence of critical installations and silos in the continental USA; and a space-based system that could reduce the number of Soviet missiles hitting the United States. The report argued that such measures "can increase stability" and strengthen deterrence by convincing an adversary that a pre-emptive first strike would not be successful.[66]

Both reports were passed to their respective review groups within the Pentagon and then on to the Senior Interagency Group on Defense Policy at the NSC. Here they were combined and reduced into a common set of recommendations for the President.[67] This report, which was also leaked to the journal *Aviation Week*, recommended *inter alia* that the United States embark on early demonstrations of BMD technologies. A detailed research and development programme was also laid out according to four possible funding levels ranging from $18-27 billion in the period FY 1985 to 1989. As the interagency report predicted: "with vigorous technology development programs, the potential for ballistic missile defense can be demonstrated by the early 1990s". This research, however, was to be conducted within the limits of existing treaties, notably the 1972 ABM treaty.

The need for further BMD research and development was justified on the basis of the following putative benefits. In the short run, "prior to deployment, the demonstration of US technology would strengthen military and negotiating stances, and options for immediate deployment would play a significant role in deterrence".[68] It would also provide a hedge against a possible Soviet BMD "breakout". Over the long run, "by constraining or eliminating effective counterforce options [by BMD], the utility of strategic and theater nuclear weapons are reduced and the threshold of nuclear war is raised. It undermines the confidence that an attack will succeed".[69]

Despite the positive tone of these recommendations, it became clear that parts of the Reagan administration did not share the technological optimism or believe the strategic arguments for a vigorous, high-profile BMD effort. For example, Richard DeLauer testified to Congress in November 1983 that the technological challenges involved in the BMD effort would be greater than those faced by the Manhattan and Apollo projects. In fact, he stated: "This is a multiple of Apollo programs which, if the systems were deployed, the Congress would be staggered at the cost."[70] Others in the administration were also concerned at the repercussions on US-Soviet and Alliance relations, both of which had reached their lowest point for many years.

The final presidential decision authorizing further research and development appears to have been a carefully crafted compromise between BMD proponents and critics within the administration. On 6 January 1984, Reagan signed NSDD 119, which formally set down the guidelines for the BMD research. It called for the "initiation of a focussed program to demonstrate the technical feasibility of enhancing deterrence and thereby reducing the risk of nuclear war through greater reliance on defensive strategic capability".[71] The directive was also deliberately cautious in referring to this effort as the "strategic defense initiative" rather than ABM research so as to avoid criticism that it was undermining the ABM treaty. Moreover, it only referred to "research" rather than "development" and to "demonstrations" rather than "tests" for the same reason.

This low-profile approach was repeated when the Strategic Defense Initiative was publicly unveiled in the FY 1985 defense budget. Total funding for FY 1985 was put at $1.7 billion, which administration officials claimed was only $250 million above the expected figure before Reagan's March 23rd speech.[72] Similarly, they argued that the technical demonstrations had also been planned beforehand. These included the acquisition and tracking of re-entry vehicles from high-altitude aircraft using infra-red sensors, short-wave chemical and nuclear explosion-generated x-ray laser tests, pointing and tracking laser experiments, and missile test intercepts using "conventional" systems.

While the rhetoric had undoubtedly been toned down, the administration' arguments were largely disingenuous. Not only was the total amount of funding for BMD-related research in FY 1985 much higher than the administration admitted—more like $2.5 billion—but the subsequent five year programme, amounting to $26 billion, was a substantial increase over what would have been expected for these years. Furthermore, it could be argued that the future of the planned demonstrations was not so certain before this latest surge of official interest in BMD. It remains to be seen whether the necessary long-term interest in ballistic missile defence can be maintained in the face of continuing technical scepticism and concern over its financial and strategic consequences.

A FAREWELL TO SPACE ARMS CONTROL?

On 20 August 1981, Foreign Minister Gromyko presented a "Draft Treaty on the Prohibition of the Stationing of Weapons of Any Kind in Outer Space" to the 36th Session of the UN General Assembly. In a covering letter to UN Secretary-General Kurt Waldheim, Gromyko explained that the purpose of the draft treaty was to supplement the provisions of the Outer Space Treaty so as to preclude "the possibility of the stationing in outer space of those kinds of weapons which are not covered by the definition of weapons of mass destruction".[73] The two key provisions of the draft are contained in Articles 1 and 3. The former proposes that "States Parties undertake not to place in orbit around the earth objects carrying weapons of any kind, install such weapons on celestial bodies, or station such weapons in outer space in any other manner, including on reusable manned space vehicles . . ."[74] In reality, this did not go much further than the Soviet position at the bilateral negotiations. Both the current Soviet and expected US ASAT systems would not have been prohibited (as both are land-based and do not necessarily have to go into orbit). There was also no satisfactory definition of the term "weapon". One positive aspect, however, is that the Soviet Union had dropped its objection to the use of the space shuttle except as a weapons platform.

Article 3 apparently uses very similar language to the Soviet draft treaty put forward at the bilateral talks:

> Each state Party undertakes not to destroy, damage, disturb the normal functioning or change the flight trajectory of space objects of other States Parties, if such objects were placed in orbit in strict accordance with article 1, paragraph 1 of this treaty.[75]

As Rebecca Strode has pointed out, the key caveat linking it to

Article 1 provides an important loophole that would permit states to use force legitimately against objects they believe to be weapons systems.[76]

The Reagan administration dismissed the Soviet draft treaty as a hypocritical propaganda ploy. In listing its major shortcomings, however, the US gave no indication of a counter proposal or even an interest in resuming bilateral discussions. After two UN General Assembly resolutions in November 1981, it was decided that the Geneva-based Committee on Disarmament would discuss the Soviet proposal at its next session.[77]

While NSDD 42 had not been entirely dismissive of ASAT arms control, it was clear from the administration's response to the Soviet proposal that it had no intention of pursuing it in the immediate future. This view was further reinforced during the Senate Foreign Relations Committee hearings on the subject of "Arms Control and the Militarization of Space" in September 1982. These hearings had been called at the behest of Senator Larry Pressler who, initially at least, had been leading a one-man campaign within the Senate to pressure the White House to resume ASAT negotiations. At the hearings Eugene Rostow, the Director of ACDA, stated that, while the whole question of ASAT arms control was still under study, there were significant "obstacles to progress". These included "assuring effective verification, minimizing so-called residual ASAT capabilities, and countering the space components of threats to U.S. forces".[78] In fact, Rostow raised the fundamental question of whether such a venture would be equitable and even desirable. In particular, he argued that ASAT arms control would remove the ability of the United States to counter threatening space systems and leave the Soviets with a demonstrated ASAT capability. As Rostow stated:

> Any interruption in the U.S. [ASAT] program at this time would leave the Soviets with both a ready ASAT capability and a considerable body of ASAT test experience from more than a decade of testing. Additionally, the threat to our national security from advances in Soviet space programs will grow. Finally, the knowledge that the United States is making steady progress toward an operational ASAT of its own could be an important inducement for the Soviet Union to explore constructive limits on space weapons.[79]

The latter argument that the United States needed an ASAT programme to bargain with at negotiations was later totally contradicted at the same hearings by DeLauer when he stated that, "The U.S. ASAT program is not a bargaining chip and never was."[80]

Growing congressional concern led to further hearings on this subject in April and May 1983. Although Rostow had now been replaced by Kenneth Adelman, the official US position had not changed. Adelman maintained that studies were still under way, although he promised to accelerate them. Again the problems of verification and residual Soviet ASAT capabilities were highlighted as the main hindrance to progress.[81]

In stark contrast to the negative US position, the Soviets continued to call for arms control in space.[82] This reached a climax in the summer of 1983, when the Soviet Union announced two significant additions to its 1981 position. The first came on August 19th in a private meeting between Andropov and a group of US senators on a visit to Moscow. Here Andropov informed the senators of an "exceptionally important decision" that:

> The USSR assumes the commitment not to be the first to put into outer space any type of antisatellite weapon, that is, imposes a unilateral moratorium on such launchings for the entire period during which other countries, including the USA, will refrain from stationing in outer space antisatellite weapons of any type.[83]

This offer, which appeared to be deliberately ambiguous to allow the United States to begin ASAT testing without necessarily provoking the Soviets to call off their moratorium, was considered by some to be a significant concession.[84]

The following day, Gromyko introduced a second draft treaty at the United Nations. This was a considerable improvement over the first in that the language was more precise and the coverage more extensive. Apart from Article 1, which outlaws the use of force in space, the heart of the draft treaty is contained in the provisions of Article 2. This calls for *inter alia*: a prohibition on the testing and deployment of "any space based weapons intended to hit targets on the Earth, in the atmosphere, or in space"; the testing and development of "new antisatellite systems"; the elimination of "such systems already in their possession"; and the testing and use "for military, including antisatellite purposes, any manned spacecraft".[85] The most significant part of this new proposal was the offer to dismantle the existing Soviet satellite interceptor. One major drawback, however, was that it would prohibit the use of the space shuttle for military purposes. While the same would also apply to the Salyut space station, the trade-off was hardly equitable.

Although this latest proposal addressed many of the United States' criticisms of the earlier draft treaty, the Reagan administration's response was still negative. "Inadequate verification" was identified by a State Department offical as one its "major weaknesses", as "it would be nearly impossible to verify through national technical means alone the dismantling and destruction of the Soviet ASAT system".[86] Also, in what appeared to be a reference to laser attacks, the State Department spokesman noted that ASAT activities" could occur without positive attribution to the perpetrator". Reference was also made to Soviet residual ASAT capabilities in systems such as the Galosh ABM system and also the unacceptable limitation on shuttle missions.

The official US response drew sharp criticism both at home and abroad. As John Pike of the Federation of American Scientists noted: "The overall impression one draws from these responses is of, not reasons, but excuses, for not entering into negotiations. It is as though the administration were waiting for the Soviets to spontaneously offer a treaty that would totally conform to American interests, hardly a realistic prospect."[87] The London *Economist* also stated that this was not an offer that the United States should pass up.[88]

Leaks to the press suggested that the administration was not entirely unified in its opposition to space arms control. Apparently, ACDA put forward a proposal to the White House in August 1983 for two parallel agreements. One consisted of a general pact among current and prospective space powers that would ban any kind of interference with satellites. The other would be confined to the United States and Soviet Union with the object of limiting ASAT systems to one apiece.[89] However, even this relatively modest arrangement was not to the liking of the White House.

Continuing frustration in the Congress at the Reagan administration's resistance to its calls for a resumption in ASAT negotiations prompted some members to pressure the White House by amending the space defence budget. On 18 July 1983 Senator Paul Tsongas succeeded in gaining acceptance of his amendment to the FY 1984 DoD Authorization Act, which denied the use of funds for the testing of the US antisatellite system against targets in space until the President certified that the administration was both negotiating in "good faith" on this matter with the Soviet Union and that testing was in the interest of US national security.[90] The House Committee on Appropriations went even further in deleting $19.4 million for advance procurement funds for the US ASAT system. After much lobbying by administration officials the money was reinstated in conference with the Senate, but with the proviso that the funds could not be expended until Congress was provided, no later than 31 March

1984, with a report on US ASAT policy.

To prepare the way for the report, administration officials began to brief Congress on the pitfalls of ASAT arms control. In testimony before the Senate Armed Services Committee, Richard Perle, Assistant Secretary of Defense for International Security Policy, stated that it was the view of the Defense Department and the intelligence community that it would be "extremely difficult, if not impossible" to verify an ASAT treaty.[91] Perle was later identified in press reports as having been the key player in blocking an internal attempt to reconsider the official US position.[92]

The report was true to form. In a covering letter to Congress, President Reagan stated that:

> . . . no arrangements or agreements beyond those already governing military activities in outer space have been found to date that are judged to be in the overall interest of the United States and its Allies. The factors that impede the identification of effective ASAT arms control measures include significant difficulties of verification, diverse sources of threats to U.S. and Allied satellites, and threats posed by Soviet targeting and reconnaissance satellites that undermine conventional and nuclear deterrence.[93]

Therefore, while the United States would continue "to study space arms control", it would not be "productive to engage in formal international negotiations".[94]

By June, however, growing congressional concern over the implications of ASAT weapons and the deterioration in US-Soviet relations caused the White House to reconsider its objections to arms control. During the passage of the FY 1985 Defense Aurthorization Bill, the House of Representatives voted to block US ASAT testing for a year, providing the Soviets continued with their unilateral moratorium.[95] After this setback, the administration redoubled its efforts to prevent the Senate from imposing further constraints on the US programme.

The Soviets were also aware of the importance of the Senate debate, as on June 11th the new President, Konstantin Chernenko, called for a ban on antisatellite weapons and reiterated his predecessor's declaration of a testing moratorium.[96] After considerable lobbying, including a specially arranged classified briefing on the Soviet space threat, the administration's supporters within the Senate succeeded in preventing further limitations on US ASAT testing and also managed to water down the restrictive language of the Tsongas Amendment.

While the administration still has to certify that it is willing to negotiate ASAT arms control, it can continue ASAT testing, providing that it demonstrates this is "necessary to avert a clear and irrevocable harm to national security" and that it would not "constitute an irreversible step [which would] gravely impair the prospects for negotiation".[97]

Although the White House had managed to avert a serious defeat on this issue, the level of the congressional opposition and, moreover, growing apprehension among Republican leaders that Reagan's poor record on arms control could affect his re-election chances, caused the White House to modify its opposition to ASAT negotiations. The first indication of this came on June 14th, when President Reagan stated in a news conference that he hadn't "slammed the door" on talks with the Soviets on this subject. A few days later, leaked reports indicated that the administration was now considering a range of options for a limited ASAT agreement short of a complete ban.[98]

In what was apparently an attempt to test the administration's interest in the subject, the Soviet Ambassador to the United States, Anatoliy Dobrynin, proposed in a note to Secretary of State George Schultz on the morning of June 19th, that the two countries engage in negotiations with the intention of "preventing the militarization of space".[99] It seems likely that the Soviets expected the United States to decline their offer, as a few hours later *Tass* published full details of the proposal. After Schultz had briefed Reagan on the situation, Robert McFarlane, the President's National Security Advisor, convened a high-level interagency group to discuss the US response. In a carefully worded statement on Saturday evening, McFarlane reported that the United States was willing to meet with the Soviets to work out "feasible negotiating approaches" to "verifiable and effective" limitations.[100] However, McFarlane also suggested that the meeting be used to discuss INF and strategic arms control issues that had been stalled after the Soviets walked out of the negotiations in December 1983.

Behind the facade of flexibility, very little had changed in the US position. A discussion of the "approaches" to ASAT limitations committed the United States to nothing, while domestically it could only reflect well on Reagan. As Leslie Gelb of the *New York Times* noted, "there was something in it for everyone: for the State Department, willingness to meet with the Russians; for the Pentagon, nothing to jeopardize the new antisatellite testing program; and for White House aides, prospects of a good domestic political reaction".[101]

The following day, *Tass* announed that any linkage between space weapon negotiations and other arms control issues was "unacceptable".[102] However, in the subsequent diplomatic exchanges the United States indicated that this was not a "precondition" for discussions on the topic of antisatellite weapons.[103] Finally, on July 27th, Deputy Foreign Minister Viktor G. Komplektov stated during a press conference in Moscow that "the current American position makes impossible the conducting of the kinds of negotiations we are talking about".[104] While talks may yet be held on this issue, the fundamental differences that currently separate the two sides cast serious doubts as to whether meaningful progress let alone an agreement can ever be reached.

CONCLUSION

In hindsight, the early part of the 1980s will most probably be viewed as a fundamental watershed in the militarization of space. During this period the chance for a significant antisatellite arms control agreement was lost—possibly for ever. The Reagan administration squandered an opportunity to take advantage of unprecedented Soviet flexibility on this issue. By the time it had begun to reconsider US policy, the Soviet position had hardened. The forthcoming tests of the US ASAT system and the commitment to the SDI will only complicate further negotiations on this topic, no doubt leaving future administrations to regret bitterly this lost opportunity.

12

Conclusion

The cold reaches of the universe must not become the new arena of an even colder war.

President John F. Kennedy
Address before the UN, 25 September 1961

INTRODUCTION

Despite President Kennedy's appeal before the UN General Assembly in September 1961, it was inevitable that the exploitation of space would become embroiled in the politics of the cold war. The adversarial relationship between the United States and the Soviet Union not only provided the incentives to exploit this medium but also the means to carry it out. Thus, the first US space programme had the goal of collecting strategic intelligence and relied on rocket boosters designed originally for long-range bombardment. Space also added a new dimension to the politics of the cold war. In the period immediately following the launch of Sputnik, space activities were divined for their military significance, used to buttress diplomatic initiatives and generally viewed as an indispensable attribute of superpower status. Sputnik also heralded a new round of the arms race by stimulating a massive US strategic weapons build-up, the consequences of which are still being felt today. Yet, in a more fundamental regard, Kennedy's appeal was fulfilled. After the dispute over satellite reconnaissance had subsided and the fear of a nuclear arms race in space had been dealt with, superpower relations in the exploitation of space remained remarkably stable and uncompetitive until the late 1970s.

The purpose of this concluding chapter is to return to the questions that motivated this study; why has there been no arms race in space and what changed in the late 1970s to reverse this trend? More importantly, what, if anything, have we to learn from this case study in US-Soviet relations? Lastly, what are the likely consequences of an arms race in space?

EXPLAINING THE US-SOVIET MILITARIZATION OF SPACE

The most widely accepted explanation for the absence of an arms race in space is that both the United States and the Soviet Union recognized the mutual benefits of reconnaissance satellites and reached a "tacit" agreement to refrain from developing weapons to counter them. As Arthur Schlesinger, Jr observed, both countries agreed to this because "satellites provided mutual reassurance and thus strengthened the system of stable nuclear deterrence".[1] Philip Klass in his study *Secret Sentries in Space* also refers to a "tacit understanding that neither the United States nor the Soviet Union would interfere with the other's space systems".[2]

The most thorough examination of this "common interests" hypothesis can be found in the work of Gerald Steinberg.[3] Steinberg, however, considers this to be an inadequate explanation as it has been the exception rather than the rule in international relations for common interests to dictate restraint in weapons procurement. Rather, the key to explaining US and Soviet behaviour in this case lies in the subtle way in which the dispute over the right to conduct satellite reconnaissance was resolved.[4] In particular, Steinberg argues persuasively that it was the success of an "informal process" of bargaining that led to the mutually beneficial outcome. More specifically:

> The establishment of the blackout on U.S. space reconnaissance programs, the cancellation of the SAINT and anti-satellite program, and Soviet introduction and subsequent modification of sections of its draft resolution dealing with espionage satellites were the major behavioral moves.[5]

In other words, it took a series of tacit signals by the key actors for them to realize their common interest in avoiding military conflict and competition in space.

However, on the basis of the findings of this study, it is highly questionable whether the tacit acceptance of satellite reconnaissance was the result of a series of carefully orchestrated "behavioral moves" and moreover whether this was instrumental in curbing subsequent antisatellite development. Rather, it is the contention here that the resolution of the dispute over satellite reconnaissance and the absence of an arms race in space were the result of a convergence of national interests, military disincentives and technical constraints, which were buttressed at important times by *formal* agreements. This is not to suggest that Soviet actions had nothing to do with US policy and vice versa but rather the outcome was the result of a broader set indigenous factors.

As the early chapters of the study illustrate, US policymakers from the outset wanted to avoid an arms race in space but not at the price of limiting their freedom of action to use space for military purposes, particularly satellite reconnaissance. These two goals were reconciled by the argument that space could and should be used for "peaceful", that is nonaggressive, purposes. To avoid any confusion and contradiction, however, the public profile of the US military space programme was deliberately de-emphasized. While this was undoubtedly conducive to resolving the dispute over satellite reconnaissance, was it critical to influencing the change in Soviet policy? A further examination of Soviet behaviour at this time suggests otherwise.

Although the United States was clearly the main beneficiary of satellite reconnaissance, the Soviets were also attracted by its potential military applications. In fact, their own Kosmos reconnaissance satellite programme was almost certainly under way when the diplomatic offensive against US activities began in 1962. By the time the Soviets withdrew their objections to "espionage satellites" in the UN in September 1963, they had already launched *nine* low resolution reconnaissance satellites. Two months later, on November 16th, the first of the second generation satellites (Kosmos 22) with higher resolution cameras was also launched.[6] The Soviet photo interpreters may have been genuinely surprised at the detail they could see from space, thus confirming the potential military advantages of this activity. An apparent change in strategic targeting doctrine at this time could also have influenced Soviet decision-makers to desist with their diplomatic objections. The movement away from targeting cities to US ICBM fields may have increased their requirement for the type of information that only satellites could provide.[7] The need for intelligence on China after the Sino-Soviet split should also not be dismissed when explaining the change in policy.

The general context of US-Soviet relations at this time is another important factor. After stepping back from the brink of war over Cuba in October 1962, both superpowers realized that co-operation was better than confrontation on a whole range of issues. In June 1963 the "hot line" agreement was signed establishing facilities in both capitals, followed by the Partial Test Ban Treaty in July and the UN resolution banning "bombs in orbit" in October. The cessation of Soviet opposition to observation satellites can thus be interpreted as a recognition of the futility of continuing with an unsupported policy in the UN and a conciliatory gesture in keeping with this early period of détente. Overall, then, while the United States might have led the Soviet Union to the water's edge in accepting satellite reconnaissance, it is just as likely that they drank for their own reasons.

To be fair to Steinberg, had the United States publicly embarrassed the Soviet Union with information gained from reconnaissance satellites, the Soviets would most probably have continued with their diplomatic opposition and even resorted to force. Yet, while the United States did not want to give the Soviet Union further incentives to shoot down reconnaissance satellites, this was not the sole reason for the "blackout", as Steinberg suggests. Kennedy and his advisors were also sensitive to the possible objections of other countries, particularly in the Third World. Maintaining a low profile is, moreover, standard practice in all intelligence operations regardless of the "sources and methods". This prevents the target country from knowing the true quality of the intelligence and reduces the likelihood of compensatory measures, such as greater concealment.

The subsequent effect of the "blackout" on Soviet statements about the US space programme also needs to be put into proper perspective. Although Soviet protests of US satellite reconnaissance diminished after 1963, this did not extend to the US space programme as a whole. Despite US efforts to reduce the provocative face of its military programme, Soviet commentators continued to criticize such US space projects as Gemini and MOL for being aggressive. The Soviets were also fully cognizant of the two US ASAT systems in the Pacific. Thus, subsequent US space activities appear to have nullified whatever positive effect that the cancellation of SAINT might have had on Soviet perceptions of US intentions.

Although the tacit acceptance of satellite reconnaissance undoubtedly removed a major source of conflict, it did not preclude the development of space weaponry altogether. After all, the *modus vivendi* applied only to the *peacetime* use of space. Contrary to Steinberg's assertions, the real reason appears to have been a combination of military, technical, economic and organizational factors which, as noted above, provided a web of disincentives and constraints.

At the same time as US policymakers wanted to project the "peaceful" image of its military space programme, they also recognized that space weapons in general offered few if any military advantages. For example, there was little incentive to use space systems for strategic bombardment. As Alton Frye observed in 1968:

> Enjoying preponderance in present types of strategic weapons, the United States has not had a strong incentive to shift the arms race to a new environment and to novel technologies.
> Furthermore, although either the United States or the Soviet Union could place thermonuclear weapons in orbit, there have appeared to be no decisive military advantages which would make deployment of bombardment satellites a national strategy.[8]

While there were some who initially thought otherwise, the logic of this argument was eventually accepted and codified in the 1963 UN resolution and later the 1967 Outer Space Treaty.

With the exception of insuring against the potential deployment of Soviet orbital bombs, antisatellite weapons also offered negligible benefits to the United States for the simple reason that Soviet military satellites posed little threat at that time. Countering them in wartime would have made little difference to the effectiveness of their terrestrial forces. This was implicitly acknowledged by Robert McNamara in 1963 when he stated: "I do not believe the Soviets are utilizing space for military purposes to nearly the extent that we are today. I say that because our operations in space for military purposes are truly quite extensive."[9] Once the likelihood of orbital bombs diminished after 1963, so too did the rationale for the two precautionary US ASAT systems that were deployed. Even when the Soviets began testing a satellite interceptor in 1968, little had changed, as a US ASAT system could neither directly counter this capability nor, as was recognized at the time, deter its use while the United States remained more dependent on military space assets.

Although the increasing Soviet military use of space from the beginning of the 1970s indicated that a US antisatellite weapon might have some utility, US policymakers still believed that the costs of developing such a system—particularly the stimulus it might give to the Soviet satellite interceptor programme—outweighed what were still only marginal military benefits.

In contrast, the Soviet Union had every reason to develop antisatellite weapons. The acceptance of satellite reconnaissance did nothing to change the military requirement to negate US space assets in wartime. In fact, the failure of the diplomatic strategy in the UN may even have stimulated the Soviet ASAT programme. Other reasons are required, therefore, to explain the low level of Soviet

ASAT development in the 1960s and 1970s.

One possibility is the technical difficulties associated with non-nuclear interception in space. The Soviets, after all, had barely managed to shoot down Gary Powers' U-2 aircraft. Khrushchev's claims that the Soviet Union could "hit a fly" in space in the early 1960s were certainly bluffs, as he later admitted.[10] By the beginning of the 1970s, the Soviets were still having difficulty perfecting their co-orbital satellite interceptor. Although they might have considered this system operational by 1971, its inherent limitations and low reliability may have encouraged the Soviets to cease further testing until a better guidance system could be developed. This, the climate of détente, and possibly more pressing priorities in the military research and development programme probably explain the hiatus in testing between 1971 and 1976.

Technical constraints should also not be discounted when explaining US behaviour. Where a military rationale for a space weapon could be demonstrated, this was often technically infeasible and prohibitively expensive. For example, while there was considerable interest in defending against a Soviet missile attack, space-based BMD proposals such as BAMBI (a "mad scientist's dream", as Herbert York described it)[11] never made it off the drawing board. Even relatively unambitious space projects such as SAINT, Dynasoar, and MOL presented enough technical problems to outweigh their advertised uses. Although more money could have solved many of these problems, their marginal contribution to politico-military needs did not make this worthwhile. The opportunity costs and the pressing demands of other programmes meant that they were always vulnerable to cancellation. This was particularly evident with the cutbacks of the Johnson and Nixon administrations following the escalation of the war in Southeast Asia.

The choice of the Thor and Nike Zeus missiles for the ASAT mission also reflects technical and economic considerations. As these boosters did not require significant modifications, they were immediately attractive to OSD. Also, while considerable research was carried out on non-nuclear interceptors, the difficulty of perfecting an ASAT system of this kind (greater demands on target prediction, accuracy of interception, homing technologies, and so forth) made it an expensive proposition that was out of proportion to the threat at that time. Although fitting nuclear warheads meant limiting the system's operational flexibility, these were in essence for emergency use only and therefore such considerations had a lower priority than cost.

While the above factors provided the main framework of disincentives and constraints in the military use of space, these were often reinforced by organizational factors. The role of the Air Force in the United States provides an example of this. After the surge of activity following Sputnik, the Air Force's interest in space declined in the late 1960s and early 1970s. This was a direct result of being rebuffed in its efforts to gain sole responsibility among the services for the space mission and also in its main goal—a manned space programme. As Leitenberg argues:

> . . . satellite intercept was just one mission harnessed to that goal. Reduced to its crudest form, one can say that the Air Force wanted a manned space mission first, and an ASAT second, not the other way round.[12]

This almost certainly explains why the SAINT project was never adequately funded by the Air Force. With the cancellation of such projects as Dynasoar and MOL, many believed in the Air Force that they had made their "pitch" and failed. This in turn reduced the incentives to try again and reinforced the bias towards the traditional mission of the Air Force, namely flying. As a result, the Air Force's space activities remained a poor relation to tactical and strategic air power in its organizational hierarchy and inevitably in its funding priorities. This undoubtedly influenced the Air Force's negative attitude towards the various ASAT modernization proposals put forward by Air Defense Command and others in the early 1970s. The provision of satellite survivability measures also suffered because the Air Force was reluctant to propose initiatives that would require the use of its own budget to defend the space assets of other services and agencies.

By the late 1970s, the factors that had stymied the development of the arms race in space began to change. In particular, the incentives for both sides to develop antisatellite weapons (and with them the costs) increased. Despite the decline in the US space budget and the ambivalence of the services, the military use of space had none the less steadily expanded. In part, this was accomplished as the services began to appreciate the "force multiplier" effect of space systems for their *traditional* missions. Whereas for the first ten to fifteen years of the space age, military space systems had predominantly supported strategic forces (that is, with early warning, communication, targeting and navigation information), satellite services have been progressively extended to general purpose military operations. Thus satellites began to facilitate battlefield surveillance, tactical targeting and communication. They offered the chance of improving the lethality of weapons systems and the effectiveness of military forces generally.

The net effect was twofold: the dependence on space systems increased, as did the threat they posed to terrestrial forces. Because satellites were both important to an adversary and threatening to one's own forces, they became doubly attractive as military targets. This undoubtedly contributed to the Soviet decision to resume satellite interceptor testing in 1976. Similarly, the expansion of the Soviet military space programme throughout the 1970s caused the United States to re-evaluate the threat posed by Soviet satellites and with it the utility of antisatellite weapons.

However, while this may have provided the ostensible rationale for the US ASAT programme in the late 1970s, it was the resumption in Soviet testing that had the most impact on US policymakers. Whereas the early phase of testing had caused relatively little concern, the tests in 1976 were perceived in a totally different light. The decline of détente, the Soviet Union's attainment of strategic parity, and an apparent aggressive interventionist policy in the Third World made US policymakers especially sensitive to any "new" development. Thus, while the satellite interceptor appeared to be very similar to the system tested between 1968-71 (and if anything had an inferior testing record), the general context of US-Soviet relations and a growing awareness that the United States was becoming increasingly dependent on a small number of vulnerable satellites caused the United States to react in a different way. Even though many still believed that the United States could not deter a Soviet ASAT attack by the threat of reciprocal action because of the prevailing asymmetry of dependency on space systems, others felt that an imbalance in ASAT capabilities could not be tolerated.

The maturation of technology in the general area of nonnuclear kill techniques and terminal homing devices that had been part of the BMD and ASAT research effort funded during the 1960s also provided decision-makers with a broader and more practicable range of options than had been previously available. However, this should not be interpreted to have provided a "technological imperative". Where the promise of technology appears to have had the most impact is in the development of directed energy weapons. Although this technology has been "pulled" as much as it "pushed", its *potential* applications undoubtedly influenced President Reagan's "Star Wars" initiative in 1983.

LESSONS FOR ARMS CONTROL

The discussion thus far suggests that diplomacy and arms control played no part in averting an arms race in space. Moreover, it also suggests that once the prevailing disincentives and constraints changed in the 1970s, an arms race became inevitable. This would be wrong in both cases. What then are the lessons for arms control?

To Steinberg, the main lesson is that "informal bargaining" offers an alternative way to bring about meaningful arms control. The above critique, however, casts considerable doubt on the role of informal bargaining in both resolving the dispute over satellite reconnaissance and preventing the subsequent development of antisatellite weapons. The tacit acceptance of satellite reconnaissance may not be a representative model for successful arms control in another regard. Because it was essentially a debate about the right to conduct espionage in space, it is difficult to conceive of an alternative agreement that would have been reached then or at any other time. As Raymond Cohen argues, clandestine activity "naturally lends itself to tacit rather than formal regulation".[13]

A more representative case study of the benefits of informal bargaining for arms control is the achievement of the 1963 UN resolution banning weapons of mass destruction from space. Here, through the use of unilateral offers of restraint and informal private exchanges, many of the problems that Steinberg recounts as bedeviling open and formal negotiations were avoided.[14] However, in the final analysis it was a formal agreement that provided the tangible conclusion to the informal bargaining. Although tacit arrangements are useful when a formal agreement is out of the question, and while informal and private contacts can be helpful in initiating discussions and reducing misperceptions, there is no substitute for the contractual obligations of formal agreements. As Bertram has argued, to

> . . . *replace* negotiated agreements by unilateral acts, reciprocal or not, seems unrealistic. The penalties for misunderstanding could be severe. It might do away with some of the inadequacies of negotiated arms control, but only at the price of creating new inadequacies. Formal negotiation and mutual agreement will have to remain the basis of East-West arms control.[15]

More cynical observers would argue that, regardless of the means by which it was achieved, "banning the bomb" from space was easy because it was never a weapon system that either side seriously considered deploying. This interpretation not only ignores the obstacles that needed to be overcome, but also devalues the benefits of the 1963 UN agreement. For example, the proposal for a separate ban on weapons of mass destruction in space was initially opposed due to the prevailing prerequisite for on–site inspection. In this sense, the UN resolution helped establish the convention of relying on national technical means of verification for arms control. The agreement was also a substantial confidence-builder at a time of great uncertainty. While the United States never seriously entertained the idea of deploying nuclear weapons in space, it was none the less concerned that the Soviets might do so.

Furthermore, as the attractiveness of technology and potential military options is constantly changing, treaties that appear superficially meaningless should only be assessed after a lengthy period has elapsed. The arms control value of many of the space-related agreements can thus be viewed in a more charitable light. For example, the drafters of the Partial Test Ban Treaty could not have foreseen that Article I also prohibits the testing of space-based nuclear-pumped x-ray lasers, which are currently being considered. Similarly, the prohibition of space-based BMD systems in Article V of the 1972 ABM treaty was not especially significant when it was drafted, but this now poses a major obstacle to the Strategic Defense Initiative. The lesson from this case study is therefore not a new one, but rather reaffirms what many have said before: arms control is important for political reasons and if conducted in a realistic way and at the right time, it can head off wasteful avenues of the arms race.

This raises the question of whether a competition in antisatellite weapons could have been avoided. In short, was there a propitious moment for an arms control agreement? With the exception of the most recent opportunity during the Reagan administration, there appears in hindsight to have been two other chances. During the 1960s both the United States and the Soviet Union were sufficiently uncertain about the future exploitation of space to forgo the option of developing antisatellite weapons. The United States was also worried that discussion of this topic might reopen the whole debate about the legitimacy of reconnaissance satellites. But by the time SALT I was signed in 1972 these concerns had diminished. Moreover, both the United States and the Soviets had rough parity in ASAT testing experience and seemingly little interest in further development. The time was therefore ripe for an ASAT arms control initiative either during the SALT I negotiations or just after. This would have been entirely consistent with the principle underlying the ABM treaty and the NTM clauses of the agreement on strategic offensive missiles. While it would be easy to say that the United States squandered an opportunity, the option was seemingly not even considered. The United States did not appear overly worried about the Soviet ASAT system at this time and later probably felt that the Soviets had ended their ASAT programme unilaterally. The Soviets also appear not to have been interested in ASAT arms control at this time, although Raymond Garthoff was later told by responsible Soviet officials that they would have "jumped at a ban in 1972".[16] This, if true, does not relieve them of blame for failing to initiate a discussion of this topic at that time.

By the late 1970s, the transformation in disincentives and constraints outlined earlier were already conspiring to make ASAT arms control more difficult, and to many an "unnatural act".[17] The period spanned by the Carter administration from 1977 to 1981 represents the second major opportunity for at least some limitations. Unfortunately, as discussed earlier, a combination of factors—an overly long delay before the ASAT talks started, Soviet intransigence towards comprehensive limits, internal US divisions, the SALT II debate and finally the crises over Cuba, Iran and Afghanistan—prevented a successful conclusion to the negotiations.

Since then the superpowers' interest in space arms control has been out of phase. By the time the Soviet Union was prepared to make major concessions, the United States was no longer committed to arms control. Now that the United States has declared an interest in ASAT negotiation, the Soviet Union shows every sign of becoming inflexible again due to the US ASAT test programme and the new ballistic missile defence effort.

From Cold War to Hot War: Implications for the Future

It now appears certain that dedicated antisatellite weapons are here to stay. Unless there is a sharp reversal in the current state of arms control both superpowers will possess operational ASAT systems by the end of this decade. This stark fact raises some necessary questions: what will be the likely course of future ASAT development? Can arms control play a role in curbing the more wasteful or dangerous aspects of this trend? What will be the impact of the possession and use of ASAT weapons on crisis stability, and what will be the effect of their use during wartime? While the answers to these questions require more thorough analysis than is possible here, some tentative assessments can be made.

It does not take much foresight or imagination to predict the probable direction of future ASAT development. While the action-reaction dynamic of arms competitions is more complex than the way it is often depicted, it does nevertheless serve as a general guideline for predicting behaviour. Soviet planning to meet the threat posed by the deployment of current and potential US ASAT systems has undoubtedly already begun. The Soviets almost certainly consider the US ASAT system to be superior to their own both in its current and potential configurations. Thus, one of the first priorities of the Soviet Union will be to improve the survivability and redundancy of its space assets. To some extent the Soviets are already highly redundant in some of the missions that satellites perform. Also, the current practice of launching a large number of satellites because of their relatively short lifetime leaves the Soviet Union in a better position to reconstitute space assets in the event of war. Furthermore, by the time the US ASAT system becomes operational, more Soviet military satellites will have been moved to higher orbits beyond the reach of the F-15/MHV.

If official US estimates are indeed correct, the Soviet Union is already developing a more advanced ASAT system based on the use of lasers. In the interim before this becomes operational, which is likely to be many years yet, the Soviets can be expected to resume testing of their satellite interceptor to improve the new guidance system and the system's overall reliability. Alternatively, they may feel that its effectiveness has already been undermined by US countermeasures. In this case they may pursue an air-launched capability similar to the US F-15/MHV system. Whatever avenue of development is followed, the net result will be the same: more US and allied satellites will be put at risk.

US officials have already expressed their intention to develop a more advanced ASAT weapon that will extend the range of the air-launched system. This programme will be further encouraged by the Soviets' survivability plans. Similarly, the United States will continue to develop its own directed energy ASAT systems and take the necessary precautions to meet the growing Soviet ASAT capability.

The cost in financial terms will be staggering. The "tens of billions" of dollars price tag that the General Accounting Office attached to the current US ASAT programme may have been exaggerated but it will surely pale in comparison to the aggregate cost of an arms race in space. This will include further generations of ASAT weapons and also the necessary survivability measures to meet the expanded threat. While it would have been prudent to fund the satellite measures regardless of whether there had been an arms control agreement, the price would have been a fraction of the figure that will now be imposed by an unconstrained threat environment. Indeed, the cost of providing survivable satellites or, alternatively, timely replacements may become so prohibitive so as to encourage a reversal of the current trend of relying more heavily on space systems. The use of cheaper, more survivable but perhaps less effective terrestrial alternatives may be the shape of things to come.

The drain on funding caused by the higher military space expenditures may also impose opportunity costs on the civil/commercial exploitation of space. Furthermore, while the operation of civil/commercial satellites is unlikely to be affected during peacetime, under wartime conditions they will certainly be considered "fair game" for ASAT attacks.

Over the long run, the most important development, if left unchecked, will be the deployment of ballistic missile defence systems. Due to the dual capability of BMD systems against satellites, the incentives to negotiate ASAT limitations will be sharply reduced. It will also undermine the prospects for a meaningful agreement if indeed talks are ever resumed. The operational overlap is just too great. However, if a long-term shift to strategic defence is considered mutually desirable, there may be common incentives to seek at least some limitations on dedicated antisatellite weapons, as the space-based components of a BMD system will be the most vulnerable.[18]

This raises the question of whether arms control can play any part in curbing future ASAT development. Given the potential effectiveness of even small numbers of ASAT weapons, it will be extremely difficult, if not impossible, to reverse the current trend in the same way that ABMs were curtailed. However, the situation is not hopeless. Depending on the stage that US ASAT testing has reached and whether Soviet tests have also resumed, it may still be possible in the next few years to reach a meaningful moratorium on further development. Over the long run this might provide a springboard for a more comprehensive ASAT limitation agreement. Although the United States would have to live with the fact that the Soviets possessed a tested, albeit limited ASAT system, the significance of this can be mitigated by further survivability measures. Moreover, Soviet confidence in the reliability of its satellite interceptor would presumably diminish the longer it was left untested. This should also go some way to assuage US fears. The timing of an initiative of this kind will clearly determine the chances of its success. While some initial tests of the US ASAT system against objects in space may provide a rough parity in testing experience and possibly even improve the chances for a mutual moratorium, an extended run of successful US tests will undoubtedly undermine Soviet incentives to agree to such an arrangement.[19]

If US deployment goes ahead and the Soviets respond in the ways described earlier, ASAT arms control will obviously be more difficult to achieve and distinctly less meaningful. Quantitative limits on the number of deployed ASAT systems would be virtually impossible to verify. The dual capability of both the current US and Soviet launch systems and the small size of the interceptor (more so in the US case) make covert deployment very easy.

However, the value of further *qualitative* restrictions should not be overlooked. The possibility of "capping" an ASAT race by limiting both sides to one (preferably their current) system has already been proposed and apparently gained some currency within ACDA.[20] As both the current systems are effective at relatively low altitudes, an agreement of this kind could be refined by prohibiting testing beyond, say, 5000 km. This would effectively "sanctuarize" the semisynchronous and geosynchronous orbits. The Soviet Union, however, may feel this to be a lopsided agreement, as more of its military satellites are in low earth orbit. This problem may recede as additional Soviet satellites migrate to higher orbits, leaving a balance of vulnerable satellites in the permitted ASAT development zone. The most significant drawback to this kind of agreement is that it may be relatively easy to convert a low altitude interceptor to one with greater reach. For example, confidence in the reliability and effectiveness of an untested high altitude interceptor could be gained from other permitted or seemingly innocuous space activities at this altitude.

Altitude restrictions may also be rendered meaningless if the development of directed energy weapons goes unchecked. Projecting a beam of light to high altitudes presents problems of its own, but it may not be as difficult as building special interceptors. Moreover, it may be impossible to detect. Thus, any restriction of kinetic kill ASAT devices will have to be augmented by equivalent limitations on DEW development if it is to be meaningful.

Some of the shortcomings of these qualitative restrictions may be alleviated by agreeing to certain co-operative measures in space, commonly known as "rules of the road". The closest analogue is the US-Soviet Incidents at Sea Agreement, which establishes rules of behaviour for naval activities and also consultative channels for resolving disputes. A similar agreement for space activities could include launch constraints (advance notification of launches with expected orbital parameters), minimum separation distances between spacecraft, inspection procedures, and consultative mechanisms to reduce misperceptions arising from ambiguous activities or accidents in space. While the benefits of such a "rules of the road" arrangement would be greater if linked with qualitative restrictions, they may be better than nothing if the latter proves impossible to achieve.

Although agreements prohibiting the use of weapons systems are generally dismissed as being next to worthless, there may be some value in providing a distinct legal threshold for antisatellite "acts" where one only exists ambiguously at the moment.[21] This might, for example, deter interference short of destruction in peacetime and in the early stages of a war.

The lessons from this brief discussion of the prospects for space arms control are twofold. First, there is still some scope to regulate and stabilize any ASAT competition. Second, this will become progressively more difficult the longer it is left unchecked. This raises the third question on the likely consequences of an unconstrained arms competition in space.

Antisatellite weapons are generally considered to be destabilizing as they threaten satellites that support deterrence.[22] For example, the destruction or disruption of early warning and strategic communication satellites could, in theory at least, facilitate a first strike and reduce the retaliatory options of the attacked state. Moreover, the fear of losing key space assets may in turn encourage states possessing ASAT weapons to pre-empt to avoid pre-emption—the "use them or lose them" argument.

While the incentive to use ASAT weapons to gain an advantage when war is considered inevitable cannot be discounted, a "decapitating" strike in space will not be as straightforward as many assume. The location of the critical early warning and communication satellites in geosynchronous and "Molniya" type orbits will make it extremely difficult to execute such an attack by the conventional direct ascent method. The lengthy flight time of the interceptor would by itself provide early warning of an impeding attack. Furthermore, the need to disable a significant part of the target satellite constellaton to be effective places a considerable demand on the attacker. For this to be fully co-ordinated with a massive launch of strategic missiles would not be easy. Long-term attack strategies using prepositioned space mines or powerful and near instantaneous beam weapons would, however, make this task considerably easier. But under these circumstances one can reasonably expect both superpowers to reduce their dependence on these vulnerable satellites.

The mutual possession of ASAT weapons could be destablizing in another regard. Ambiguous activities or an accident in space could be attributed to the use of antisatellite weapons where before this would not have been such a definite possibility. Suspicion of this kind could lead to a major crisis or, worse still, hostilities in space. The probability of this occurring will depend on the circumstances of the incident(s), the nature of the event, and the importance of the affected satellite(s). For instance, in periods of high tension the malfunction of the data transmission system on a photoreconnaissance satellite or the collision between a navigation satellite and a piece of space debris would be the cause of some alarm. Whether suspicious of this kind would be subsequently translated into retaliatory action is impossible to predict, but it is not implausible. The discussion so far is indicative of the problems of gauging the effect of ASAT deployment and use on crisis stability. This uncertainly increases considerably when assessing the impact of ASAT use during a conflict.

Apart from the actual success of an antisatellite attack, the most important determinant of its impact derives from the role and function of the satellite. This is measured in terms of the level of dependency, or conversely the degree of redundancy, in the satellite system. Redundancy is a function of the ability to either reconstitute a lost satellite or rely on alternative systems without a significant reduction in capability. This in turn varies according to the type, intensity and geographical location of the conflict in which antisatellite weapons are used.

At the lowest level of interstate conflict in which ASAT use is plausible, that is, war between two superpower proxies, the net effect on the course of the conflict is likely to be minimal for the simple reason that space systems are unlikely to play a significant role in the fighting. While the product gained from reconnaissance satellites was alleged to have been provided to the warring states in the 1973 Middle East war and more recently in the Falkland Islands conflict, its impact on the course of hostilities appears to have been negligible. Although the incentives to use ASATs could change in future wars of this type, there are likely to be more telling ways of aiding client states than engaging in ASAT warfare with the risks that this entails. In short, the fear of escalation is likely to outweigh the marginal utility of ASAT attacks at this level of conflict.

This concern is just as applicable in a conflict involving one of the superpowers against a client of the other. In both cases, however, the incentives to use ASAT weapons might increase depending on what is at stake in the conflict. One factor that will determine the utility of an ASAT attack will be the geographical location of the conflict. For example, the Soviet Union's dependency on satellites would be considerably lower in a conflict close to its borders than in, say, Central America or the Caribbean basin. The loss, temporary or otherwise, of vital communication links or photographic intelligence would be compounded by the relative lack of alternative back-ups in geographically distant contingencies.

The impact of satellite loss will also depend on the ability of the respective superpowers to reconstitute their space assets. For the reasons outlined earlier, the Soviet Union is likely to be in a better position to respond to the demands of wartime reconstitution than the United States. While the United States could eventually replace a lost satellite or use an alternative system, the delay could be crucial.

On the face of it, ASAT attacks in the context of a European war would seem to make little difference. An initial attack against a US reconnaissance satellite would in itself provide warning of an attack. Also, space systems are currently not relied on to perform vital battlefield missions, as local military commanders still use ground- and aircraft-based systems for communication, intelligence gathering and targeting. Moreover, the availability of civil alternatives is—on paper at least—immense. The loss of use of NATO's satellites, for example, would not be great because of the abundance of other civil satellites and submarine cables between the United States and Europe. Over an extended period, however, the relative dependency on space and terrestrial assets could reverse itself. The attrition of terrestrial systems would presumably be high and as a result place the onus on satellites to provide and channel information.

In the future, the level of dependency on space assets for war-fighting is likely to be much higher for NATO. The use of space systems for tactical communication, navigation, air defence and possibly even antitactical ballistic missile defence will progressively raise the utility of Soviet ASAT attacks in a European conflict. If the "deep strike" defence doctrine is adopted by NATO, the use of reconnaissance satellites for tactical/theatre targeting purposes will also increase.

If a conventional war on Europe escalates to the use of nuclear weapons, satellites will play an important part in its initial stages. Satellites would provide timely early warning, targeting data, and strategic communications. Later they would provide the means for post-attack assessment. While this provides additional incentives to attack satellites, the net effect may be to reduce the likelihood of the war remaining limited. Once a nuclear war turns into a "spasm" exchange of strategic systems, antisatellite attacks would be largely irrelevant. The mass destruction of ground stations would make attacks on any remaining satellites, if not impossible, then meaningless.

So far the utility and therefore impact of antisatellite weapons has been discussed within specific thresholds of conflict. Perhaps the most important repercussion is the propensity of ASAT weapons to escalate a conflict to higher levels of violence. Possible scenarios are not difficult to find. For instance, the use of ASAT systems for a show of force as suggested by General Stafford could be totally misconstrued by an adversary.[23] Similarly, the belief that distinctions can be made between satellites that support conventional and nuclear forces may encourage supposedly discrete ASAT attacks but be essentially erroneous and lead to further escalation. The different methods of disrupting satellites may also encourage false expectations of the likely response. For example, the belief that temporary non-destructive interference during low levels of conflict would not be so inflammatory as attacks leading to permanent damage may be entirely wrong and even difficult to achieve. As noted above, attacks on satellites during the early stages of a nuclear war could render the conflict less controllable.

The advent of antisatellite and other space weapons will be akin to opening the mythical Pandora's box. The putative benefits of such weapons will be short lived or, more likely, illusory. Instead, the superpowers will become locked into a never ending, ever demanding search for security in space that will leave them worse off than before. The opportunity costs both in financial and operational terms will be immense. More worrisome is that it will add yet another potential source of conflict to an already overtaxed international system. In short, outer space will never be the same again.

Appendices

Appendix I: US Space Programme Expenditures

Table 1. Space Activities of the US Government, FY 1959-84 Historical Budget Summary—Budget Authority (in millions of dollars)

Fiscal Year	NASA Space[a]	NASA % Total Space	Defense	% Total Space	Energy	Commerce	Interior	Agriculture	NSF	Total Space
1959	260.9	33.2	489.5	62.3	34.3	...	...	...	...	784.7
1960	461.5	43.3	560.9	52.6	43.5	...	...	...	0.1	1065.8
1961	926.0	51.2	813.9	45.0	67.7	...	...	...	.6	1808.2
1962	1796.8	54.5	1298.2	39.4	147.8	50.7	...	...	1.3	3294.5
1963	3626.0	66.7	1549.9	28.5	213.9	43.2	...	...	1.5	5434.5
1964	5016.3	73.4	1599.3	23.4	210.0	2.8	...	...	3.0	6831.4
1965	5137.6	73.8	1573.9	22.6	228.6	12.2	...	...	3.2	6955.5
1966	5064.5	72.6	1688.8	24.2	186.8	26.5	...	...	3.2	6969.8
1967	4830.2	71.9	1663.6	24.7	183.6	29.3	...	...	2.8	6709.5
1968	4430.0	67.7	1921.8	30.3	145.1	28.1	0.2	0.5	3.2	6528.9
1969	3822.0	63.9	2013.0	33.6	118.0	20.0	.2	.7	1.9	5975.8
1970	3547.0	66.4	1678.4	31.4	102.8	8.0	1.1	.8	2.4	5340.5
1971	3101.3	65.4	1512.5	31.8	94.8	27.4	1.9	.8	2.4	4740.9
1972	3071.0	67.1	1407.0	30.7	55.2	31.3	5.8	1.6	2.8	4574.7
1973	3093.2	64.1	1623.0	33.6	54.2	39.7	10.3	1.9	2.6	4824.9
1974	2758.5	59.4	1766.0	38.0	41.7	60.2	9.0	3.1	1.8	4640.3
1975	2915.3	59.3	1892.4	38.5	29.6	64.4	8.3	2.3	2.0	4914.3
1976	3225.4	60.6	1983.3	37.2	23.3	71.5	10.4	3.6	2.4	5319.9
T.Q.[b]	849.2	63.3	460.4	34.3	4.6	22.2	2.6	.9	.6	1340.5
1977	3440.2	57.5	2411.9	40.3	21.7	90.8	9.5	6.3	2.4	5982.8
1978	3622.9	55.6	2728.8	43.2	34.4	102.8	9.7	7.7	2.4	6508.7
1979	4030.4	54.3	3211.3	43.2	58.6	98.4	9.9	8.2	2.4	7419.2
1980	4680.4	53.8	3848.4	44.2	59.6	92.6	11.7	13.7	2.4	8688.8
1981	4992.4	50.0	4827.7	48.0	40.5	87.0	12.3	15.5	2.4	9977.8
1982	5527.6	44.0	6678.7	54.0	60.6	144.5	12.1	15.2	2.0	12440.7
1983	6327.9	41.0	8490.9	54.0	38.9	177.8	4.6	20.4	—	15588.5
1984 Est	6590.4	38.0	10590.3	61.0	34.1	234.8	4.7	23.0	—	17477.3

Note: a. Excludes amounts for air transportation; b. T.Q. = Transitional Quarter.
Source: Aeronautics and Space Report of the President 1982 Activities (NASA, Washington, DC, 1983), p. 96, with updated figures from OMB and DoD.

Table 2. Funding of US Defense Department (DoD) Space Programmes by Mission, FY 1961-84 (in millions of current dollars)

Mission[a]	FY 1961[f]	1962	1963	1964	1965	1966	1967	1968	1969	1970	1971	1972	1973
Manned space flight	58.0	100.0	131.8	89.7	47.0	151.4	237.1	431.0	515.0	121.8	—	—	—
Communications	55.2	104.6	35.4	80.2	25.7	61.1	58.6	60.6	62.0	126.8	84.6	54.2	182.8
Navigation	23.6	22.0	42.1	27.9	27.6	15.5	22.0	24.7	11.5	5.1	7.4	6.5	10.8
Early warning	122.5	180.8	102.4	61.5	50.4	58.6	45.6	68.0	(159.0)[h]	(211.8)	(266.5)	(228.4)	233.2
Weather	—	—	—	—	—	—	—	—	—	—	—	—	21.7
Geodesy	—	—	—	—	—	—	—	—	12.1	8.6	5.8	5.8	4.9
Space Defense[b]	8.2	33.0	40.9	66.9	39.1	8.0	16.6	15.4	16.5	9.8	3.9	3.4	3.2
Vehicle and Engine Devlop.	3.7	68.3	286.1	389.8	274.4	190.0	98.9	72.5	72.3	74.6	46.0	36.2	32.9
Space Ground Support[c]	57.8	102.6	167.7	171.9	235.1	240.1	316.9	260.2	183.9	169.3	150.5	152.3	143.6
Research and Development[d]	74.2	155.7	174.3	157.9	167.1	156.9	131.1	124.1	79.9	82.6	84.6	118.0	116.4
General Support[e]	420.7	531.2	569.2	553.5	713.0	807.2	732.8	864.5	900.9	868.0	863.0	801.7	873.0
Total Mil	813.9	1298.2	1549.9	1599.3	1579.4[g]	1688.8	1663.6	1921.8	2013.1	1678.4	1512.3	1406.5	1622.5

Table 2 (continued)

Mission[a]	1974	1975	1976	1976 (TQ)[j]	1977	1978	1979	1980	1981	1982	1983	1984
Manned space flight	—	—	—	—	—	—	—	—	—	—	—	—
Communications	275.3	361.5	361.4	57.29	720.9	574.2	458.6	506.2	687.4	986.1	1329.6	1406.0
Navigation	38.1	47.6	104.8	23.1	104.9	93.8	117.7	185.6	167.0	431.2	295.4	460.0
Early warning	103.7	136.5	88.7	16.0	87.9	150.0	214.3	207.3	267.3	565.7	707.2	756.9
Weather	24.5	29.1	54.7	8.1	67.9	78.7	61.2	67.9	86.5	109.7	232.4	110.1
Geodesy	6.6	7.7	46.3	2.2	7.7	7.3	8.5	10.3	11.6	28.1	60.8	79.9
Space Defense[b]	(3.5)[i]	(3.2)	—	—	—	—	—	—	—	—	—	—
Vehicle and Engine Develop.	26.0	36.8	54.7	18.5	106.3	289.6	509.6	661.0	758.5	842.3	1072.5	1214.2
Space Ground Support[c]	189.9	91.9	111.7	24.9	123.6	173.7	210.9	242.3	337.8	472.0	614.4	806.9
Research and Development[d]	133.8	137.1	159.3	50.1	209.5	296.0	434.0	427.7	573.7	759.6	848.7	1098.7
General Support[e]	968.1	1044.2	1047.9	260.3	981.2	1065.5	1196.5	11540.1	1938.0	2586.7	3329.3	4657.6
Total Mil	1766.0	1892.4	1983.3	460.4	2411.1	2728.8	3211.3	3848.4	4827.7	6681.1	8490.9	10590.3

Notes:

a. The missions listed exclude satellite reconnaissance due to the March 1962 DoD Directive that ended all official reference to reconnaissance activities. However as the official totals for FY 1959 and FY 1960 (see Table I) include satellite reconnaissance projects it is likely that subsequent funding for this activity has been hidden in the above totals. Funding for CIA satellite intelligence gathering activities is excluded from this table. These figures are not publicly available. The mission elements have also varied in subsequent years as reflected in the table.
b. Space Defense includes such projects as SAINT, Program 505, 437.
c. Defined as including range support, instrumentation, satellite detection, tracking and control.
d. Defined as including basic and applied research and component development.
e. Defined as including laboratory and research centre in house programmes, development support organisations, general operational support, and space related military construction not otherwise charged to specific space projects.
f. Figures for FY 1961 are the first available using this mission break down.
g. This total is different from Table 1 where the account is given as $1573.9 million.
h. For FY 1969-72 there are no separate figures for the Early Warning Mission. However the entry 'miscellaneous' which is in parenthesis includes Early Warning.
i. Space Defense funding for FY 1974 and FY 1975 is added in parenthesis but it is excluded from subsequent listings for these years.
j. TQ = Transitional Quarter.

Source: NASA Authorization Hearings, FY 1961-84, and information supplied by the DoD.

Table 3: DoD and NASA Space Programme Expenditures, FY 1959-84 (in constant dollars)[a]

Fiscal Year	Department of Defense ($ millions)	NASA (Space) ($ millions)
1959	1815.0	967.4
1960	2086.7	1716.9
1961	2971.5	3380.8
1962	4720.7	6533.8
1963	5673.1	13272.3
1964	5738.4	17999.0
1965	5603.1	18289.7
1966	5695.7	17080.9
1967	5313.3	15427.0
1968	5839.6	13461.0
1969	6010.7	11412.4
1970	4818.8	10183.7
1971	4136.0	8480.4
1972	3711.4	8100.7
1973	4139.2	7889.0
1974	4132.9	6455.6
1975	3981.5	6133.6
1976	3888.8	6324.3
TQ[b]	902.7	1665.1
1977	4413.3	6295.0
1978	4647.1	6169.8
1979	5030.2	6313.3
1980	5170.5	6288.3
1981	5806.0	6004.1
1982	7381.4	6109.2
1983	8966.1	6682.1
1984	10590.3	6590.4

Note: (a) 1984=100. Using deflators provided in *National Defense Budget Estimates for FY 1984* (Office of the Assistant Secretary of Defense (Comptroller), March 1983; (b) TQ=Transitional Quarter.

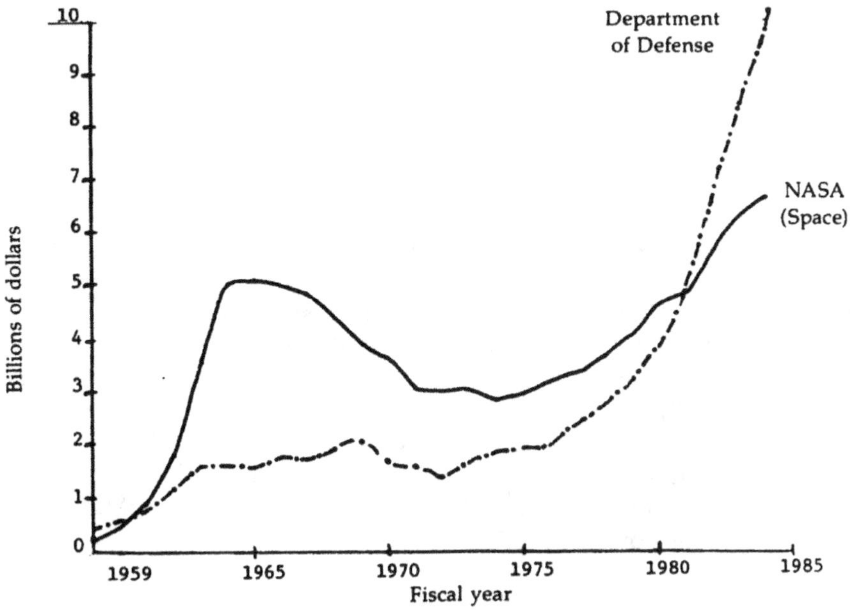

Figure 1: DoD and NASA Space Programme Expenditures, FY 1959-84 (billions of current dollars)

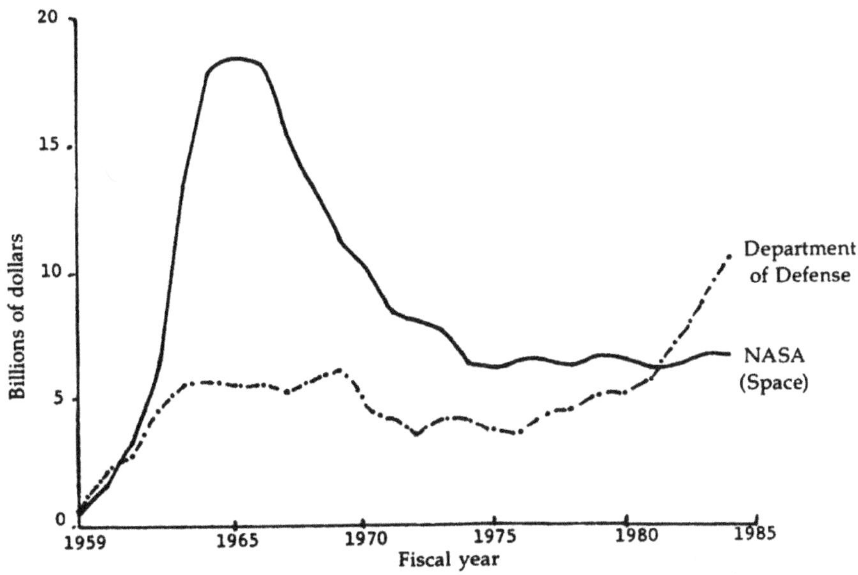

Figure 2: DoD and NASA Space Programme Expenditures, FY 1959-84 (billions of constant dollars)

Appendix II: US-Soviet Antisatellite Tests and Space Launches

Table 1: US Antisatellite and Related Tests, 1959-84

Number	Date	Programme Title	Service	Location	Outcome
1	13/10/59	Bold Orion	Air Force	Eastern Test Range	Success
2	6/4/62	Hi Ho	Navy	Pacific Test Range	Failure
3	26/7/62	Hi Ho	Navy	Pacific Test Range	Success
4	17/12/62	Program 505 (Mudflap)	Army	White Sands N.M.	Success
5	15/2/63	Program 505	Army	White Sands N.M.	Success
6	21/3/63	Program 505	Army	Kwajalein Atoll	Failure
7	19/4/63	Program 505	Army	Kwajalein Atoll	Failure
8	24/5/63	Program 505	Army	Kwajalein Atoll	Success
9	6/1/64	Program 505	Army	Kwajalein Atoll	Success
10	14/12/64	Program 437	Air Force	Johnston Island	Success
11	2/3/64	Program 437	Air Force	Johnston Island	Success
12	?/4/64	Program 505	Army	Kwajalein Atoll	Success
13	21/4/64	Program 437	Air Force	Johnston Island	Success
14	28/5/64	Program 437	Air Force	Johnston Island	Failure
15	16/11/64	Program 437 (CTL)[(a)]	Air Force	Johnston Island	Success
16	5/4/65	Program 437 (CTL)	Air Force	Johnston Island	Success
17	?/6-7/65	Program 505	Army	Kwajalein Atoll	Success
18	?/6-7/65	Program 505	Army	Kwajalein Atoll	Success
19	?/6-7/65	Program 505	Army	Kwajalein Atoll	Success
20	?/6-7/65	Program 505	Army	Kwajalein Atoll	Failure
21	7/12/65	Advanced Program 437	Air Force	Johnston Island	N/A
22	13/1/66	Program 505	Army	Kwajalein Atoll	Success
23	18/1/66	Advanced Program 437	Air Force	Johnston Island	N/A
24	12/3/66	Advanced Program 437	Air Force	Johnston Island	N/A
25	2/7/66	Advanced Program 437	Air Force	Johnston Island	N/A
26	30/3/67	Program 437 (CEL)[(b)]	Air Force	Johnston Island	Success
27	15/5/68	Program 437 (CEL)	Air Force	Johnston Island	Success
28	21/11/68	Program 437 (CEL)	Air Force	Johnston Island	Success
29	28/3/70	Program 437 (CEL)	Air Force	Johnston Island	Success
30	25/4/70	Special Defense Program	Air Force	Johnston Island	Failure
31	24/9/70	High Altitude Program	Air Force	Johnston Island	Success
32	21/1/84	PMALS[(c)]	Air Force	Western Test Range	Success
33	13/11/84	PMALS	Air Force	Western Test Range	Success

(a) CTL = Combat Test Launch
(b) CEL = Combat Evaluation Launch
(c) PMALS = Prototype Miniature Air Launched System

Table 2. Soviet Antisatellite Tests, 1968-82

	Test Number	Date	Target	Target Orbit (km) Incl	Target Orbit (km) Perigee	Target Orbit (km) Apogee	Interceptor	Intercept Orbit (km) Incl	Intercept Orbit (km) Perigee	Intercept Orbit (km) Apogee	Attempted Intercept Altitude	Mission Type	Probable Outcome
	1	20 Oct 68	K248	62.25	475	542	K249	62.23	502	1639	525	2 Rev	Failure
	2	1 Nov 68	K248	62.25	473	543	K252	62.34	535?	1640?	535	2 Rev	Success
	3	23 Oct 70	K373	62.93	473	543	K374	62.96	530	1053	530	2 Rev	Failure
Phase I	4	30 Oct 70	K373	62.92	466	555	K375	62.86	565	994	535	2 Rev	Success
	5	25 Feb 71	K394	65.84	572	614	K397	65.76	575?	1000?	585	2 Rev	Success
	6	4 Apr 71	K400	65.82	982	1006	K404	65.74	802	1009	1005	2 Rev	Success
	7	3 Dec 71	K459	65.83	222	259	K462	65.88	231	2654	230	2 Rev	Success
	8	16 Feb 76	K803	65.85	547	621	K804	65.86	561	618	575	1 Rev	Failure
	9	13 Apr 76	K803	65.86	549	621	K814	65.9?	556?	615?	590	1 Rev	Success
	10	21 Jul 76	K839	65.88	983	2097	K843	—	—	—	1630?	2 Rev	Failure[a]
	11	27 Dec 76	K880	65.85	559	617	K886	65.85	532	1266	570	2 Rev	Failure[b]
	12	23 May 77	K909	65.87	993	2104	K910	65.86	465?	1775?	1710	1 Rev	Failure
Phase II	13	17 Jun 77	K909	65.87	991	2106	K918	65.9?	245?	1630?	1575?	1 Rev	Success[c]
	14	26 Oct 77	K959	65.83	144	834	K961	65.8?	125?	302?	150	2 Rev	Success
	15	21 Dec 77	K967	65.83	963	1004	K970	65.85	949	1148	995	2 Rev	Failure[b]
	16	19 May 78	K967	65.83	963	1004	K1009	65.87	965	1362	985	2 Rev	Failure[b]
	17	18 Apr 80	K1171	65.85	966	1010	K1174	65.83	362	1025	1000	2 Rev	Failure[b]
	18	2 Feb 81	K1241	65.82	975	1011	K1243	65.82	296	1015	1005	2 Rev	Failure[b]
	19	14 Mar 81	K1241	65.82	976	1011	K1258	65.83	301	1024	1005	2 Rev	Success
	20	18 Jun 82	K1375	65.84	979	1012	K1379	65.84	537	1019	1005	2 Rev	Failure[b]

K = Kosmos
Incl = Inclination°
Notes: a. Apparently failed to enter intercept orbit; b. Reportedly used new optical sensor; c. Conflicting data exist for intercept orbit.
Source: Nicholas Johnson, *The Soviet Year in Space 1983* (Teledyne Brown Engineering, Colorado Springs, 1984), p. 39.

Table 3. Summary of US and Soviet Space Launches/Payloads, 1957-81

Summary of USA & USSR Announced Launches

	Calendar Year											Number of Successful Launches														
	57	58	59	60	61	62	63	64	65	66	67	68	69	70	71	72	73	74	75	76	77	78	79	80	81	Total
NASA	0	0	8	10	16	20	11	24	23	29	18	12	13	7	7	9	9	3	11	2	3	8	3	1	4	251
NASA/USA Gov't	0	0	0	0	0	0	2	1	1	4	3	3	1	1	1	2	2	1	2	3	2	2	3	3	4	41
NASA/Commercial	0	0	0	0	0	1	1	0	1	1	3	1	2	3	2	2	1	3	3	7	1	3	2	2	5	44
NASA/International	0	0	0	0	0	2	0	2	1	0	2	3	4	2	6	5	1	8	3	4	7	7	1	0	0	58
TOTAL NASA	0	0	8	10	16	23	14	27	26	34	26	19	20	13	16	18	13	15	19	16	13	20	9	6	13	394
Air Force	0	1	5	8	16	31	24	31	34	39	27	25	18	16	17	13	10	8	9	11	10	13	7	6	5	384
Navy	0	1	0	2	3	3	4	4	5	4	4	1	1	1	0	0	0	0	0	0	0	0	0	0	0	33
Army	0	3	0	1	0	0	0	0	1	0	1	0	0	0	0	0	0	0	0	0	0	0	0	0	0	6
TOTAL DoD	0	5	5	11	19	34	28	35	40	43	32	26	19	17	17	13	10	8	9	11	10	13	7	6	5	423
TOTAL USA SUCCESSES	0	5	13	21	35	57	42	62	66	77	58	45	39	30	33	31	23	23	28	27	23	33	16	12	18	817
TOTAL USSR	2	1	3	3	6	20	17	30	48	44	66	74	70	81	83	74	86	81	89	99	98	88	87	89	98	1437
	Number of Unsuccessful Launches (Not included in numbers above)																									
NASA	0	4	6	7	8	4	1	3	4	2	1	3	1	1	1	0	0	1	1	0	0	0	0	0	0	48
NASA/USA Gov't	0	0	0	0	0	0	0	0	0	0	0	0	0	0	1	0	1	0	0	0	0	0	0	1	0	3
NASA/USA Commercial	0	0	0	0	0	0	0	0	0	0	0	1	1	0	0	0	0	0	1	0	1	0	0	0	0	4
NASA/International	0	0	0	0	0	0	0	0	0	0	1	0	0	0	0	0	0	1	0	0	2	0	0	0	0	4
Total NASA Unsuccessful	0	4	6	7	8	4	1	3	4	2	2	4	2	1	2	0	1	2	2	0	3	0	0	1	0	59
Total DoD Unsuccessful	1	8	4	8	7	6	8	5	4	3	2	1	0	0	2	2	0	0	1	0	0	1	0	2	1	66

Table 3 (continued)

Summary of USA and USSR Announced Payloads

	Calendar Year										Number of Successful Missions or Payloads															
	57	58	59	60	61	62	63	64	65	66	67	68	69	70	71	72	73	74	75	76	77	78	79	80	81	Total
NASA	0	0	8	9	15	17	10	23	22	19	17	13	11	5	7	9	8	3	11	2	3	8	3	1	4	228
NASA/USA Gov't	0	0	0	0	0	0	2	1	1	4	3	3	1	1	1	2	2	1	2	3	2	2	3	3	4	41
NASA/USA Commercial	0	0	0	0	0	1	1	0	1	0	3	1	2	2	2	2	1	3	3	7	1	3	1	2	5	41
NASA/International	0	0	0	0	0	2	0	2	2	0	2	3	4	2	6	5	1	9	3	4	8	7	1	0	0	61
TOTAL NASA	0	0	8	9	15	20	13	26	26	23	25	20	18	10	16	18	12	16	19	16	14	20	8	6	13	371
Air Force	0	1	5	8	18	33	39	39	49	63	48	42	29	20	31	17	12	7	11	18	14	14	9	9	5	541
Navy	0	1	0	3	7	7	10	11	15	4	12	1	10	1	0	0	0	1	0	0	0	0	0	0	0	83
Army	0	3	0	1	0	0	0	0	4	3	1	0	1	1	0	0	0	0	0	0	0	0	0	0	0	14
TOTAL DoD	0	5	5	12	25	40	49	50	68	70	61	43	40	22	31	17	12	8	11	18	14	14	9	9	5	638[a]
TOTAL USA SUCCESSES	0	5	13	21	40	60	62	76	94	93	86	63	58	32	47	35	24	24	30	34	28	34	17	15	18	1009
TOTAL USSR	2	1	3	3	6	20	17	35	64	44	66	74	70	88	97	89	107	95	111	121	105	120	102	110	125	1675
	Number of Unsuccessful Missions or Payloads (Not included in numbers above)																									
NASA	0	4	6	8	9	7	2	5	5	7	2	3	3	3	1	0	0	1	1	0	0	0	0	0	0	67
NASA/USA Gov't	0	0	0	0	0	0	0	0	0	0	0	0	0	0	1	0	1	0	0	0	0	0	0	1	0	3
NASA/USA Commercial	0	0	0	0	0	0	0	0	0	1	0	1	1	1	0	0	0	0	1	0	1	0	1	0	0	7
NASA/International	0	0	0	0	0	0	0	0	0	0	1	0	0	0	0	0	0	1	0	0	2	0	0	0	0	4
Total NASA Unsuccessful	0	4	6	8	9	7	2	5	5	8	3	4	4	4	2	0	1	2	2	0	3	0	1	1	0	81
Total DoD Unsuccessful	1	8	4	8	7	6	8	5	4	3	2	1	0	0	2	2	0	0	1	0	0	2	0	4	1	69

(a) Subject to change as DoD payloads become unclassified

Source: US Congress, Senate, Subcommittee on Science, Technology, and Space of the Committee on Commerce, Science, and Transportation Hearings, *NASA Authorization for Fiscal Year 1983*, 97th Congress, 2nd Session (1982), pp. 75-6.

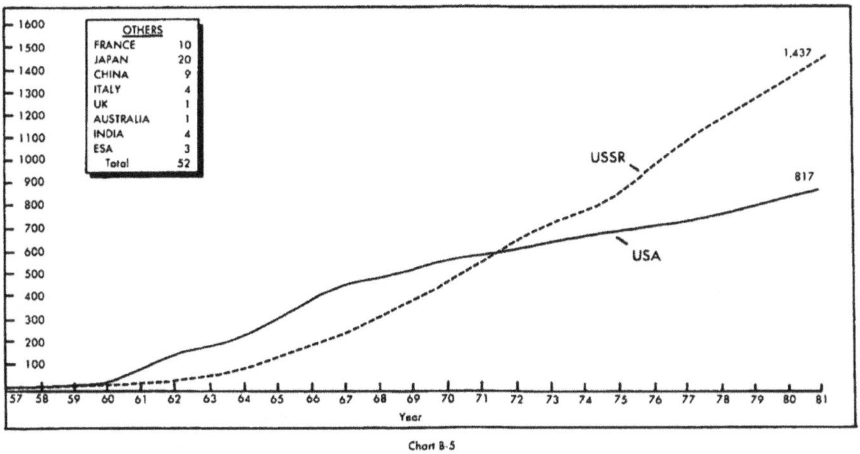

Figure 1: Successful USA & USSR Announced Launches

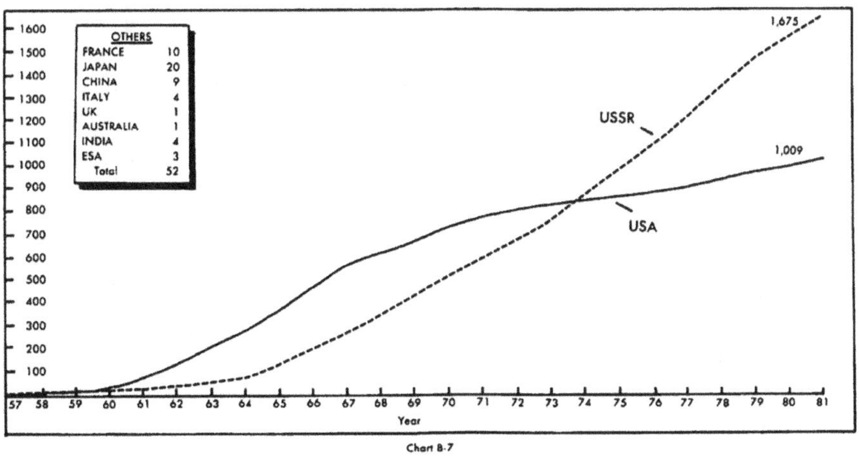

Figure 2: Successful USA & USSR Announced Payloads

Table 4: Summary of Possible US and Soviet Military Satellites by Mission, 1958-83

Year	Photographic Reconnaissance Satellites		Electronic Reconnaissance Satellites		US MIDAS and Vela Satellites		Early Warning Satellites		Ocean Surveillance Satellites		Navigation Satellites		Communication Satellites		Meteorological Satellites		Geodetic Satellites		Cumulative Total	
	USA	USSR	USA	USSR	MIDAS	Vela	USA	USSR	USA	USSR	USA	USSR	USA	USSR	USA	USSR	USA	USSR	USA	USSR
1958	—	—	[a]	—	—	—	—	—	—	—	—	—	1	—	—	—	—	—	1	—
1959	6	—	—	—	—	—	—	—	—	—	1	—	—	—	—	—	—	—	8	—
1960	6	—	—	—	2	—	—	—	—	—	2	—	2	—	2	—	—	—	22	—
1961	13	—	—	—	3	—	—	—	—	—	3	—	2	—	1	—	—	—	44	—
1962	26	5	4	—	1	—	—	—	—	—	1	—	3	—	4	—	1	—	84	5
1963	17	7	7	—	2	2	—	—	—	—	3	—	4	—	3	2	—	—	122	14
1964	24	12	8	—	—	2	—	—	—	—	3	—	3	3	3	2	2	—	167	31
1965	21	17	5	—	—	2	—	—	—	—	4	—	7	8	6	4	6	—	218	60
1966	23	21	10	—	2	—	1	—	—	—	4	—	11	2	6	2	4	—	279	85
1967	18	22	8	5	—	2	—	1	—	1	3	—	17	5	6	4	1	—	334	123
1968	16	29	7	7	—	—	1	1	—	1	1	—	11	4	4	2	1	2	375	169
1969	12	32	6	11	—	2	1	—	—	—	—	—	5	2	3	2	1	2	405	218
1970	9	29	7	10	—	2	3	—	—	1	1	1	3	14	5	6	1	—	436	279
1971	7	28	3	15	—	—	1	—	4	2	—	2	5	21	2	4	—	2	458	353
1972	8	30	3	7	—	—	2	1	—	1	1	3	3	24	4	5	—	2	479	426
1973	5	35	2	12	—	—	2	1	—	1	1	3	4	33	2	3	—	1	495	515
1974	5	28	3	10	—	—	—	1	—	2	1	4	3	24	4	6	—	2	511	592
1975	4	34	2	8	—	—	2	2	1	3	1	4	5	37	3	5	1	2	530	687
1976	4	34	1	11	—	—	1	1	4	2	1	8	11	29	3	3	1	1	554	776
1977	3	33	—	8	—	—	2	3	4	3	1	8	4	16	2	3	—	1	570	851
1978	2	35	1	6	—	—	2	2	1	—	4	8	6	42	4	—	—	1	590	945
1979	2	35	1	5	—	—	2	2	—	3	—	6	3	27	2	4	—	—	606	1027
1980	2	35	1	6	—	—	—	5	4	4	2	6	3	36	2	2	—	—	616	1121
1981	2	37	—	4	—	—	2	5	—	8	1	5	2	39	2	2	—	—	625	1221
1982	3	35	1	6	—	—	1	5	9	9	—	11	2	26	—	2	—	—	632	1305
1983	2	37	1	6	—	—	0	3	—	2	1	13	—	26	1	1	—	1	643	1394
Total	240	610	81	137	10	12	23	33	24	43	40	82	120	418	74	64	19	16	Total 2037	

Note: a. This table excludes Soviet FOBS satellite interceptors and NATO satellites.
Source: B. Jasani, *Outer Space: A New Dimension of the Arms Race*; pp. 94-5.

Glossary of Acronyms

ABM	Anti-Ballistic Missile
ABMA	Army Ballistic Missile Agency
ACDA	Arms Control and Disarmament Agency
ADC	Air Defense Command (Later ADCOM)
AEC	Atomic Energy Commission
AFB	Air Force Base
ARDC	Air Research and Development Command
ARPA	Advanced Research Projects Agency (Later DARPA)
ASAT	Antisatellite
BAMBI	Ballistic Missile Boost Intercept
BMD	Ballistic Missile Defence
CIA	Central Intelligence Agency
CinC	Commander in Chief
CSOC	Consolidated Space Operations Centre
CMLC	Civilian Military Liaison Committee
DARPA	Defense Advanced Research Projects Agency
DDR&E	Director of Defense Research and Engineering
DoD	Department of Defense
DSARC	Defense Systems Acquisition Review Committee
DSB	Defense Science Board
ELINT	Electronic Intelligence
EHF	Extremely High Frequency
EORSAT	Electronic Ocean Reconnaissance Satellite
FOBS	Fractional Orbital Bombardment System
FY	Financial year
GCD	General and Complete Disarmament
GEODSS	Ground Based Electro-Optical Deep Space Surveillance
HATV	High Altitude Test Vehicle

ICBM	Intercontinental Ballistic Missile
INF	Intermediate Nuclear Forces
INTELSAT	International Telecommunications Satellite
IOC	Initial Operating Capability
IRBM	Intermediate Range Ballistic Missile
ISA	International Security Affairs (DoD)
ITV	Instrumented Test Vehicle
JCS	Joint Chiefs of Staff
JRDB	Joint Research and Development Board
KH	Keyhole (satellite series)
LTV	Ling Temco Vought
LWIR	Long Wave Infrared
MHV	Miniature Homing Vehicle
MIDAS	Missile Defense Alarm System
MIRV	Multiple Independently Targeted Re-entry Vehicle
MODS	Military Orbital Development System
MOL	Manned Orbital Laboratory
MX	Missile Experimental
NASA	National Aeronautics and Space Administration
NASC	National Aeronautics and Space Council
NIE	National Intelligence Estimate
NNTRP	National Nuclear Test Readiness Program
NOAA	National Oceanic and Atmospheric Administration
NORAD	North American Air Defense Command
NRL	Naval Research Laboratory
NRO	National Reconnaissance Office
NSA	National Security Agency
NSAM	National Security Action Memorandum
NSDD	National Security Decision Directive
NSDM	National Security Decision Memorandum
NSSM	National Security Study Memorandum
NTM	National Technical Means (of verification)
OMB	Office of Management and Budget
OSS	Office of Strategic Services
OST	Office of Science and Technology
PBW	Particle Beam Weapon
PD	Presidential Directive
PE	Program Element
PFIAB	President's Foreign Intelligence Advisory Board
PMALS	Prototype Miniature Air-launched System
POM	Program Objective Memorandum
PPBS	Planning Programming Budgetary System

PRM	Presidential Review Memorandum
PSAC	Presidential Science Advisory Board
PVO	Protivovozdushnaya Oborona (Soviet Air Defense)
R&D	Research and Development
RDB	Research and Development Board
RORSAT	Radar Ocean Reconnaissance Satellite
SAB	(Air Force) Science Advisory Board
SAC	Strategic Air Command
SAINT	Satellite Inspector
SALT	Strategic Arms Limitation Talks
SAM	Surface to Air Missile
SAMOS	Satellite and Missile Observation System
SAMSO	Satellite and Missile System Organization (Later SSD)
SCF	Satellite Control Facility
SDI	Strategic Defense Initiative
SDS	Satellite Data System
SECDEF	Secretary of Defense
SIG	Senior Interagency Group
SLBM	Sea Launched Ballistic Missile
SPADOC	Space Defense Operations Center
SPO	System Program Office
SSBN	Strategic Nuclear Ballistic Missile Submarine
SSD	Space Systems Division (later Space Division)
STS	Space Transportation System (Space Shuttle)
TCP	Technology Capabilities Panel
UHF	Ultra High Frequency
UN	United Nations
UNGA	United Nations General Assembly
USAAF	United States Army Air Force
USAF	United States Air Force

A Note on Research Methodology and Sources

> DoD has been quite candid with the American public regarding this subject [US ASAT R&D]. For example, in our DDR&E Posture Statements and in our testimony before the various Congressional Committees, we have provided considerable detail on our space defense programs [deleted].
>
> Hearings, *NASA Authorization for Fiscal Year 1978*, p. 1637

The military use of space especially intelligence gathering operations is, for the reasons discussed in this book, among the most sensitive and secretive areas of US defence policy. As a result, the serious scholar wishing to study this subject faces an immense if not impossible task. This, more than any other reason, accounts for why there have been so few unclassified in-depth historical accounts of US military space policy and programmes.

Even more scarce are detailed studies or even references to US antisatellite activities. There are few other facets of US military policy that rival its classified nature. As an illustration, internal documents concerning the early US ASAT programmes in the Pacific were stamped "Special Handling" or "Special Access Required". In May 1972, Deputy Secretary of Defense Kenneth Rush issued a classified memorandum which stated that even reference to the fact that the United States was conducting antisatellite research (never mind its content), including the acronym ASAT, would be classified. While this order was eventually rescinded by Deputy Secretary W. Graham Claytor, Jr, on 30 October 1979, the amount of available information remained pitifully small. Contrary to the above epigram from the FY 1978 NASA hearings, the DoD has not been candid with the American public.

None the less, for the determined researcher there are a number of standard practices and formal mechanisms that can be employed to penetrate the veil of secrecy. These were found to be especially productive during the research for this book. First, a wealth of declassified documents resides at the Eisenhower, Kennedy and Johnson Presidental Libraries. While a considerable amount is still classified, this can be gradually reduced by use of the Mandatory Review procedure for documents over 20 years old. Second, access to documents of a later period or material not stored in the presidential libraries can be sought by Freedom of Information Act (FOIA) requests. While both methods can be time-consuming and frustrating, they were especially fruitful to this study.

However, many of the documents requested were denied or partially declassified with deletions in vital places. To help bridge many of the gaps in the information, particularly in the latter period of the study, over 100 interviews were conducted with former members of the US government and armed forces. These interviews proved to be highly productive but generally only on the condition that they not be for attribution. Although this restriction did not always apply, for the purpose of maintaining uniformity the identity of the interviewee has not been given.

Material of this kind always requires some circumspection, as the interviewee's recollection of events and decisions can be biased and confused. Where possible, this material has been corroborated with documentary evidence or further interviews. Never the less, there is always an irreducible amount of uncertainty that requires subjective judgement on the reliability of the source. Where this has been necessary, the interviewee's qualifications and proximity to events is used as the primary yardstick. In fairness to the reader, the interviewee's former affiliation and a letter from the Greek alphabet has been added to distinguish the sources.

In addition to these primary sources, a survey of US congressional hearings and governmental reports was carried out. Although the sections dealing with the military use of space are among the most heavily censored, the deletions are by no means consistent within and between these official publications. When compared, they can provide a rich source of information. Similarly, the prodigious leaks in newspapers and journals like *Aviation Week & Space Technology* provide considerable supporting evidence. Where these are the only sources, the use of such qualifying terms as "apparently" and "reportedly" have been inserted in the text.

Selected Bibliography

1. Government Publications

US Congressional Hearings

Rather than list each congressional hearing individually, the majority of sources can be condensed as follows:

(a) Senate and House Subcommittee on Appropriations, *Department of Defense Appropriations*, FY 1960-84.
(b) Senate and House Armed Services Committee, *Authorizations For Military Procurement*, FY 1962-84.
(c) Senate Committee on Aeronautical and Space Sciences, subsequently, Subcommittee on Science, Technology and Space of the Committee on Commerce, Science and Transportation, *NASA Authorization Hearings*, FY 1960-84.

The following hearings have also been cited:

United States, Congress, Senate, Preparedness Investigating Subcommittee of the Committee on Armed Services. *Inquiry into Satellite and Missile Programs* 85th Congress, 1st and 2nd Sess. (1958)

United States, Congress, House, Committee on Science and Astronautics, *Defense Space Interests*, 87th Congress, 1st Sess. (1961)

United States, Congress, House, Committee on Science and Astronautics, *Space Posture*, 88th Congress, 1st Sess. (1963)

United States, Congress, House, Subcommittee of the Committee on Government Operations, *Missile and Space Ground Support Operations*, 89th Congress, 2nd Sess. (1966)

United States, Congress, Joint Committee on Atomic Energy, *Scope, Magnitude and Implications of the United States Antiballistic Missile Program*, 90th Congress, 1st Sess. (November 1967)

United States, Congress, Senate, Committee on Foreign Relations, *Treaty on Outer Space*, 90th Congress, 1st Session (1967)

United States, Congress, House, Subcommittee of the Committee on International Relations, *First Use of Nuclear Weapons*, 94th Congress, 2nd Sess. (March 1976)

United States, Congress, Senate, Committee on Foreign Relations, *Warnke Nomination*, 95th Congress, 1st Sess. (1977)

United States, Congress, Senate, Committee on Foreign Relations, *The SALT II Treaty*, 96th Congress, 1st Sess. (1979)

United States, Congress, Senate, Subcommittee on Arms Control, Oceans, International Operations and Environment of the Committee on Foreign Relations, *Arms Control and the Militarization of Space*, Hearings, 97th Congress, 2nd Sess. (20 September 1982)

United States, Congress, Senate, Committee on Foreign Relations, *Controlling Space Weapons*, Hearings, 98th Congress, 1st Session (1983)

United States, Congress, House Investigations Subcommittee on Armed Services, *Aerospace Force Act*, Hearings, 97th Congress, 2nd Session (1983)

US Congressional Reports

United States, Congress, House, Committee on Science and Astronautics, *Military Astronautics* (Preliminary Report), 87th Congress, 1st Sess. (1961)

United States, Congress, House, Committee on Science and Astronautics, Staff Report *Science, Astronautics and Defense* (The 1961 Review of Scientific and Astronautical Research and Development in the Department of Defense), 87th Congress, 1st Sess. (1961)

United States, Congress, Committee on Science and Astronautics, *A Chronology of Missile and Astronautic Events*, 87th Congress, 1st Sess. (1961)

United States, Congress, House, Committee on Government Operations *Government Operations in Space (Analysis of Civil-Military Roles and Relationships* 13th Report, 89th Congress, 1st Sess. (1965)

United States, Congress, Senate, Committee on Aeronautical and Space Sciences, *Soviet Space Programs 1962-1965: Goals and Purposes, Achievements, Plans and International Implications*, Staff Report, 1966

United States, Congress, Senate, Committee on Aeronautical and Space Sciences, *International Co-operation in Outer Space: A Symposium*, 92nd Congress, 1st Sess. (1971)

United States, Congress, Senate, Committee on Aeronautical and Space Sciences, Staff Report, *Soviet Space Programs, 1966-70*, 92nd Congress, 1st Sess. (1971)

United States, Congress, Senate, Committee on Aeronautical and Space Sciences, Staff Report, *Soviet Space Programs, 1971-1975*, vol. I (1976)

United States, Congress, House, Subcommittee on Space Science and Applications of the House Committee on Science and Technology, *United States Civilian Space Programs 1958-1978*, vol. 1, 97th Congress, 1st Sess. (1981)

Special Reports

President's Space Task Group, *The Post Apollo Space Program: Directions for the Future*, Space Task Group, Report to the President (September 1969)

The Next Decade in Space. A report of the Space Science and Technology Panel of the President's Science Advisory Committee, Executive Office of the President, Office of Science and Technology (March 1970)

US Aeronautics and Space Activities, 1958-67. Message for Congress from the President of the United States

Speeches and Press Releases

Remarks by Laurence Kavanau, Special Asst. for Space to Director for Defense Research and Engineering before Aviation/Space Writers Association (21 May 1963) DoD News Release No. 718-63

White House Press Release—Description of a Presidential Directive on National Space Policy. The White House (20 June 1978)

Defense Secretary Weinberger's Testimony Before the Senate Armed Services Committee on US Strategic Capabilities *Official Text* ICA, US Embassy, London (6 October 1981)

Pentagon Release on US Strategic Programme *Official Text* (5 October 1981)

Text, Reagan Remarks on Return of Columbia Four—*Official Text* ICA, US Embassy, London (4 July 1982)

Fact Sheet, National Space Policy, *Weekly Compilation of Presidential Documents* (12 July 1982), pp. 872-6

Miscellaneous Official Documents

Fiscal year 1979, 1980, 1981, 1982, Arms Control Impact Statements (USGPO, Washington, DC)
Documents on Disarmament, 1962-78 (USGPO, Washington, DC)
Arms Control and Disarmament Agreements: Texts and Histories of Agreements, US ACDA Publication (USGPO, Washington, DC, August 1980)
Civilian Space Policy and Options, Office of Technology Assessment Report (USGPO, Washington, DC, 1981)
DoD Space Based Laser Program and Potential, Progress and Problems, Unclassified segment of GAO Report C-MASAD-82-10 (26 February 1982)
Consolidated Space Operations Center Lacks Adequate DoD Planning, GAO Report MASAD-82-14 (29 January 1982)
Soviet Military Power, 2nd edn (Government Printing Office, Washington, DC, March 1983), p. 67

2. Secondary Sources

Books

AEROSPACE CORPORATION, *The Aerospace Corporation: Its Work 1960-1980* (The Aerospace Corporation, Los Angeles, 1980)
ARMACOST, M. H., *The Politics of Weapons Innovation: The Thor Jupiter Dispute* (Columbia University Press, New York, 1969)
BAKER, D., *The Shape of Wars to Come: The Hidden Facts behind the Arms Race in Space* (Patrick Stephens, Cambridge, 1981)
BALL, D., *Politics and Force Levels. The Strategic Missile Program of the Kennedy Administration* (Univerisity of California Press, London, 1980)
BAMFORD, J., *The Puzzle Palace: A Report on NSA, America's Most Secret Agency* (Houghton Mifflin, Boston, 1982)
BEARD, E., *Developing the ICBM: A Study in Bureaucratic Politics* (Columbia University Press, New York, 1976)
BERMAN, R. P. and BAKER, J. C., *Soviet Strategic Forces: Requirements*

and Responses (The Brookings Institution, Washington, DC, 1982)

BRZEZINSKI, Z., *Power and Principle: Memoirs of the National Security Advisor 1977-1981* (Farrar, Straus, Grove, New York, 1983)

BULL, H., *The Control of the Arms Race: Disarmament and Arms Control in the Missile Age* (Weidenfeld and Nicolson, London, 1961)

COHEN, R., *International Politics: The Rules of the Game* (Longman, London, 1981)

EISENHOWER, D. D., *The White House Years: Waging Peace 1956-61* (Doubleday, Garden City, New York, 1965)

EWALD, Jr, W. B., *Eisenhower the President: Crucial Days 1951-1960* (Prentice Hall, Englewood Cliffs, New Jersey, 1981)

FREEDMAN, L., *U.S. Intelligence and the Soviet Strategic Threat* (Macmillan, London, 1977)

FUTRELL, R. F., *Ideas, Concepts, Doctrine: A History of Basic Thinking in the United States Air Force 1907-1964* (Air University, Maxwell AFB, Alabama, 1971)

GALLAGHER, M. P. and SPIELMAN, K. F., *Soviet Decision Making for Defense: A Critique of U.S. Perspectives on the Arms Race* (Praeger, New York, 1972)

GAVIN, J., *War and Peace in the Space Age* (Hutchinson, London, 1959)

GLASSTONE, S., *Sourcebook on the Space Sciences* (Van Nostrand, Princeton, 1965)

GLASSTONE, S. and DOLAN, P. J., *The Effects of Nuclear Weapons* (US Dept. of Defense and US Dept. of Energy, 1977)

GRAY, C. S., *American Military Space Policy, Information Systems, Weapon Systems and Arms Control* (Abt Books, Cambridge, Mass., 1983)

HEIKAL, M., *The Road to Ramadan* (Ballantine Books, New York, 1975)

HORELICK, A. L. and RUSH, M., *Strategic Power and Soviet Foreign Policy* (University of Chicago Press, Chicago, 1966)

JASANI, B., *Outer Space-Battlefield of the Future?* (SIPRI, Taylor & Francis, London, 1982)

JASANI, B. (ed.), *Outer Space: A New Dimension of the Arms Race* (SIPRI, Taylor & Francis, London, 1982)

KARAS, T., *The New High Ground: Strategies and Weapons of Space Age War* (Simon and Schuster, New York, 1983)

KILLIAN, J. R., *Sputnik, Scientists and Eisenhower* (The MIT Press, Cambridge, Mass., 1977)

KISTIAKOWSKY, G., *A Scientist at the White House: The Private Diary of*

President Eisenhower's Special Assistant for Science and Technology (Harvard University Press, Cambridge, Mass., 1976)

KLASS, P., *Secret Sentries in Space* (Random House, New York, 1970)

LABRIE, R. P. (ed.), *SALT Handbook: Key Documents and Issues* (American Enterprise Institute, Washington, DC, 1979)

LAKOFF, S. (ed.), *Knowledge and Power: Essays in Science and Government* (Free Press, New York, 1966)

LASBY, C. G., *Project Paperclip: German Scientists and the Cold War* (Atheneum, New York, 1971)

LEVINE, A., *The Future of the U.S. Space Program* (Praeger, New York, 1975)

LINDSEY, T. R., *The Falcon and the Snowman* (Penguin, Harmondsworth, 1980)

LOGSDON, J. M., *The Decision to Go to the Moon: Project Apollo and the National Interest* (MIT Press, Cambridge, Mass., 1970)

LOMOV, N. A., Col. (ed.), *Scientific-Technical Progress and the Revolution in Military Affairs* (A Soviet View), USAF Studies in Soviet Military Thought, no. 3 (USGPO Washington, DC, 1973)

LONG, F. and RATHJENS, G. (ed.), *Arms, Defense Policy and Arms Control* (Norton, New York, 1976)

LONG, F. A. and REPPY, J., *The Genesis of New Weapons: Decision Making for R&D* (Pergamon, New York, 1980)

MARCHETTI, I. V. and MARKS, J., *The CIA and the Cult of Intelligence* (Jonathan Cape, London, 1974)

MOSELEY, L., *Dulles: A Biography of Eleanor, Allen and John Foster Dulles and their Family Network* (Hodder and Stoughton, London, 1978)

NEWHOUSE, J., *Cold Dawn: The Story of SALT* (Holt, Reinhart and Winston, New York, 1973)

NIEBURG, H. L., *In the Name of Science* (Quadrangle Books, Chicago, 1966)

ORDWAY, F. I. and SHARPE, M. R., *The Rocket Team* (Thomas Y. Crowell, New York, 1979)

PEEBLES, C., *Battle for Space* (Beaufort Books, New York, 1983)

POTTER, W. (ed.), *SALT and Verification* (Westview, Boulder, Colorado, 1980)

POWERS, T., *The Man who Kept the Secrets: Richard Helms and the CIA* (Pocket Books, New York, 1979)

PRADOS, J., *The Soviet Estimate: U.S. Intelligence Analysis and the Russian Military Strength* (The Dial Press, New York, 1982)

ROSTOW, W., *Open Skies: Eisenhower's Proposal of July 21, 1955* (Univ. of Texas Press, Austin, 1982)

SCHAUER, W. H., *The Politics of Space* (Holmes and Meir, New York, London, 1976)
SCHLESINGER, Jr, A. M., *A Thousand Days* (Houghton Mifflin, Boston, 1965)
SMITH, G., *Doubletalk: The Story of SALT I* (Doubleday, New York, 1980)
SOKOLOVSKIY, V. D., *Military Strategy*, 3rd edn, edited by Harriet Fast Scott (Macdonald and Jane's, London, 1975)
SOKOLOVSKIY, V. D., *Military Strategy* (Prentice Hall, New Jersey, 1963)
SPIELMAN, K. F., *Analyzing Soviet Strategic Arms Decisions* (West View, Boulder, Colorado, 1978)
STEINBERG, G. M., *Satellite Reconnaissance: The Role of Informal Bargaining* (Praeger, New York, 1983)
STOIKO, M., *Soviet Rocketry: The First Decade of Achievement* (David and Charles, Newton Abbot, 1970)
TALBOTT, S. *Khrushchev Remembers Vol. 2: The Last Testament* (Penguin, London, 1977)
VAN DYKE, V., *Pride and Power: The Rationale of the Space Program* (University of Illinois Press, Urbana, 1964)
The War Reports of General George C. Marshall, General H. H. Arnold, Admiral Ernest J. King (Lippincott, New York, 1947)
WOLFE, T., *The Right Stuff* (Bantam, New York, 1980)
WOLFE, T. W., *The SALT Experience* (Ballinger, Cambridge, 1979)
WOLFE, T. W., *Soviet Strategy at the Crossroads* (Harvard University Press, Cambridge, Mass., 1964)
YORK, H., *Race to Oblivion* (Simon and Schuster, New York, 1970)

3. Articles and Monographs

ALEXANDER, A., "Decision Making in Soviet Weapons Procurement", *Adelphi Paper*, No. 147/8 (IISS, London, Winter 1978-9)
ANUREEV, I. I., "Anti-Missile and Space Defense Weapons", Joint Publications Research Service, Arlington, Virginia, 1972
BALL, D., "Can Nuclear War Be Controlled?", *Adelphi Paper*, No. 169 (IISS, London, 1981)
BERTRAM, C., "The Future of Arms Control: Part II: Arms Control and Technological Change: Elements of a New Approach", *Adelphi Paper*, No. 141 (IISS, London, 1978)
BOWERS, R., Lt. Col. *et al.*, "A Turning Point in Space: Has the time arrived for a National Review of our Space Policy?". Study Project,

US Army War College, Carlisle Barracks (13 May 1977)

BUCHHEIM, R., "Arms Control as Applied to Satellites and Other Space Objects", Paper presented at Los Alamos Nuclear Laboratory (27 May 1981)

CARTER, A., "Directed Energy Missile Defense in Space", Background Paper (Office of Technology Assessment, Washington, DC, 1984)

CLARKE, P., "The Polyot Missions", *Spaceflight,* vol. 22, no. 9-10 (Sept-Oct. 1980)

CLEMENS, W. C., "Outer Space and Arms Control", Center for Space Research, MIT CSR-TR-66-14 (October 1966)

DMS Market Intelligence Report: "Space Intercept Programs" (January 1968); "Thor/Delta" (November 1971); "Burner 2"; "Survivable Satellite" (January 1976) (DMS, Connecticut)

GARDNER, R. N., "Co-operation in Outer Space", *Foreign Affairs,* vol. 41, no. 2 (January 1963)

GARTHOFF, R. L., "Banning the Bomb in Outer Space", *International Security* vol. 5, no. 3 (Winter 1980-1)

FREEDMAN, L., "The Soviet Union and Anti-Space Defense", *Survival,* vol. 19, no. 1 (January 1977)

HAFNER, D., "Averting a Brobdingnagian Skeet Shoot: Arms Control Measures for Antisatellite Weapons", *International Security,* vol. 5, no. 3 (Winter, 1980-1)

HAFNER, D., "Outer Space Arms Control: Unverified Practices, Unnatural Acts?", *Survival,* vol. XXV, no. 6

HALL, R. C., "Early U.S. Satellite Proposals", *Technology and Culture,* vol. IV (Autumn 1963)

HOLLOWAY, D., "Decision Making in Soviet Defense Policies" in "Prospects for Soviet Power in the 1980s", Part II, *Adelphi Paper,* No. 152 (IISS, London, 1979)

JOHNSON, N., *The Soviet Year in Space 1981; The Soviet Year in Space 1982; The Soviet Year in Space 1983;* (Teledyne Brown Engineering, Colorado Springs, Colorado)

KATZ, A., *The Soviets and the U2 Photos—An Heuristic Argument,* Rand Memo RM 3584-PR (March 1963)

MAY, M., "War or Peace in Space", *The California Seminar on International Security and Foreign Policy,* Discussion Paper no. 93 (Santa Monica, California, March 1981)

MEYER, S. M., "Anti-satellite Weapons and Arms Control: Incentives and Disincentives from the Soviet and American Perspectives", *International Journal* (Autumn 1981)

MEYER, S. M. "Soviet Military Programmes and the New High

Ground", *Survival*, vol. 25, no. 5 (Sept.-Oct. 1983)
MIHALKA, M., "Soviet Strategic Deception", *Journal of Strategic Studies*, vol. 5, no. 1 (March 1982), pp. 40-3
"Military Race in Space", *The Defense Monitor*, vol. 9, no. 9 (1980)
PEEBLES, C. L., "The Guardians: A History of the 'Big Bird' Reconnaissance Satellites", *Spacecraft*, vol. 20, no. 11 (November 1978)
PERRY, G. E., "Russian Ocean Surveillance Satellites", *Royal Air Force Quarterly*, vol. 18 (Spring 1975)
PERRY, G. E., "Russian Hunter-Killer Satellite Experiments", *Royal Air Force Quarterly*, vol. 17 (Winter 1977)
PIKE, J., "Limits on Space Weapons: The Soviet Initiative and the American Response", *FAS Staff Study* (12 September 1983)
RICHELSON, J. T., "United States Strategic Reconnaissance: Photographic and Imaging Satellites", *ACIS Working Paper*, no. 38, UCLA (Los Angeles, May 1983)
RICHELSON, J. T.,"PD-59, NSSD-13 and the Reagan Strategic Modernization Program", *Journal of Strategic Studies*, vol. 6, no. 2 (June 1983)
STARES, P. B., "Arms Control in Outer Space: On Trying to Close the Stable Door Before the Horse Bolts", *Arms Control*, vol. 1, no. 3 (December 1980)
STARES, P. B., "Reagan's BMD Plan: The Ultimate Defence?", *ADIU Report*, vol. 5, no. 3 (May/June 1983)
STARES, P. B., "Space and US National Security", *Journal of Strategic Studies*, vol. 6, no. 4 (December 1983)
TSIPIS, K., "Laser Weapons", *Scientific American*, vol. 245, no. 5 (December 1981)
WILKES, O., "Spacetracking and Space Warfare", *PRIO Publication S-1/77*, International Peace Research Institute (Oslo, December 1978)
WILSON, M. "Killer Satellites—The 'Seeds' of Space War", *Command* (November/December 1978)
WORTHMAN, P. E., "The Promise of Space", *Air University Review*, vol. XX, no. 2 (Jan.-Feb. 1969)
YORK, H. and GREB, G. H., "Strategic Reconnaissance", *Bulletin of the Atomic Scientists*, vol. 33, no. 4 (April 1977)

In addition articles have been cited from the following newspapers and periodicals: *Aerospace Daily; Air Force Magazine; Astronautics, Aviation Week and Space Technology* (formerly *Aviation Week*); *Baltimore Sun; Defense/Space Daily; Defense Week; Flight International; International*

Herald Tribune; Los Angeles Times; Military Thought; Missiles and Rockets; New Scientist; Newsweek; New York Times; Ordnance; Science; Science News; Space Astronautics; Time; U.S. News and World Report; Wall Street Journal; Washington Post.

Unpublished Sources

GREENWOOD, J. T., "A Short History of the Air Force Ballistic Missile and Space Program 1954-1974. Unpublished Monograph

LEITENBERG, M., "The History of the U.S Antisatellite Program". Unpublished Monograph for Cornell Peace Studies Program 1979

SAWYER, H. L., "The Soviet Space Controversy, 1961-1963". Doctoral Thesis for Fletcher School of Law and Diplomacy, Massachusetts (May 1969)

STEINBERG, G., "The Legitimization of Satellite Reconnaissance: An Example of Informal Arms Control Negotiations". Doctoral Thesis, Cornell University, New York, 1981

Notes

1. *Introduction*

1. See, for example, Bhupendra Jasani, *Outer Space: Battlefield of the Future?* (Stockholm International Peace Research Institute [SIPRI], Taylor and Francis, 1978); Bhupendra Jasani (ed.), *Outer Space: A New Dimension of the Arms Race* (SIPRI, Taylor and Francis, 1982); Thomas Karas, *The New High Ground: Strategies and Weapons of Space-Age War* (Simon and Schuster, New York, 1983); James Canan, *War in Space* (Harper and Row, New York, 1982); David Ritchie, *Space War* (Atheneum, New York, 1982); Keith B. Payne (ed.), *Laser Weapons in Space: Policy and Doctrine* (Westview Press, Boulder, Co., 1983); Colin Gray, *American Military Space Policy: Information Systems, Weapons Systems and Arms Control* (Abt Books, Cambridge, Mass., 1983); and Curtis Peebles, *Battle for Space* (Beaufort Books, New York, 1983).

2. See Jasani, *Outer Space: A New Dimension of the Arms Race*, pp. 41 and 112. The total excludes satellites launched for and by other countries as well as manned flights, deep space probes, and moon flights. See also Table 1, Appendix I.

3. Ibid. This is just the publicly acknowledged amount. Other estimates of the classified US military space programmes put a further $2-3 billion on the total quoted. See "Military Race in Space", *Defense Monitor*, vol. 9, no. 9 (1980).

4. The orbital characteristics of a satellite which are most commonly used throughout this study include: "orbital period", which is the time taken to complete one full revolution of the earth; "orbital inclination", which is the angle between the orbital plane of the satellite and the equatorial plane of the earth; the "perigee", which is the shortest distance between the satellite and the earth; the "apogee", which is the longest distance between the satellite and the earth; and the "ground track", which is the estimated path that the satellite travels over the surface of the globe. For further details see Jasani, *Outer Space: A New Dimension of the Arms Race*, pp. 5-10; and Samuel Glasstone, *Sourcebook on the Space Sciences* (Van Nostrand, Princeton, NJ, 1965), pp. 41- 82.

5. Walter Clemens, Jr, *Outer Space and Arms Control* (Center for Space Research, Massachusetts Institute of Technology, 1966), p. 112.

6. The term "sanctuary" is used here with caution, as it refers to the absence of space-based weapons, not the absence of weapons that could be used in space. In addition to the limited antisatellite systems deployed in the 1960s, both superpowers

presumably had some capability to jam satellites and also to use available ICBMs and ABMs for this purpose.

7. The term antisatellite is used hereafter without a hyphen. It is also used interchangeably with its acronym ASAT.

8. See Paul Stares, "Reagan's BMD Plan: The Ultimate Defence?" *ADIU Report*, vol. 5, no. 3 (May/June 1983).

9. Tom Wolfe, *The Right Stuff* (Bantam, New York, 1980), p. 57.

10. Ibid. (his emphasis).

11. For other examples, see Chapter 4 of Vernon Van Dyke, *Pride and Power: The Rationale of the Space Program* (University of Illinois Press, Urbana, 1964).

2. *The Origins of the US Military Space Programme, 1945–1957*

1. See Clarence G. Lasby, *Project Paperclip: German Scientists and the Cold War* (Atheneum, New York, 1971). See also Frederick I. Ordway III and Mitchell R. Sharpe, *The Rocket Team* (Thomas Y. Crowell, New York, 1979).

2. See Edmund Beard, *Developing the ICBM: A Study in Bureaucratic Politics* (Columbia University Press, New York, 1976), pp. 68-9; and Vernon van Dyke, *Pride and Power: The Rationale of the Space Program* (University of Illinois Press, Urbana, 1964), p. 10.

3. *The War Reports of General George C. Marshall, General H.H. Arnold, Admiral Ernest J. King* (Lippincott, New York, 1947), pp. 452-6. Quoted in R.J. Perry, *Origins of the USAF Space Program 1945-1956*, AFSC Historical Publications Series 62-24-10, vol. V to History of OCAS (1961), p. 9. Due to the paucity of information of this period and the official nature of this publication, Perry's work is relied on heavily during this section.

4. US Congress, House, Committee on Science and Technology, *United States Civilian Space Programs 1958-1978*, vol. 1, 97th Congress, 1st Session (Government Printing Office, Washington, DC, 1981), p. 35. In fact, the idea of using rockets for reconnaissance purposes goes back at least to the work of Alfred Maul, who first devised and tested rocket-borne camera devices in 1904. See R. Cargill Hall, "Early U.S. Satellite Proposals," *Technology and Culture*, vol. IV (Autumn 1963), pp. 425-6. The United States later started using the V-2 rockets to test photographic equipment. See the US Congress, House, Committee on Science and Astronautics, *A Chronology of Missile and Astronautic Events*, 87th Congress, 1st Session (Government Printing Office, Washington, DC, 1961), p. 10.

5. House Committee on Science and Technology, *U.S. Civilian Space Programs*, p. 36.

6. "Preliminary Design of an Experimental World Circling Spaceship", Report No. SE: 11827, Douglas Aircraft Company, Inc., Santa Monica Plant Engineering Division, Contract WBB-038 (2 May 1946).

7. See Hall, *Early U.S. Satellite Proposals*, p. 414.

8. Ibid.

9. Perry, *Origins of the USAF Space Program*, p. 11. According to Hall's account, Dr Harvey Hall was summoned to General LeMay's office in mid-March to be told that the Army would not support the Navy project. See Hall, *Early U.S. Satellite Proposals*, p. 414.

10. "Preliminary Design of an Experimental Earth Circling Spaceship," p. 9.

11. Ibid.

12. Ibid, p. 10.

13. Ibid. The report also stated that a satellite was virtually undetectable from the ground by means of present-day radar.
14. Ibid., p. 11.
15. Ibid., p. 14.
16. Ibid., p. 11.
17. To avoid being outmanoeuvred by the Army, the Chief of Naval Operations directed the Bureau of Aeronautics to proceed with further preliminary studies. In January 1947 the Navy also appealed to the JRDB for it to create a special *ad hoc* committee to determine which of the services should have jurisdiction over the space programme, but this never materialised. Project RAND also carried out a second study in mid-1946 which was completed in April 1947. This essentially refined the earlier study.
18. Perry, *Origins of the USAF Space Program*, p. 20.
19. Ibid.
20. Hall, *Early Satellite Proposals*, p. 427.
21. Perry, *Origins of the USAF Space Program*, pp. 21-3.
22. Ibid., p. 23 (emphasis added).
23. James Forrestal, *First Annual Report of the Secretary of Defense* (Department of Defense, Washington, DC, 1948), p. 129. This represented the first public statement that the United States was giving any consideration to the development of satellites which, significantly, was condemned at the time by Soviet commentators. See House Committee on Science and Technology, *U.S. Civilian Space Programs*, p. 39. This was also the last mention of space R&D by a US official before Secretary of Defense Charles Wilson's press release in 1956.
24. The Navy even tried to get the support of the National Advisory Committee for Aeronautics (NACA) for its High Altitude Test Vehicle (HATV). See Hall, *Early U.S. Satellite Proposals*, p. 429.
25. Perry, *Origins of the USAF Space Program*, p. 31.
26. Hall, *Early U.S. Satellite Proposals*, pp. 430-1.
27. For example, see "Utility of a Satellite Vehicle for Reconnaissance", RAND Report R-218, April 1951. Both of these reports remain classified. See also H. York and G.H. Greb, "Strategic Reconnaissance", *Bulletin of the Atomic Scientists*, vol. 33, no. 4 (April 1977), pp. 33-42, for an examination of similar activities using aircraft.
28. J.E. Lipp and R.M. Salter (eds), "Project Feedback Summary Report", RAND Report R-262, 1 March 1954, quoted in Perry, *Origins of the USAF Space Program*, pp. 36-7. According to Perry (p. 32) this report demonstrated that it would be feasible to develop a system with a 40-foot resolution, which was a marked improvement on the 1946 estimate of a 200-foot resolution.
29. Perry, *Origins of the USAF Space Programme*, p. 41; York and Greb, "Strategic Reconnaissance", p. 38; and "Chronology of Air Force Space Activities", Air Force Document (full reference unknown), p. 2. GOR No. 80 apparently called for a 20-foot resolution.
30. "Space and Missile Systems Organization: A Chronology 1954-1976", AFCS Historical Publications (SAMSO, Chief of Staff, History Office), pp. 29, 32, and 35.
31. "Chronology of Air Force Space Activities", p. 2; and York and Greb, "Strategic Reconnaissance", p. 38. According to Philip Klass, *Secret Sentries in Space* (Random House, New York, 1971), p. 83, Lockheed was notified on 30 June 1956. The development of its associated Hustler (later redesignated Agena) upper-stage vehicle was also included in the contract. Lockheed apparently gave the code name Pied Piper

to this project. See "SAMSO: A Chronology", p. 36.

32. York and Greb, "Strategic Reconnaissance", p. 38.

33. James R. Killian, Jr, *Sputnik, Scientists and Eisenhower: A Memoir of the First Special Assistant to the President for Science and Technology* (The MIT Press, Cambridge, Mass., 1977), p. 83.

34. See Perry, *Origins of the USAF Space Program*, p. 44. For comprehensive accounts of booster developments see Beard, *Developing the ICBM*; John T. Greenwood, "A Short History of the Air Force Ballistic Missile and Space Program, 1954-1974", unpublished monograph; M.H. Armacost, *The Politics of Weapons Innovation: The Thor Jupiter Controversy* (Columbia University Press, New York, 1969); and R.L. Perry, "System Development Strategies: A Comparative Study of Doctrine, Technology and Organization in USAF Ballistic and Cruise Missile Programs, 1950-1960", RAND Memo RM-485B PR (August 1966).

35. See the information supplied by Colonel Paul Worthman in W. W. Rostow, *Open Skies: Eisenhower's Proposal of July 21, 1955* (University of Texas Press, Austin, 1982), pp. 189-93. See also "Final Report Project 119L" (1st Air Division, Meteorological Survey), n.d.

36. Rostow, *Open Skies*, p. 193.

37. See Leonard Mosley, *Dulles: A Biography of Eleanor, Allen and John Foster Dulles and their Family Network* (Hodder and Stoughton, London, 1978), p. 432; and Klass, *Secret Sentries in Space*, p. 77. Perry's account of Project Feedback does include the rather cryptic statement that "most of this work was financed under a special supplement to the working contract with RAND. . ." Perry, *Origins of the USAF Space Program*, p. 34. Worthman also states that the camera developed for WS-119L became the basis for the Discoverer system, Rostow, *Open Skies*, p. 192.

38. Victor Marchetti and John Marks, *The CIA and the Cult of Intelligence* (Jonathan Cape, London, 1974), p. 35. See also House Committee on Science and Technology, *United States Civilian Space Programs*, p. 159.

39. Perry, *Origins of the USAF Space Program*, p. 56. According to Perry, WS-117L was funded at rather less than 10 per cent of the requirements level. However, there may have been funds channelled via the CIA.

40. Quoted in Klass, *Secret Sentries in Space*, p. 78.

41. Perry, *Origins of the USAF Space Program*, p. 56.

42. Ibid., p. 45.

43. See Samuel Glasstone, *Sourcebook on the Space Sciences* (Van Nostrand, Princeton, NJ, 1965), pp. 30-1.

44. House Committee on Science and Technology, *United States Civilian Space Programs*, p. 40. The acronym MOUSE was derived from a 1951 study by the British Interplanetary Society. For further information on the early scientific satellite proposals see Ordway and Sharpe, *The Rocket Team*, pp. 374-5; and Enid Curtis Bok Schoettle, "The Establishment of NASA", in Sanford Lakoff (ed.), *Knowledge and Power: Essays in Science and Government* (Free Press, New York, 1966), pp. 164-73.

45. Ordway and Sharpe, *The Rocket Team*, p. 376. This was to use the Army's Redstone missile as the first-stage launch vehicle.

46. House Committee on Science and Technology, *United States Civilian Space Programs*, p. 41.

47. Perry, *Origins of USAF Space Program*, p. 48. See also van Dyke, *Pride and Power*, p. 14. The "peaceful" intent of the US space programme was later codified in NSC Action No. 1553 of 21 November 1956 (see Chapter 3).

48. House Committee on Science and Technology, *United States Civilian Space*

Programs, p. 43. On 5 August 1955, the Soviet Union announced its plan to orbit a satellite.

49. "NSC 5520, U.S. Scientific Satellite Program", Records of the White House Office of the Special Assistant for National Security Affairs, NSC Series, Policy Matters Subseries, Dwight D. Eisenhower Library (May 1955), pp. 2 and 9.

50. Ibid., p. 3. This observation was drawn from the TCP's report "Meeting the Threat of Surprise Attack", pp. 146-7.

51. Ibid., p. 6.

52. US Congress, House, Committee on Science and Astronautics, *Military Astronautics*, Preliminary Report, 87th Congress, 1st Session (1961), p. 2.

53. See remarks by Laurence Kavanau, Special Assistant for Space to the Director for Defense Research and Engineering, before the Aviation/Space Writers Association, 21 May 1963, DoD News Release No. 718-63, p. 4.

54. Ibid.

55. York and Greb, "Strategic Reconnaissance", p. 40.

3. *Eisenhower and the Space Challenge*

1. J. Killian, *Sputnik, Scientists and Eisenhower*, p. 7. for an equally evocative description see T. Wolfe, *The Right Stuff* (Bantam, New York, 1980), p. 57.

2. Eisenhower was briefed in the NSC on 24 January and 10 May, 1957 of the likelihood of a Soviet space launch. See Williams Bragg Ewald, Jr, *Eisenhower and the President: Crucial Days 1951-1960* (Prentice-Hall, Englewood Cliffs, NJ, 1981). Eisenhower admits in his memoirs that he was told as early as November 1956 of an estimate from "our intelligence people" that the Soviet Union would be able to launch a satellite after November 1957. Dwight D. Eisenhower, *The White House Years: Waging Peace 1956-61* (Doubleday, Garden City, New York, 1965), p. 206. The Soviet Union had also announced that it was on the verge of launching an artificial satellite.

3. The disclosure of the existence of U-2 aircraft was considered at the time, but it was rejected by the President. See Eisenhower, *The White House Years*, p. 225.

4. In particular, the choice of Vanguard over Redstone.

5. Memorandum of Conference with the President, 8 October 1957, 8.30 a.m. and at 5.00 p.m.; and Minutes of Cabinet Meetings, 18 October 1957, Dwight D. Eisenhower Diary Series, October 1957, Staff Notes, Eisenhower Library.

6. Summary of Important Facts in the Development by the United States of an Earth Satellite. Statement by the President, the White House, 9 October 1957.

7. Enid Curtis Bok Schoettle, *The Establishment of NASA*, p. 187.

8. See *Military Astronautics*, Preliminary Report of the Committee on Science and Astronautics, US House of Representatives, 87th Congress, 1st Session (Washington, DC, Government Printing Office, 1961), p. 3; *Sputnik Scientists and Eisenhower*, p. 3; and R. F. Futrell, *Ideas, Concepts, Doctrine: A History of Basic Thinking in the United States Air Force 1907-1964* (Air University, Maxwell AFB, Alabama, 1971), p. 293.

9. See Killian, *Sputnik, Scientists and Eisenhower*, pp. 128-9.

10. Legislative Leadership Meeting, Supplementary Notes, 4 February 1958, Dwight D. Eisenhower Diary Series, Staff Notes, February 1958, p. 1, Eisenhower Library.

11. Memorandum of Conference with the President, 6 February 1958, Dwight D. Eisenhower Diary Series, Staff Notes, February 1958, p. 1-2, Eisenhower Library.

12. Ibid., p. 4. See also Minutes of Cabinet Meeting, 7 February 1958,

9.00-11.00 a.m., Dwight D. Eisenhower Diary Series, Staff Notes, February 1958, p. 2, Eisenhower Library.

13. Killian, *Sputnik, Scientists and Eisenhower,* p. 132 and Memorandum of Conference with the President, 5 March 1958, Dwight D. Eisenhower Diary Series, Staff Notes, March 1958, Eisenhower Library; and also Memorandum of Conference with the President, 12 March 1958, 10.30 a.m., DDE Diary Series, Staff Notes, March 1958, Eisenhower Library.

14. DoD was apparently unhappy before then with the wording of the Bill and the haste in which it was presented to Congress. See Killian, *Sputnik, Scientists and Eisenhower,* p. 134.

15. Logsdon, *The Decision to Go to the Moon: Project Apollo and the National Interest* (Massachusetts Institute of Technology Press, Cambridge, Mass., 1970), p. 23. See also Carroll Kilpatrick, "Democrats Charge Ike's Space Plan Gives Military Too Much Control", *Washington Post* (7 June 1958); and J. W. Finney, "Senate Critical on Space Agency", *New York Times* (7 May 1958).

16. National Aeronautics and Space Act 1958. Quoted in Logsdon, *The Decision to Go to the Moon,* p. 23. The Space Act was finally signed by Eisenhower on 29 July.

17. For a comprehensive discussion of many of these developments, for example the "battles" over the transfer of ABMA and the Saturn project to NASA, see Futrell, *Ideas, Concepts, Doctrine,* pp. 292-302.

18. George Kistiakowsky, *A Scientist at the White House: The Private Diary of President Eisenhower's Special Assistant for Science and Technology* (Harvard University Press, Cambridge, Mass., 1976), pp. 39 and 57. See also York, *Race to Oblivion,* pp. 138-9.

19. Memorandum for General Goodpaster, 15 September 1959, Ann Whitman Series, Administration Series, Kistiakowsky File (2), Eisenhower Library.

20. Memorandum for the Chairman, Joint Chiefs of Staff, "Subject: Coordination of Satellite and Space Vehicle Operations", attached to Memorandum for General Goodpaster, ibid. (emphasis in original). This document was eventually made public in a press release from the DoD on 18 September 1959.

21. *Military Astronautics,* p. 6. For a full account of this episode see Max Rosenberg, *The Air Force in Space 1959-1960,* USAF Division Liaison Office (June 1962), pp. 18-21.

22. Futrell, *Ideas, Concepts, Doctrine,* p. 295.

23. Van Dyke, *Pride and Power,* p. 192.

24. See John Prados, *The Soviet Estimate: U.S. Intelligence Analysis and the Russian Military Strength* (The Dial Press, New York, 1982), p. 106.

25. Thomas Powers, *The Man Who Kept the Secrets: Richard Helms and the CIA* (Pocket Books, New York, 1979), p. 121.

26. For further discussion of the deception surrounding the Corona/Discoverer programme and its progress see Mosley, *Dulles,* p. 432; and Jeffrey T. Richelson, *United States Strategic Reconnaissance: Photographic Imaging Satellites,* ACIS Working Paper No. 38 (Center for International and Strategic Affairs, UCLA, May 1983), pp. 6-7.

27. Kistiakowsky, *Diary,* p. 192.

28. Ibid., p. 244. See also Rosenberg, *The Air Force in Space,* pp. 31-5.

29. Kistiakowsky, *Diary,* p. 192. The polar satellite was for data relay purposes.

30. See Rosenberg, *The Air Force in Space,* p. 33; and Carl Berger, *The Air Force in Space, Fiscal Year 1961* (USAF Historical Division Liaison Office, 1966), p. 35. It is possible that Congress was not aware of the Corona programme.

31. Berger, *The Air Force in Space, Fiscal Year 1961,* p. 34.

32. Interview with former member of PSAC (Beta), Harvard, Cambridge, Mass., 17 September 1980. See also Kistiakowsky, *Diary,* pp. 339, 347, and 378. Dr Land had

been a central figure in the intelligence recommendations of the Killian Technological Capabilities Panel.

33. Kistiakowsky, *Diary*, pp. 382 and 387.

34. Ibid., p. 395.

35. Prados, *The Soviet Estimate*, p. 122. According to James Bamford, *The Puzzle Palace: A Report on NSA, America's Most Secret Agency* (Houghton Mifflin, Boston, 1982), p. 189, the panel known as the Executive Committee, or EXCOM, did not come into existence until 1965. Bamford also states that the EXCOM comprises the Director of the CIA, the Assistant Secretary of Defense for Intelligence and the Scientific Advisor to the President.

36. White House Press Release, 9 October 1957.

37. Killian, *Sputnik, Scientists and Eisenhower*, p. 124. This point was reiterated in a private interview with former PSAC member (Alpha), Cambridge, Massachusetts, 4 May 1981.

38. Ibid., p. 297.

39. NSC 5814/1, "Preliminary U.S. Policy on Outer Space", White House Office, Office of the Special Assistant for National Security Affairs, NSC Series, Policy Papers Subseries, Eisenhower Library; and NASC, "U.S. Policy on Outer Space", 26 January 1960, the White House Office, Office of the Special Assistant for National Security Affairs, NSC Series, Policy Papers Subseries, Eisenhower Library. The NASC directive was also referred to as NSC 5918 in its early stages of drafting.

40. Quoted from Futrell, *Ideas, Concepts, Doctrine*, p. 280. After this speech, Schriever was given instructions not to use the word "space" in future speeches. Similarly, General Power, Commander of ARDC, learned that it was "inappropriate" for an officer to speak of the military potential of space.

41. Ibid. The moon also figured prominently in these early discussions on the military utility of space. See the statements of Lt. Gen. Putt, Air Force Deputy Chief of Staff for Research and Development to Congress. These views, however, were also not universally held by other sections of the Air Force, both for technical and conceptual reasons. For example, studies made by the office of the Air Force Deputy Chief of Staff for Plans and Programs indicated that control of space would be a far more complex matter than the control of the air because manoeuvrable space vehicles were still some way from being developed.

42. Ibid., p. 280.

43. James Gavin, *War and Peace in the Space Age* (Hutchinson, London, 1959), pp. 215-16. This view was echoed by Senator Keating, who declared in 1959 that if the Soviet Union put a satellite in orbit "for the purpose of viewing what was going on in this country, we should try to shoot it down and any other country would". Quoted in Van Dyke, *Pride and Power*, p. 38.

44. Briefing on Army Satellite Program, 11 November 1957, White House Office of Special Assistant for Science and Technology, November 1957. Eisenhower Library (diagrams were included).

45. NSC 5802/1, "U.S. Policy on Continental Defense", 19 February 1958, p. 5 (Eisenhower Library).

46. NSC 5814/1, pp. 26-7. See also note 56.

47. Letter, James Douglas to Gordon Gray, 23 May 1960, White House, Office of the Special Assistant for National Security Affairs, 1952-61, Reconnaissance Satellites, Eisenhower Library.

48. NSC 5814/1, p. 8 (my emphasis).

49. See Bamford, *The Puzzle Palace*, p. 138; and Mumey Marder, "Testimony on Spy

Planes Reopens Shrouded Chapter of Cold War", *Washington Post* (13 December 1982), p. A2.

50. See Oral History Interview with Robert Amory, Jr (Washington, DC, 9 February 1966), John F. Kennedy Library.

51. York, *Race to Oblivion*, p. 131.

52. Interview with former senior official of the Department of Defense (Alpha), San Diego, California, December 1980.

53. Kistiakowsky, *Diary*, p. 229. As a footnote to his incident, Kistiakowsky states that "the recurrent proposals to develop and test a missile for destroying or 'inspecting' satellites was an outgrowth of a rather widespread fear that the Soviet Union would launch satellites with nuclear warheads to be released on targets in an all out war. It took several years' arguments to convince everybody but the ultra-space cadets that nuclear warheads in satellites were not the way to prepare for war". Ibid., pp. 229-30.

54. Ibid., p. 230.

55. Ibid.

56. Ibid., pp. 245-6 (my parentheses). Kistiakowsky later states in his diary:

> During the period covered in this diary there was considerable occasional pressure to develop antisatellite missiles. I opposed these proposals, successfully using the arguments that for us the satellites were far more important than for the Soviet Union which could get most of the information from the open press, such as *Aviation Week*.

Ibid., p. 224.

57. Interview with former official of ARPA (Alpha), Los Angeles, California, 12 June 1981. See also Steinberg, *The Legitimization of Satellite Reconnaissance*, Chapter 2.

58. "Satellite Countermeasures", *Time*, vol. 63, no. 8 (3 May 1954).

59. See Futrell, *Ideas, Concepts, Doctrine*, p. 279.

60. *Summary of Foreign Policy Aspects of the U.S. Outer Space Program*, attached to Memorandum from McGeorge Bundy, *Chronology of Development of U.S. Policy with Respect to Outer Space*, 7 June 1962, Johnson Library, p. 8.

61. For example, President Eisenhower's State of the Union Address on 10 January 1957; and two days later a Memorandum on Disarmament was submitted to the First Committee of the 12th UNGA: as well as Eisenhower's letters to Bulganin on 12 January 1958, and 15 February 1958. See *Chronology of Development of U.S. Policy with Respect to Outer Space*, pp. 8-10; and Futrell, *Ideas, Concepts, Doctrine*, p. 296, for the text of relevant sections of these proposals.

62. NSC 5814/1, "U.S. Policy on Outer Space", White House Office, Office of the Special Assistant for National Security Affairs, NSC Series, Policy Papers Subseries, Eisenhower Library.

63. NASC, "U.S. Policy on Outer Space", 26 January 1960, White House Office, Office of the Special Assistant for National Security Affairs, NSC Series, Policy Paper Subseries, Eisenhower Library, p. 12. Kistiakowsky recounts that in the discussions leading up to the codification of this document the NASA Director, T. Keith Glennan, urged him not to accept the DoD-JCS version of the section on "foreign policy" because "it would totally prevent the possibility of international agreements". Kistiakowsky, *Diary*, p. 167.

64. *Summary of Foreign Policy Aspects of the U.S. Outer Space Program*, p. 13. The

proposal to include a prohibition on weapons of mass destruction as part of a GCD package was initially recommended by the Joint Disarmament Study headed by Charles Coolidge. This study also noted the importance of communications and reconnaissance satellites to the United States. It states: "Our ability to use them must not be compromised, in spite of the probable opposition of the Soviet Union to permitting them". See Memorandum for George B. Kistiakowsky, Subject: Proposed Prohibition of Weapons in Space Vehicles, 2 February 1960, Eisenhower Library.

65. Eisenhower, *Waging Peace*, p. 556.

66. Kistiakowsky, *Diary*, p. 334.

67. It is difficult to ascertain the extent of the intelligence gathered by military satellites before the Kennedy administration. Although the capsule from Discoverer XIII was successfully recovered for the first time in August 1960, Discoverer XIV launched on 18 August was apparently the first to involve photography. There were only two other successful Discoverer satellites (XVII and XVIII) before Kennedy entered office. Although SAMOS II was a success, it is unlikely to have provided the same resolution images as the Discoverer satellites. In *Politics and Force Levels: The Strategic Missile Program of the Kennedy Administration* (University of California Press, London, 1980), p. 101, Desmond Ball recalls an interview with Herbert Scoville in which Scoville says "forget about the early SAMOS satellites—they never produced anything useful at all".

68. Any scepticism that the Soviet Union would not appreciate the potential value of space reconnaissance to the United States would have been removed once they examined the sophisticated U-2 camera. See also A. Katz, *The Soviets and the U-2 Photos—An Heuristic Argument*, RAND Memorandum RM 3584-PR (March 1963). Katz argues that the Soviets would have been able to gauge the capability of U-2 reconnaissance cameras before the U-2 incident from published material on Second World War reconnaissance cameras. He also gives an account of the Soviet analysis of the U-2 camera presented at the trial of Gary Francis Powers.

4. *Kennedy and the Years of Uncertainty*

1. Logsdon, *The Decision to Go to the Moon: Project Apollo and the National Interest* (Massachusetts Institute of Technology Press, Cambridge, Mass., 1970), p. 72. Wiesner was also appointed Presidential Science Advisor.

2. Ibid., p. 73. See also R. F. Futrell, *Ideas, Concepts, Doctrine: A History of Basic Thinking in the United States Air Force 1907-1964* (Air University, Maxwell, AFB, Alabama, 1971), pp. 385-6.

3. US Congress, House, Committee on Science and Astronautics, *Military Astronautics*, preliminary report, 87th Congress, 1st Session (1961), p. 11. Just after the Kennedy administration took office, Deputy Secretary of Defense Roswell Gilpatric apparently told Secretary of the Air Force Zuckert that the Air Force would receive the military mission in space provided it "put its house in order"—a reference to the fact that the Air Force responsibility for research, development, test and procurement was still not centralized in one command. As a result, Air Force Systems Command was created on 1 April 1961. See Futrell, *Ideas, Concepts, Doctrine*, p. 402.

4. Futrell, *Ideas, Concepts, Doctrine*, p. 388. Department of Defense Directive No. 5160.34, Subject: Reconnaissance, Mapping and Geodetic Programs.

5. H. L. Nieburg, *In the Name of Science* (Quadrangle Books, Chicago, 1966), pp. 40, 50.

6. Futrell, *Ideas, Concepts, Doctrine*, pp. 388-9 and Logsdon, *The Decision to Go to the Moon*, pp. 77-80.

7. U-2 flights continued over China, however, after Gary Powers had been shot down in May 1960. See also Desmond Ball, *Politics and Force Levels: The Strategic Missile Programme of the Kennedy Administration* (University of California Press, London, 1980), p. 102, for information on the impact of satellite reconnaissance on National Intelligence Estimates (NIEs) of Soviet missile strength.

8. Transcript, Abram Chayes Oral History Interview, pp. 215-16, John F. Kennedy Library (my emphasis). After the U-2 incident, Keith Glenman of NASA proposed a joint US–Soviet undertaking to develop a meteorological satellite. This was opposed by CIA Director McCone because it would "end SAMOS and other intelligence satellites". See George Kistiakowsky, *A Scientist at the White House: The Private Diary of President Eisenhower's Special Assistant for Science and Technology* (Harvard University Press, Cambridge, Mass., 1976), p. 321.

9. Memorandum from Assistant Secretary of Defense for Public Affairs to the President, 26 January 1961, John F. Kennedy Library.

10. Ibid. President Kennedy was also "especially concerned" about unauthorized statements on the use of nuclear reactors aboard the Transit navigation satellites as he issued NSAM 50 on 12 May 1961, reserving for himself all first official announcements. National Security Action Memorandum No. 50, Subject: Official Announcements of Launching into Space of Systems Involving Nuclear Power in Any Form, 12 May 1961, NSC Series, John F. Kennedy Library.

11. Transcript, Oral History Interview with Solis Horwitz, John F. Kennedy Library, p. 22. See also George C. Wilson, "U.S. is Formulating New Space Policy", *Aviation Week & Space Technology*, vol. 76, no. 25 (18 June 1962), p. 26, for further evidence of Kennedy's anger following the disclosure of Air Force briefings on a manned SAINT programme.

12. Transcript, Oral History Interview with Eugene Zuckert, p. 124, John F. Kennedy Library.

13. "Space Secrecy Rule Stirs Fear, Confusion", *Aviation Week & Space Technology*, vol. 76, no. 21 (21 May 1962), p. 26. The term "blackout" is adopted from Gerald M. Steinberg, *Satellite Reconnaissance: The Role of Informal Bargaining* (Praeger, New York, 1983), p. 40.

14. Ibid., p. 27. See also "Space Secrecy Muddle", *Aviation Week & Space Technology*, vol. 76, no. 17 (23 April 1962).

15. Transcript, Oral History Interview, Solis Horwitz, p. 23.

16. Memorandum for the President from the Secretary of State, Subject: United Nations Outer Space Activities, 2 February 1961, John F. Kennedy Library.

17. Memorandum for the Secretary of State from Special Assistant for National Security Affairs, Subject: United Nations Outer Space Activities, 28 February 1961, John F. Kennedy Library. Bundy also stated that Wiesner would be taking the lead in planning relations with the Soviet Union on this issue.

18. *Summary of Foreign Policy Aspects of the U.S. Outer Space Program*, p. 18. The text of this is used extensively in the following paragraphs.

19. Ibid., p. 19.

20. Memorandum for Mr McGeorge Bundy. Subject: *Chronology of Development of U.S. Policy with Respect to Outer Space*.

21. Letter, Bundy (ISA) to Farley (S/AE), 12 January 1962, in *Summary of Foreign Policy Aspects of the U.S. Outer Space Program*, Annex C, p. 1 (my emphasis).

22. Ibid., Annex C, p. 2.

23. Ibid., p. 3.
24. Ibid., p. 22.
25. Transcript, Oral History Interview, Eugene Zuckert, p. 133.
26. Raymond L. Garthoff, "Banning the Bomb in Outer Space", *International Security*, vol. 5, no. 3 (Winter 1980/1), p. 26. The bulk of NSAM 156 still remains classified. National Security Action Memorandum 156 (26 May 1962), John F. Kennedy Library.
27. The Committee lasted until 1969, but its most active period was between 1962 and 1967. For details see Garthoff, "Banning the Bomb in Outer Space", p. 26.
28. Department of State, *Documents on Disarmament, 1962* (Government Printing Office, Washington, DC, 1963), Part 2, pp. 871-2.
29. Transcript, Oral History Interview, Abram Chayes, pp. 871-2.
30. Memorandum for the President from the Director of the Arms Control and Disarmament Agency, Subject: Arms Control Aspects of Proposed Satellite Reconnaissance Policy, 6 July 1962, John F. Kennedy Library, pp. 1-2. Foster also argued that advance notifications would build confidence for arms control.
31. Garthoff, "Banning the Bomb in Outer Space", p. 27.
32. *Documents on Disarmament, 1962*, p. 1121.
33. Ibid., p. 1131.
34. See Ward Wright, "United Nations Committee Receives New Soviet Space Use Proposal", *Aviation Week & Space Technology*, vol. 78, no. 5 (20 April 1960), p. 29; and Ward Wright, "United Nations Group Fails to Gain Accord on Space Use Rule", *Aviation Week & Space Technology*, vol. 78, no. 20 (20 May 1963), p. 129.
35. Quoted in Futrell, *Ideas, Concepts, Doctrine*, p. 430.
36. Logsdon, *The Decision to Go to the Moon*, p. 77.
37. See Berger, *The Air Force in Space FY 1961* (USAF Historical Division Liaison Office, April 1966), p. 12. According to a former senior Air Force officer (Alpha), the development of an ASAT system was "one of the most important recommendations of the Gardner Committee's Report that was submitted on 20 March 1961". Interview, Washington, DC, 10 October 1981.
38. *Military Astronautics*, p. 20.
39. Ibid., p. 21.
40. Quoted in Futrell, *Ideas, Concepts, Doctrine*, p. 431.
41. Quoted in Philip Klass, *Secret Sentries in Space* (Random House, New York, 1971), p. 62.
42. Arnold L. Horelick and Myron Rush, *Strategic Power and Soviet Foreign Policy* (University of Chicago Press, Chicago, 1966), p. 97.
43. Ibid., p. 98.
44. US Congress, Senate, Committee on Aeronautical and Space Sciences, *Soviet Space Programs 1962-1965: Goals and Purposes, Achievements, Plans and International Implications*, Staff Report (30 December 1966), p. 75.
45. Interview, senior member of the Aerospace Corporation, Los Angeles, 12 June 1981. Although the interviewee recalls seeing the telegram, it is uncertain why the cable went directly to the Aerospace Corporation rather than through standard Air Force channels—unless time was its primary consideration or McNamara wanted to bypass the Air Staff.
46. US Congress, House, Subcommittee of the Committee on Appropriations, *Department of Defense Appropriations for 1963*, Hearings, 87th Congress, 2nd Session (1962), Part 2, p. 28.
47. Quoted in Vernon van Dyke, *Pride and Power: The Rationale of the Space Program*

(University of Illinois Press, Urbana, 1964), p. 41.

48. US Congress, House, Committee on Aeronautical and Space Sciences, *NASA Authorization for Fiscal Year 1963*, Hearings, 87th Congress, 2nd Session (June 1962), Part 2, p. 335.

49. Transcript of News Conference at the Pentagon, Friday, 18 September 1964, p. 4.

50. Ibid. McNamara made no reference to the Air Force SAINT programme at this briefing.

51. *New York Times* (13 May 1962), p. 59.

52. "USAF Starts Manned SAINT Studies", *Aviation Week & Space Technology*, vol. 74, no. 23 (4 June 1962), p. 34.

53. *New York Times* (12 June 1962), p. 16.

54. See US Congress, House, Committee on Government Operations, *Government Operations in Space: An Analysis of Civil-Military Roles and Relationships*, 13th Report, 89th Congress, 1st Session (1965), p. 82.

55. US Congress, Senate, *NASA Authorization for Fiscal Year 1963*, p. 348.

56. Larry Booda, "Air Force Still Limited on Space Studies", *Aviation Week & Space Technology*, vol. 77, no. 5 (30 July 1962), pp. 16-17.

57. "Vostok's 3 and 4 Rendezvous, Dock in Orbit: US Re-evaluating Military Space Needs", *Aviation Week & Space Technology*, vol. 77, no. 8 (20 August 1962), pp. 26-7; and "U.S.-Soviets Disagree on Rendezvous", *Aviation Week & Space Technology*, vol. 77, no. 9 (27 August 1962), pp. 36-7.

58. Memorandum for the President from Secretary of the Air Force, 22 August 1962, John F. Kennedy Library.

59. Ibid.

60. Memorandum for Eugene Zuckert, 27 August 1962, John F. Kennedy Library.

61. Letter to the President from Secretary of the Air Force, 4 September 1962, John F. Kennedy Library.

62. Ibid.

63. US Congress, House, *Government Operations in Space*, pp. 77-8. A later report states that this was the nadir of Air Force-OSD relations as Rubel's "militant speech . . . struck the Air Force very hard". See Col. Paul E. Worthman, "The Promise of Space", *Air University Review*, vol. XX, no. 2 (January-February 1969), pp. 124-5.

64. See Futrell, *Ideas, Concepts, Doctrine*, p. 389.

65. US Congress, House, Subcommittee of the Committee on Appropriations, *Department of Defense Appropriations for FY 1964*, Hearings, 88th Congress, 1st Session (1963), Part 1, p. 476.

66. See Edward H. Kolcum, "Kennedy Stresses Peaceful Space Theme", *Aviation Week & Space Technology*, vol. 77, no. 12 (17 September 1962), pp. 26-7; and Garthoff, "Banning the Bomb in Outer Space", pp. 30-1.

67. J. W. Finney, "Air Force Shelves Plan to Intercept and Inspect Spacecraft", *New York Times* (4 December 1962).

68. Philip Klass, "USAF Halts SAINT Work: Shifts to Gemini", *Aviation Week & Space Technology*, vol. 77, no. 24 (10 December 1962).

69. Ibid.

70. History of Air Defense Command, vol. 1 (July-December 1964), p. 38.

71. Quoted from US Congress, Senate, *Soviet Space Programs 1962-1965*, p. 75.

72. Ibid.

73. Garthoff, "Banning the Bomb in Outer Space", p. 33.

74. NSAM 258, "Assignment of Highest National Priority to Program 437" (6 July 1963), John F. Kennedy Library.

75. Transcript of News Conference at the Pentagon, 18 September 1964, p. 4.

76. See "Anti-satellite Weapon", *Aviation Week & Space Technology*, vol. 79, no. 16 (14 October 1963), p. 25; "Pentagon Seeks Satellite Killer", *New York Times* (21 October 1963); and Michael Getler, "Accuracy, Payload, Cost Favor Ground-Based Anti-Satellite Systems", *Missiles and Rockets*, vol. 13, no. 18 (28 October 1963), p. 19.

77. *Summary of Foreign Policy Aspects of the U.S. Outer Space Program*, p. 14.

78. Ibid., p. 15.

79. Garthoff, "Banning the Bomb in Outer Space", p. 28. Much of the text in this section is drawn from this article. Unless othewise stated the source of this information is the same.

80. The US response was without any knowledge of the inspection requirements of such a ban or whether these could be met. It was made almost automatically as a result of the ongoing debate with the Soviet Union on inspection and verification.

81. Garthoff, "Banning the Bomb in Outer Space", p. 27.

82. This was Recommendation 19 on the report and was brought to the attention of the President as "Tab C: Considerations on a Separate Agreement Banning Weapons of Mass Destruction in Satellites", 2 pp. (TOP SECRET).

83. Memorandum for the President, from the Director of the Arms Control and Disarmament Agency, Subject: Arms Control Aspects of Proposed Satellite Reconnaissance Policy, 6 July 1962, John F. Kennedy Library, pp. 3-4.

84. Ibid., p. 4.

85. Garthoff, "Banning the Bomb in Outer Space", p. 27.

86. See also "Background on U.S. Position on a Separate Ban on Weapons of Mass Destruction in Outer Space", 2 August 1962, Box 336, John F. Kennedy Library.

87. Garthoff replied that he believed this had not been the case but chose to report the conversation to the Committee. Garthoff, "Banning the Bomb in Outer Space", p. 29.

88. Ibid.

89. List of NSC Documents, John F. Kennedy Library.

90. From Entry Sheet to File: NSAM 183, Box 338, John F. Kennedy Library. The contingency papers, which are still classified, are:

(a) US Outer Space Policy in the UN General Assembly, 5 pp. (SECRET)

(b) Definitions of Peaceful Uses of Outer Space, 1 p. (CONFIDENTIAL)

(c) Photographic and Observation Satellites, 1 p. (CONFIDENTIAL)

(d) US Military Space Programs and Disarmament and Outer Space, 2 pp. (CONFIDENTIAL)

(e) Effects of High Altitude Explosions, 1 p. (CONFIDENTIAL).

91. Excerpt from Remarks by Deputy Secretary of Defense Gilpatric, South Bend, Indiana, 5 September 1962, John F. Kennedy Library (my emphasis).

92. The JCS views had also been made known to McNamara in a memorandum on 14 September from the Chairman of the JCS, General Lemnitzer. Memorandum for the Secretary of Defense from L. L. Lemnitzer, Subject: Recommendations Respecting US Approach to a Separate Arms Control Measure for Outer Space (TOP SECRET).

93. Memorandum for the President from the Director of the Arms Control and Disarmament Agency, Subject: Report on a Separate Arms Control Measure for Outer Space, 17 September 1962, John F. Kennedy Library. The report on their deliberations and the JCS document are classified.

94. Memorandum to the President, Subject: Separate Arms Control Measures for

Outer Space, 2 pp. (TOP SECRET), with "Recommendations Respecting U.S. Approach to a Separate Arms Control Measure for Outer Space", 23 pp. (TOP SECRET).

95. National Security Action Memorandum No. 192, John F. Kennedy Library.

96. Interview with a former senior member of the State Department (Alpha), Washington, DC, 22 October 1980.

97. Garthoff, "Banning the Bomb in Outer Space", p. 31.

98. Garthoff does mention a further meeting between Foster and Soviet Ambassador Tsarapkin in early 1963. Ibid., p. 32.

99. Ibid.

100. *Documents on Disarmament, 1962*, p. 1123.

101. Interview with a former senior member of the Department of State (Alpha), Washington, DC, 22 October 1980.

102. Garthoff, "Banning the Bomb in Outer Space", p. 32.

103. No reason is given in Garthoff's account as to why Rusk decided to suspend negotiations. This was probably transmitted to the White House in the Memorandum from U. Alexis Johnson to McGeorge Bundy, Subject: Further Action Under NSAM 192, 29 April 1962, 1 p. (CONFIDENTIAL).

104. Memorandum for the President from Rusk, Subject: Contingency Plan for US Reaction to Soviet Placing of a Nuclear Weapon in Space, 8 May 1963, 4 pp. (SECRET). A further document entitled "Satellite Inspection and Non-Nuclear Anti-Satellite Systems", 2 pp. (SECRET), was also included. Garthoff makes no mention of this second document in his account, although it probably refers to the ability of the United States to determine beyond all doubt that a satellite was hostile before it launched a nuclear ASAT system. The discussion of nonnuclear ASAT systems was most probably in anticipation of the results of the Partial Test Ban Negotiations that were reaching their conclusion in the summer of 1963.

105. Interview with a former senior member of the Department of State (Alpha).

106. Garthoff, "Banning the Bomb in Outer Space", p. 34.

107. Ibid.

108. Ibid.

109. Kennedy's sensitivity was obviously a reaction to the opposition that the Test Ban Treaty was receiving in Congress. This became apparent in a news conference the next day when he stated there "is not an agreement" but just a declaration of US intent not to station weapons of mass destruction in space. He also stated that there is "no way we can verify" a similar statement by the Soviet Union and therefore "we obviously have to take our own precautions". See "Kennedy Denies Space Arms Ban", *New York Times* (10 October 1963), p. 18. After the agreement had been reached he apparently informed Congress of the precautions that the United States was taking in the form of antisatellite weapons to pre-empt possible criticism.

110. For further discussion of the negotiation of the 1963 UN resolution see Clemens, *Outer Space and Arms Control*, pp. 62-7.

111. Garthoff, "Banning the Bomb in Outer Space", p. 36.

5. *The Johnson Years: The Consolidation of Policy*

1. "Planning Implications for National Security of Outer Space in the 1970s", Basic National Security Policy Planning Task I (1) (Department of State, 30 January 1964).

2. The study group consisted of representatives from State, Defense, NASA, ACDA, CIA, OST, and JCS.

3. "Planning Implications for National Security of Outer Space in the 1970s", p. vi.

4. Ibid., p. ix.

5. Ibid., p. 62.

6. Ibid., p. 7.

7. Ibid., pp. 48 and 44. Furthermore, the report states: "We would undoubtedly desire to be able to take defensive action, but it is not clear that there would be any reason for the United States to deploy a similar offensive force in space simply because the Soviet Union might have done so." (p. 48.)

8. Ibid., p. 61.

9. Ibid., p. 30.

10. Harold Brown would later state, in congressional testimony in 1965, that it is "clear that weapons of this sort [i.e., orbital bombs] are not very great threats to us in the near future, and they are unlikely to ever be". See US Congress, House, Committee on Appropriations, *Department of Defense Appropriations for 1965*, Hearings, 88th Congress, 2nd Session (1964), Part 5, p. 12.

11. "Politics and New U.S. Weapons", *Los Angeles Times* (20 September 1964), p. 25.

12. Remarks in Sacramento, on the steps of the State Capitol, *Public Papers of Lyndon B. Johnson*, vol. II, no. 57a (1963-4), pp. 1086-90.

13. US Congress, House, Committee on Government Operations, *Government Operations in Space: An Analysis of Civil-Military Roles and Relationships*, 13th Report, 89th Congress, 1st Session (1965), pp. 86-7.

14. Ibid., p. 87.

15. US Congress, House, *Department of Defense Appropriations for 1965*, p. 12.

16. Interview with former senior official of the Department of the Air Force (Alpha), Washington, DC, 18 May 1981.

17. See Clemens, *Outer Space and Arms Control*, pp. 77-8.

18. Ibid., p. 78.

19. Interview with former senior official of the Department of the Air Force (Alpha).

20. Clemens, *Outer Space and Arms Control*, p. 81.

21. Raymond L. Garthoff, "Banning the Bomb in Outer Space", *International Security*, vol. 5, no. 3 (Winter 1980/1), p. 38. See also Clemens, *Outer Space and Arms Control*, pp. 70-4.

22. Kosmos 139, 160, 169, 170, 171, 178, 179, 183, 187. There is the possibility that some earlier tests relating to this system were carried out in 1965. Interview with Charles Sheldon III of the Congressional Research Service, Washington, DC, 21 October 1980.

23. US Congress, Senate, Committee on Aeronautical and Space Sciences, *Soviet Space Programs, 1966-1970*, Staff Report, 92nd Congress, 1st Session (1971), p. 336.

24. US Congress, House, Subcommittee of the Committee on Appropriations, *Department of Defense Appropriations for 1969*, Hearings, 90th Congress, 2nd Session (1968), Part 1, p. 252.

25. "Nike Gets New Antisatellite Role", *Aviation Week & Space Technology*, vol. 87, no. 17 (2 October 1967), p. 62.

26. US Congress, House, *Department of Defense Appropriations for 1969*, Part 1, p. 456.

27. Letter from Commissioner of the US Atomic Energy Authority to Leonard Meeker, Legal Advisor, Department of State, 30 June 1965. Legislative Background: Outer Space Treaty. L.B. Johnson Library.

28. Outer Space Treaty Chronology, Office File of J. Califano. Celestial Bodies Treaty History: Negotiations. L.B. Johnson Library.

29. Ibid.

30. Memorandum for the Secretary of Defense from Joint Chiefs of Staff, Subject: Principles for Inclusion in a Treaty on the Exploitation of Celestial Bodies, JCSM-839-65, 23 November 1965. L.B. Johnson Library.

31. See the following: letter from Robert McNamara to Dean Rusk, 11 January 1966; letter from Leonard Meeker to Arthur Barber, 27 January 1966; letter from Arthur Barber to Leonard Meeker, 14 February 1966. All found in Legislative Background: Outer Space Treaty: The Outer Space Committee becomes the Arena. L.B. Johnson Library.

32. Memorandum for the President from W. Rostow, 5 April 1966. Legislative Background: Outer Space Treaty: The Turning Point. L.B. Johnson Library.

33. Intelligence Note No. 390. Subject: "Soviets Table Space-Exploration Treaty Designed to be Acceptable to U.S.", 17 June 1966, US Department of State, Director of Intelligence and Research. L.B. Johnson Library.

34. Talking Paper for the Chairman, JCS, for use at a meeting of the National Security Council on 15 September 1966. Subject: Outer Space, 14 September 1966. Legislative Background: Outer Space Treaty: The Second and Final Negotiations. L.B. Johnson Library.

35. Outer Space Treaty Chronology, p. 3.

36. *Arms Control and Disarmament Agreements: Texts and Histories of Agreements* (US Arms Control and Disarmament Agency, Washington, DC, 1980), p. 52.

37. Ibid, p. 52.

38. Interview with former senior official of the State Department (Beta), Washington, DC, 30 October 1980. Later at the Senate Hearings on the Outer Space Treaty, General Wheeler, Chairman of the JCS, stated: "We gave up nothing that was currently attractive to us from a military point of view in stating our intention not to orbit weapons of mass destruction." US Congress, Senate, Committee on Foreign Relations, *Treaty on Outer Space*, Hearings, 90th Congress, 1st Session (1967), p. 94.

39. Interview with former senior member of the State Department (Gamma), Washington, DC, 21 May 1981. Also, US Congress, House, *Department of Defense Appropriations for 1969*, Part 9, p. 456.

40. Interview with former senior official of the State Department (Gamma), Washington, DC, 21 May 1981.

41. US Congress, Senate, *Treaty on Outer Space*, p. 84.

42. Quoted in US Congress, Senate, Committee on Aeronautical and Space Sciences, *International Cooperation in Outer Space: A Symposium*, 92nd Congress (1971).

6. *US Antisatellite Research and Development, 1957-1970*

1. Interview with former senior official of the Defense Department (Beta), Los Angeles, California, December 1980.

2. George Kistiakowsky, *Scientist at the White House: The Private Diary of President Eisenhower's Special Assistant for Science and Technology* (Harvard University Press, Cambridge, Mass., 1976), pp. 30-1.

3. See "Consolidated List of Military Departmental Contracts for Space Activities under ARPA Cognizance within $500,000" (Office of the Special Assistant for National Security Affairs, White House, 1 June 1959), OCB Series, Eisenhower Library.

4. US Congress, House, Subcommittee of the Committee on Appropriations, *Department of Defense Appropriations for 1960*, 86th Congress, 1st Session (1959), p. 23.

5. Ibid., p. 97. For further details on space weapon proposals funded by ARPA see Lee Bowen, *The Threshold of Space: The Air Force in the National Space Program 1945-1959* (USAF Historical Division Liaison Office, September 1960), pp. 26-7; and Max Rosenberg, *The Air Force in Space 1959-1960* (USAF Historical Division Liaison Office, June 1982), pp. 38-40.

6. See the testimony of David G. Stone, Head of the Stability and Controls Branch, Pilotless Aircraft Research Division, NASL, Langley Research Center in the United States. US Congress, Senate, Committee on Aeronautical and Space Sciences, *NASA Authorization for Fiscal Year 1960*, 86th Congress, 1st Session (1959), pp. 646-50.

7. The Hardtack series was in fact the first set of high-altitude nuclear tests (Teak and Orange being detonated on 1 August 1958 and 12 August 1958, respectively), but Argus was the first nuclear test in space. See Samuel Glasstone and Philip J. Dolan, *The Effects of Nuclear Weapons* (US Department of Defense and US Department of Energy, Washington, DC, 1977), p. 45.

8. Memorandum of Conference with the President, 6 March 1958, 3.00 p.m., Dwight D. Eisenhower Diary Series (Ann Whitman), Staff Notes, March 1958. Eisenhower Library. According to Killian's account of Argus, the decision was given individually by the President on 1 May 1958. James R. Killian, Jr, *Sputnik, Scientists and Eisenhower: A Memoir of the First Special Assistant to the President for Science and Technology* (The MIT Press, Cambridge, Mass., 1977), p. 188.

9. Memorandum for the President, Subject: Preliminary Results of the Argus Experiment, 3 November 1958. Ann Whitman, Administration Series, James Killian, 1957, Eisenhower Library. For Killian's own account see Killian, *Sputnik, Scientists and Eisenhower*, pp. 186-91 and p. 221.

10. Glasstone and Dolan, *The Effects of Nuclear Weapons*, p. 45. See also US Defense Nuclear Agency, "Operation Argus 1958", Doc. No. 6039F (April 1982).

11. "Preliminary Results of the Argus Experiment", pp. 3-4.

12. US Congress, House, *Department of Defense Appropriations for 1960*, Part 6, p. 112.

13. Glasstone and Dolan, *The Effects of Nuclear Weapons*, p. 45. These were also known as the DOMINIC series.

14. Ibid., p. 47; and US Congress, Senate, Preparedness Investigating Subcommittee of the Committee on Armed Services, Hearings, 88th Congress, 1st Session (1963), Part 1, p. 172.

15. Ibid., p. 179.

16. It is interesting to note that Kosmos V, launched on 28 May 1962, was reported to have taken "detailed photos" (though most probably radiation measures) of the Starfish Prime test. See "Manned Orbiting Laboratory", *Astronautics and Aeronautics*, vol. 2, no. 10 (October 1965).

17. Private communication with member of Martin Marietta Aerospace, 15 September 1981. In a report in *Astronautics*, vol. 5, no. 1 (January 1960), the missile was designated as ALBM 199B.

18. "ALBM Comes Close to Satellite Path", *Aviation Week & Space Technology*, vol.

71, no. 18 (2 November 1959), p. 33. Interview with former senior Air Force officer (Alpha), Los Angeles, California, 16 June 1981. *Aviation Week* also reported an earlier ALBM antisatellite test from a B-58 aircraft at the Discoverer V satellite, but this apparently failed due to telemetry problems. See "B-58 Launches ALBM in Trajectory Test", *Aviation Week*, vol. 71, no. 15 (5 October 1959), p. 34.

19. This test had been largely forgotten by the Air Force and only during the drafting of a press release on the current Air Force air-launched ASAT system did the Air Force rediscover the project and prevent themselves from stating that the F-15/MHV system was the first of its kind.

20. US Congress, House, Committee on Science and Astronautics, *Science, Astronautics and Defense (The 1961 Review of Scientific and Astronautical Research and Development in the Department of Defense)*, Staff Report, 87th Congress, 1st Session (1961), p. 54.

21. Interview with former aerospace industry contractor (Beta), Washington, DC, 30 October 1980.

22. US Congress, House, Committee on Science and Technology, *United States Civilian Space Programs 1958-1978*, vol. 1, 97th Congress, 1st Session (Government Printing Office, Washington, DC, 1981), p. 197.

23. James Trainor, "Non-Nuclear Anti-Satellite System in Making", *Missiles and Rockets* (11 May 1964), p. 13.

24. US Congress, House, *U.S. Civilian Space Programs 1958-1978*, p. 183.

25. See Barry Miller, "U.S. Begins Laser Weapons Programs", *Aviation Week & Space Technology*, vol. 76, no. 13 (26 March 1962), pp. 41-4; Barry Miller, "Fiscal 1962 Laser Funding Completed", *Aviation Week & Space Technology*, vol. 77, no. 2 (9 July 1962), pp. 42-5; "DoD Funds $15 Million for Laser R&D", *Aviation Week & Space Technology*, vol. 78, no. 14 (8 April 1963), p. 75; and Barry Miller, "Aerospace Military Laser Uses Explored", *Aviation Week & Space Technology*, vol. 78, no. 16 (22 April 1963), pp. 54-69.

26. "Black Eye Funding Sought", *Aviation Week & Space Technology*, vol. 76, no. 26 (25 June 1962), p. 23.

27. Milton Leitenberg, "The History of the U.S. Antisatellite Weapon Systems", unpublished monograph for Cornell Peace Studies Program (1979), p. 15.

28. Quoted from R. F. Futrell, *Ideas, Concepts, Doctrine: A History of Basic Thinking in the United States Air Force 1907-1964* (Air University, Maxwell AFB, Alabama, 1971), pp. 432-3.

29. Rosenberg, *The Air Force in Space 1959-1960*, p. 40.

30. See "Study O.K.'d on Satellite Interception", *Baltimore Sun* (12 June 1959); and "Satellite Interceptor", *Ordnance* (September-October 1959).

31. Interview with former senior official of the Department of the Air Force (Gamma), Cambridge, Massachusetts, 4 May 1981.

32. Rosenberg, *The Air Force in Space 1959-1960*, p. 41. See also "USAF Launches Anti-Satellite Program", *Aviation Week & Space Technology*, vol. 73, no. 20 (14 November 1960), pp. 26-7.

33. Interview with former official of ARPA (Alpha), Los Angeles, California, 12 June 1981. Kistiakowsky had received confirmation of its detection on 3 February from the Assistant Director of the CIA, Herbert "Pete" Scoville. See also Curtis Peebles, *Battle For Space* (Beaufort Books, New York, 1982), p. 60.

34. Kistiakowsky, *Diary*, p. 239.

35. Ibid., p. 246 (my emphasis).

36. Ibid., p. 251.

37. See Peebles, *Battle for Space*, p. 60.

38. Memorandum for Assistant Secretary of the Air Force (Research and Development) from York (DDR&E). Subject: Satellite Inspection, 16 June 1960.

39. Interview with former official of ARPA (Beta), Cambridge, Massachusetts, 23 September 1980. York was apparently sceptical of the SAINT proposal at the time.

40. Memorandum for ASAF (R&D).

41. Covering Brief to Director, Defense Research and Engineering. From Assistant Director, Defense Research and Engineering (Air Defense) (no date attached). At this time the Guidance and Control Panel of the Air Force's Scientific Advisory Board (SAB) was also conducting its own review of satellite interception. In their final report, dated July 1960, they endorsed the original proposal of the Air Force and suggested that the "project include a research and experimental objective aimed at the effective and sequential gathering of engineering design information and the evolution of techniques necessary for the ultimate construction of a military operational weapon system". See Report of the Scientific Advisory Board Guidance and Control Panel on Satellite Interception, July 1960, p. 3.

42. These reservations had been outlined by Jack Ruina in the Covering Brief to DDR&E, ibid.

43. Memorandum for Assistant Secretary of the Air Force (Research and Development), Subject: Satellite Inspection, From: Deputy Director DDR&E, 25 August 1960.

44. Ibid.

45. "USAF Launches Anti-Satellite Program", *Aviation Week & Space Technology*, p. 26.

46. *The Aerospace Corporation: Its Work, 1960-1980* (The Aerospace Corporation, Los Angeles, California, 1980), p. 23.

47. "USAF Developing Wide Space Capabilities", *Aviation Week & Space Technology*, vol. 75, no. 13 (25 September 1961), pp. 94-5.

48. "USAF Launches Anti-Satellite Program", *Aviation Week & Space Technology*, p. 26.

49. Berger, *The Air Force in Space FY 1961* (USAF Historical Division Liaison Office, April 1966), p. 70. According to this account, on 21 September 1960 the NSC directed that any tests involving the destruction of a satellite or space vehicle could not be carried out without explicit presidential approval.

50. "National Space Program", 3 June 1961, NSC, Vice President's Security File, Space: National Program, Box 15, L. B. Johnson Library.

51. See Larry Booda, "NASA Defense Seek More Space Funds", *Aviation Week & Space Technology*, vol. 74, no. 1 (2 January 1961), p. 16. See also Berger, *The Air Force in Space FY 1961*, p. 71.

52. Quoted in Vern Haugland, "Destroy That Satellite", *Ordnance*, vol. 46, no. 247 (July-August 1961), p. 52.

53. "USAF Starts Manned Saint Studies', *Aviation Week & Space Technology*, vol. 76, no. 23 (4 June 1962), p. 34.

54. Interview with Charles Sheldon III, Congressional Research Services, Washington, DC, 21 October 1980.

55. Philip Klass, "USAF Halts Saint Work: Shifts to Gemini", *Aviation Week & Space Technology*, vol. 77, no. 24 (10 December 1962), p. 36.

56. Gerald T. Cantwell, *The Air Force in Space Fiscal year 1964* (USAF Historical Division Liaison Office, June 1967), p. 62.

57. Ibid. A later report also cited the results of studies completed by RCA, which

stated that an orbital inspection system would be "frightfully expensive" if the Soviets put low-cost decoys into orbit. See Philip Klass, "Doubts are Growing About Value of Satellite Inspection in Orbit", *Aviation Week & Space Technology*, vol. 77, no. 25 (17 December 1962), p. 33.

58. Klass, "USAF Halts Saint Work", p. 36.

59. See Edward H. Kolcum, "USAF Keys Space Plan to Three Programs", *Aviation Week & Space Technology*, vol. 78, no. 4 (28 January 1963), p. 26.

60. See "Industry Observer", *Aviation Week & Space Technology*, vol. 78, no. 7 (14 January 1963); and "Satellite Inspection Study Proposals Asked", *Aviation Week & Space Technology*, vol. 78, no. 22 (3 June 1963), p. 33. See also Cantwell, *The Air Force in Space FY 1964*, pp. 62-3.

61. For further information on the DoD experiment on the NASA Gemini spaceflights see Klass, *Secret Sentries in Space*, p. 148 and pp. 177-8; US Congress, House, *U.S. Civilian Space Programs 1958-1978*, pp. 366-9; and *United State Aeronautics and Space Activities, 1962, Message From the President of the United States* (27 January 1964), p. 43.

62. Kistiakowsky, *Diary*, p. 2.

63. US Congress, House, Committee on Science and Astronautics, *Military Astronautics*, preliminary report, 87th Congress, 1st Session (1961), p. 26.

64. "Zeus Seen as Anti-Satellite Weapon", *Missiles and Rockets* (20 March 1961), p. 14.

65. The Bell Laboratories' own history of their ABM R&D programme states that: "At the request of the Secretary of Defense, the Army was asked in early 1962 to prepare the Zeus system on Kwajalein for the eventuality of having to intercept and destroy satellites." ABM Research and Development, "Bell Laboratories: Project History" (October 1975), pp. 1-31.

66. Interview with former senior official of the Department of the Air Force (Alpha), Washington, DC, 18 May 1981.

67. Later in congressional testimony the range is given as 200 miles.

68. This and subsequent data on the initial test programme is taken from "Bell Laboratories: Project History", pp. 1-31.

69. The master log of US space flights in US Congress, House, *U.S. Civilian Space Programs 1958-1978*, p. 1, 278, reports that a "target balloon" was launched on 21 May 1963 in connection with Project Mudflap.

70. "Bell Laboratories: Project History", pp. 1-32.

71. Minutes of Meeting: Briefing for Secretary McNamara on Satellite Detection, Inspection and Negation by John Whitman, DDR&E (AD), 27 June 1963, p. 2.

72. Memorandum for Mr William Moyers, Special Assistant to the President, Subject: Weekly Report for the President (14 April 1964), LBJ Confidential File, Department of Defense, April 1964, Box 111, L. B. Johnson Library.

73. Memorandum for Mr Jack Valenti, Special Assistant to the President, Subject: Weekly Report to the President (25 May 1965), LBJ Confidential File, Department of Defense, May 1965, Box 112, LBJ Library. It is not certain whether this was an ASAT-related Nike Zeus test.

74. Memorandum for Mr Jack Valenti, Special Assistant to the President, Subject: Weekly Report to the President (24 August 1965), LBJ Confidential File, Department of Defense, August 1965, Box 113, L. B. Johnson Library.

75. Memorandum for Mr Jack Valenti, Special Assistant to the President, Subject: Weekly Report to the President (23 February 1966), LBJ Confidential File, Department of Defense, February 1966, Box 115, L. B. Johnson Library.

76. JCS Message 2414392 (23 May 1966).

77. Minutes of Briefing (27 June 1967), p. 3.

78. US Congress, Senate, Committee on Aeronautical and Space Sciences, *NASA Authorization for Fiscal Year 1970*, Hearings, Part 1, 91st Congress, 1st Session (1969), p. 856.

79. US Congress, House, Subcommittee of the Committee on Appropriations, *Department of Defense Appropriations for 1971*, Hearings, 91st Congress, 2nd Session (1970), Part 6, p. 87.

80. "History of Air Defense Command", vol. 1 (July-December 1964), p. 32; and "Advanced Development Objective for an Anti-Satellite Program", ADO-40 (9 February 1963). Interestingly, ADO-40 states that "consideration should be given to any unique advantages that might accrue from an air-launch capability".

81. A participant in the discussion of ASAT requirements in this period also recalls that an Air Force intelligence briefing on the potential Soviet orbital bombs "led to a request being sent via General Schriever at Systems to the Air Force's 'West Coast' establishment for them to review a range of launch vehicles to meet this threat". Interview with retired senior officer of the Air Force (Alpha), Los Angeles, 16 June 1981.

82. Qualitative Operational Requirement (QOR) for a Satellite Interceptor System (SIS), 1 May 1962. ADC was informed by AFHQ that this requirement was being pursued under Phase 1 of ADO-40.

83. Supporters of the SAINT programme apparently saw Program 437 as a threat to their "grand plan" and criticized the choice of the Thor missile on the basis that it was not "robust" or "credible" enough to meet the threat. Interview with retired senior officer of the Air Force (Delta), Washington, DC, October 1980.

84. Memorandum for Secretary of the Air Force, Subject: Proposed Early Satellite Intercept Capability (13 December 1962).

85. Memorandum for Secretaries of the Army, Navy, Air Force and Director of ARPA, Subject: Anti-Satellite Programs (7 February 1963).

86. "History of Air Defense Command", vol. 1 (July-December 1964), p. 38; and Memorandum From Major-General Heuter to AFSC, Subject: Early Satellite Intercept Capability, Program 437 (15 February 1963).

87. Memorandum for the Chief of Staff, Subject: Program 437, 20 March 1963. This request was later reaffirmed in a separate memorandum from Brockway McMillan, Assistant Secretary for Research and Development, to the Deputy Chief of Staff, Research and Development (28 March 1963).

88. Memorandum from McMillan to the Deputy Chief of Staff.

89. Minutes of Meeting, Briefing for Secretary McNamara (27 June 1963), p. 3.

90. "History of Air Defense Command", vol. 1 (July-December 1964), pp. 53-8.

91. Cantwell, *The Air Force in Space FY 1964*, p. 61.

92. Memorandum for Jack Valenti, Special Assistant to the President (20 April 1965), LBJ Confidential File, Department of Defense, April 1965, Box 112, L. B. Johnson Library.

93. "History of Air Defense Command", vol. 1 (July-December 1964), p. 58.

94. Interview with retired senior Air Force officer (Epsilon), Albuquerque, New Mexico, November 1980. James Trainor, "Non-Nuclear Anti-Satellite System in Making", pp. 12-13.

95. Memorandum for Mr William Moyers, Special Assistant to the President, Subject: Weekly Report to the President, 18 February 1964. LBJ Confidential File, February 1964, Box 110, L. B. Johnson Library.

96. "History of Air Defense Command", vol. 1 (July-December 1964), pp. 59-60.

97. Cantwell, *The Air Force in Space FY 1964*, p. 60. The warhead would have been detonated from the ground by a "AEC-developed weapon discrete burst command system".

98. "History of Air Defense Command", vol. 1 (July-December 1964), pp. 60-1.

99. "History of Air Defense Command", vol. 1 (January-June 1966), p. 296.

100. Ibid.

101. Interview with former aerospace industry contractor (Alpha), Washington, DC, 20 October 1980. According to the interviewee, the decision to proceed with Program 437 X (AP) was taken before the West Coast establishment knew about it.

102. Cantwell, *The Air Force in Space FY 1964*, p. 65.

103. Ibid., pp. 65-6.

104. Interview with former official of the Department of the Air Force (Alpha), 18 May 1981.

105. Information on the characteristics of this system is taken from Cantwell, *The Air Force in Space FY 1964*, p. 66.

106. Interview with former aerospace industry contractor (Alpha), Washington, DC, 20 October 1980.

107. Interview with a former official of the Department of the Air Force (Alpha), Washington, DC, 18 May 1981.

108. "History of Air Defense Command", vol. 1 (January-June 1967), pp. 342-3. A submarine cable was also reportedly laid between Johnston Island and Hawaii to improve communication links. See DMS Market Intelligence Report, *Space Intercept Programs* (January 1968).

109. Memorandum for R. E. Kintner, Special Assistant to the President, Subject: Weekly Report to the President, 4 April 1967, and 11 April 1967. LBJ Confidential File: DoD, April 1967, LBJ Library. The lethal range was judged to be about five nautical miles.

110. "History of Air Defense Command", vol. 1 (January-June 1967), p. 344.

111. "History of Air Defense Command" (1969), p. 453.

112. CSAF Message 3020207. Subject: Program 437 Phasedown. July 1970 and Program Management Directive for Program 437, Thor Missile Launch Support. 10 August 1974.

113. Memorandum to Secretary of the Air Force, Subject: Anti-Satellite System, 4 May 1970.

114. "History of Air Defense Command" (1971), p. 388.

115. "History of Air Defense Command" (1972), p. 143. See also Message from ADC to Chief of Staff Air Force, Subject: Program 437 Concept of Operations, 21 January 1971.

116. Memorandum for Secretaries of the Army, Navy, Air Force and Director of ARPA, 7 February 1963.

117. Cantwell, *The Air Force in Space FY 1964*, p. 64.

118. Ibid.

119. Memorandum for the Chief of Staff, Subject: Satellite Interception, 9 December 1963.

120. Cantwell, *The Air Force in Space FY 1964*, p. 67.

121. *Aviation Week & Space Technology*, vol. 81, no. 19 (November 1964), p. 19.

122. *Report to the Congress from the President of the United States, US Aeronautical and Space Activities (1964)*, p. 45.

123. *Aviation Week & Space Technology*, vol. 82, no. 12 (22 March 1965), p. 13.

124. "History of Air Defense Command", vol. 1 (January-June 1967), pp. 344-50.

125. "History of Air Defense Command" (1969), p. 456.

126. Ibid., p. 458.

127 "History of Air Defense Command" (1970), p. 346.

128. Review and Summary of X20 Military Applications Studies, 14 December 1963, Aeronautical Systems Division, AFSC Wright Patterson Air Force Base. See also Futrell, *Ideas, Concepts, Doctrines*, p. 276.

129. Ibid.

130. Berger, *The Air Force in Space FY 1961*, p. 51.

131. Ibid.

132. William E. Howard, "Dynasoar Streamlining Urged", *Missiles and Rockets*, vol. 9, no. 1 (July 1961), p. 10.

133. For a thorough account of the cancellation of Dynasoar (X20) see Cantwell, *The Air Force in Space FY 1964*, pp. 24-30.

134. Owen Wilkes, "Space Tracking and Space Warfare", PRIO Publication S-1/77 (International Peace Research Institute, Oslo, December 1978), p. 23.

135. US Congress, Senate, Committee on Armed Services, *Military Procurement Authorization Fiscal Year 1964*, Hearings, 88th Congress, 1st Session (1963), pp. 74-5.

136. Hal Gellings, "Shepherd Touching Off Interservice Row", *Missiles and Rockets*, vol. 6, no. 10 (7 March 1960), pp. 21-5. Military Space Projects: Report of Progress for June-July, August 1960" (DDR&E, 20 October 1960), Eisenhower Library.

137. Use here is made of Wilkes' "Space Tracking and Space Warfare", categorization of space object identification systems.

138. Ibid., p. 39.

139. US Congress, Senate, Committee on Appropriations, *Military Procurement Authorization Fiscal Year 1966*, 89th Congress, 1st Session (1965), p. 68.

140. Memorandum for Robert Kintner, Subject: Weekly Report to the President, 28 June 1966; LBJ Confidential File, Department of Defense, June 1966, Box 118, L. B. Johnson Library.

7 *The New Soviet Space Challenge, 1968–1977*

1. The Polet or Polyot launches on 1 November 1963 and 12 April 1964, as well as the Kosmos 102 and 125 flights on 27 December 1965 and 20 July 1966, respectively, all demonstrated a capacity to manoeuvre in orbit, which may also have contributed to the satellite interceptor programme. See "Polyot 'manoeuvring' Claim Questioned", *Aviation Week & Space Technology*, vol. 79, no. 20 (11 November 1963), pp. 28-9; Philip S. Clark, "The Polyot Missions", *Spaceflight*, vol. 22, no. 9-10 (September-October 1980), pp. 312-15; and US Congress, Senate, Committee on Aeronautical and Space Sciences, *Soviet Space Programs, 1971-1975*, vol. 1 (Government Printing Office, Washington, DC, 30 August 1976), p. 61.

2. Quoted from G. E. Perry, "Russian Hunter-Killer Satellite Experiments", *Royal Air Force Quarterly*, vol. 17 (Winter 1977), p. 329.

3. Ibid.

4. US Congress, Senate, *Soviet Space Programs, 1971-1975*, p. 426.

5. See George C. Wilson, "Russia May be Testing Anti-Satellite Vehicle", *International Herald Tribune* (6 December 1968), p. 4; George C. Wilson, "Soviet Target Shots in Space May be Anti-Satellite Arms", *International Herald Tribune* (5/6 April

1969); and *Newsweek* (10 February 1969).

6. "Launching the Killer Cosmos", *Newsweek* (16 February 1970).

7. Perry, "Russian Hunter-Killer Satellite Experiments", p. 331.

8. Ibid., p. 332.

9. Ibid.

10. *Aviation Week & Space Technology* also suggested that Kosmos 520, launched on 19 September 1972, may have been involved in an intercept test with Kosmos 516, but this is unlikely given their orbital elements. This view was later revised. See "Cosmos 520 May Be Orbital Destruct Test", *Aviation Week & Space Technology*, vol. 97, no. 14 (2 October 1972), p. 17; and "Russians Continuing Complex Military Satellite Exercise", *Aviation Week & Space Technology*, vol. 97, no. 15 (9 October 1972), p. 20. Of more interest, however, is a reference to the Soviet ASAT test programme in the FY 1978 NASA authorization hearings. Here, in a written reply to a question, it is stated: "Although check-out of both ASAT and target systems was conducted in 1973 and 1974, the Soviets did not launch the boosters". See US Congress, Senate, Subcommittee on Science, Technology and Space of the Committee on Commerce, Science and Transportation, *NASA Authorization for Fiscal Year 1978*, Hearings, 95th Congress, 1st Session (1977), Part 3, p. 1636.

11. Perry, "Russian Hunter-Killer Satellite Experiments", p. 333.

12. Performance data are taken from N. L. Johnson, *The Soviet Year in Space: 1982* (Teledyne Brown Engineering, January 1983), p. 26.

13. See R. P. Berman and J. C. Baker, *Soviet Strategic Forces: Requirements and Responses* (The Brookings Institution, Washington, DC, 1982), pp. 136-57, for examples.

14. See Philip Klass, "USSR Accelerates Recon Satellite Race", *Aviation Week & Space Technology*, vol. 92, no. 14 (6 April 1970), pp. 72-9. Berman and Baker, *Soviet Strategic Forces*, however, argue that the Soviet invasion of Czechoslovakia was probably the first extensive use of reconnaissance satellites during a military operation involving Soviet forces.

15. See US Congress, Senate, *Soviet Space Programs, 1971-1975*, pp. 464-5.

16. "New Soviet Satellite Observes NATO Work from Unusual Orbit", *Aviation Week & Space Technology*, vol. 97, no. 13 (25 September 1972), p. 19.

17. See US Congress, Senate, *Soviet Space Programs, 1971-1975*, pp. 466-73; and "Soviets Launch Five Spy Satellites in Two Weeks", *New Scientist*, vol. 60, no. 869 (25 October 1973), p. 260.

18. See US Congress, Senate, *Soviet Space Programs, 1971-1975*, p. 473; and David Baker, *The Shape of Wars to Come* (Patrick Stephens, Cambridge, 1981), pp. 77-9.

19. Anwar el-Sadat, *In Search of Identity: An Autobiography* (Harper and Row, New York, 1977), p. 260. Mohammed Heikal, in his book *The Road to Ramadan* (Ballantine Books, New York, 1975), p. 241, states that Sadat was shown *aerial* photographs by the Soviet Union.

20. Berman and Baker, *Soviet Strategic Forces*, pp. 161-2.

21. Quoted in Philip Klass, "Soviets Push Ocean Surveillance", *Aviation Week & Space Technology*, vol. 99, no. 11 (10 September 1973), pp. 12-13.

22. Quoted in Philip Klass, "Soviets Appear to be Operational with Ocean Surveillance Satellite", *Aviation Week & Space Technology*, vol. 102, no. 25 (23 June 1975), p. 18.

23. For details of the Soviet ocean reconnaissance programme see G. E. Perry, "Russian Ocean Surveillance Satellites", *The Royal Air Force Quarterly*, vol. 18 (Spring 1975), pp. 60-7; US Congress, Senate, *Soviet Space Programs, 1971-1975*, pp. 430-2; and

"Soviets Change Ocean Missions", *Aviation Week & Space Technology*, vol. 110, no. 18 (30 April 1979), p. 33.

24. For details of its coverage see Berman and Baker, *Soviet Strategic Forces*, pp. 162-5; and S. M. Meyer, "Soviet Military Programmes and the 'New High Ground'", *Survival*, vol. 25, no. 5 (September/October 1983), p. 210.

25. See Philip Klass, "Soviets Spur Military Space Net", *Aviation Week & Space Technology*, vol. 101, no. 1 (8 July 1974), pp. 12-13; and Craig Covault, "Geodetic Launches and Soviet Targetting", *Aviation Week & Space Technology*, vol. 104, no. 23 (7 June 1976), pp. 23-4.

26. See "Soviets Resume Satellite Intercept Tests", *Aviation Week & Space Technology*, vol. 104, no. 17 (26 April 1976), p. 21; and "Satellite Killer Question", *Aviation Week & Space Technology*, vol. 104, no. 14 (5 April 1976), p. 9.

27. "Another Soviet Space Intercept Test Conducted", *Aviation Week & Space Technology*, vol. 104, no. 17 (26 April 1976), p. 21.

28. "Soviet Anti-Satellite Mission Fails to Reach its Target", *Aviation Week & Space Technology*, vol. 105, no. 5 (2 August 1976), p. 24.

29. Perry, "Russian Hunter-Killer Satellite Experiments", p. 333.

30. See Meyer, "Soviet Military Programmes and the 'New High Ground'", p. 212; and Johnson, *The Soviet Year in Space: 1982*, p. 26.

31. "Satellite Killers", *Aviation Week & Space Technology*, vol. 104, no. 25 (21 June 1976), p. 13.

32. "Russia's Killer Satellites", *Foreign Report* (14 January 1981), p. 2.

33. Ibid.

34. *Aviation Week & Space Technology*, vol. 99, no. 6 (6 August 1973), p. 9.

35. John Douglas, "High Energy Laser Weapons", *Science News*, vol. 110, no. 1 (3 July 1976), p. 12.

36. For further details on these incidents see Philip Klass, "Anti-satellite Laser Use Suspected", *Aviation Week & Space Technology*, vol. 103, no. 23 (8 December 1975), pp. 12-13; "DoD Continues Satellite Blinding Investigation", *Aviation Week & Space Technology*, vol. 104, no. 1 (5 January 1976), p. 18; and "Pentagon Denies Soviet Anti-satellite Laser Use", *Defense/Space Daily* (24 November 1976), p. 26.

37. See Karl F. Spielman, *Analyzing Soviet Strategic Arms Decisions* (West View, Boulder, Colorado, 1978), pp. 4-6.

38. Ibid., Matthew P. Gallagher and Karl F.Spielman, *Soviet Decision-Making for Defense: A Critique of U.S. Perspectives on the Arms Race* (Praeger, New York, 1972); David Holloway, "The Soviet Style of Military R & D" in Franklin A. Long and Judith Reppy (eds), *The Genesis of New Weapons: Decision-Making for R & D* (Pergamon, New York, 1980); and Arthur Alexander, "Decision-Making in Soviet Weapons Procurement", *Adelphi Paper*, no. 147/148 (IISS, London, Winter 1978-9).

39. Alexander, "Decision-Making in Soviet Weapons Procurement", *Adelphi Paper*, p. 1.

40. US Congress, Senate, Committee on Aeronautical and Space Sciences, *Soviet Space Programs, 1966-1970*, Staff Report, 92nd Congress, 1st Session (1971), p. 339.

41. Thomas W. Wolfe, *Soviet Strategy at the Crossroads* (Harvard University Press, Cambridge, 1964), pp. 204-5.

42. "Missiles and Strategy", *Red Star* (21 March 1962), quoted in Wolfe, *Soviet Strategy at the Crossroads*.

43. "Outer Space and Strategy", *Red Star* (21 March 1962), quoted in Wolfe, *Soviet Strategy at the Crossroads*.

44. V. D. Sokolovskiy, *Soviet Military Strategy*, Harriet Fast Scott (ed.), 3rd edn. (Macdonald and Jane's, London, 1975), p. 458.

45. *Dictionary of Basic Military Terms: A Soviet View*, USAF Series on Soviet Military Thought, no. 9 (Government Printing Office, Washington, DC, 1972), p. 177.

46. See, for example, Lawrence Freedman, "The Soviet Union and Anti-Space Defence", *Survival*, vol. 19, no. 1 (January 1977), p. 18; *Soviet Aerospace Handbook*, US Air Force Pamphlet 200-21 (Government Printing Office, Washington, DC, May 1978), p. 65; and Stuart Cohen, "The Evolution of Soviet Views on SALT Verification: Implications for the Future", in William Potter (ed.), *SALT and Verification* (Westview, Boulder, Colorado, 1980), p. 62.

47. Wolfe, *Soviet Strategy at the Crossroads*, p. 206.

48. "45 Years on Guard Over the Socialist Fatherland", *Pravda* (23 February 1963), quoted in Wolfe, *Soviet Strategy at the Crossroads*, p. 206.

49. Col. A. Krasnov, "Space and Combat Readiness", *Military Thought*, no. 4 (April 1966), p. 32. Other citations include Engineer Col. A. Vasil'yev, "Development of Space Systems of Armament in the U.S.", *Military Thought*, no. 3 (March 1967), pp. 77-85; and Engineer Col. A. Vasil'yev, "The Development of Military Space Systems and Equipment in the USA", *Military Thought*, no. 88 (August 1971), pp. 77-82.

50. Manfred Otto, *National Zeitung* (East Berlin, 14 September 1968), p. 6, quoted in US Congress, Senate, *Soviet Space Programs, 1966-1970*, p. 341.

51. V. D. Sokolovskiy, *Soviet Military Strategy* (Prentice-Hall, New Jersey, 1963), p. 177.

52. *Dictionary of Basic Military Terms*, p. 109.

53. See Col. N. Maksimov, "New ABM Weapons and Methods of Destroying Ballistic Missiles", *Military Thought*, no. 5 (May 1962), p. 84; Lt. Col. V. Vasil'yev, "The Organisation of Air Defence for the North American Continent", *Military Thought*, no. 7 (July 1964), p. 86; Engineer Lt. Col. B. Aleksandrov, "Problems of Space Defence and Means of Resolving Them", *Military Thought*, no. 9 (September 1964), pp. 84-5; Major General M. Cherednichenko, "Scientific Technical Progress and the Development of Armaments and Military Technology", no. 4, *Military Thought* (April 1972), p. 35; Col. Gen. N. A. Lomov (ed.), *Scientific-Technical Progress and the Revolution in Military Affairs (A Soviet View)* (Moscow, 1973); *USAF Studies in Soviet Military Thought*, no. 3, pp. 54-6; and Ivan I. Anureev, "Anti-missile and Space Defense Weapons" (Joint Publications Research Service, Arlington, Virginia, 1972). See also Sawyer, *The Soviet Space Controversy, 1961-1963*, doctoral thesis (Fletcher School of Law and Diplomacy, Tufts University, May 1969), pp. 88-90.

54. See US Congress, Senate, *Soviet Space Programs, 1966-1970*, pp. 346-8.

55. See Freedman, "The Soviet Union and Anti-Space Defence", p. 23.

56. Private communication.

57. "Right to Destroy Satellites Sought by Soviets in UN", *Aviation Week & Space Technology*, vol. 97, no. 17 (23 October 1972), p. 20.

58. For details on the organization of the Soviet military space programme see US Congress, Senate, *Soviet Space Programs, 1966-1970*, pp. 69-105; William H. Schauer, *The Politics of Space: A Comparison of the Soviet and American Space Programs* (Holms and Meier, New York, 1976), pp. 25-30; Michael Stoiko, *Soviet Rocketry: The First Decade of Achievement* (David and Charles, Newton Abbot, 1970), pp. 191-4; and Nicholas Johnson, *The Soviet Year in Space: 1983* (Teledyne Brown, 1984), p. 2.

59. Sawyer, *The Soviet Space Controversy*, p. 116.

60. Walter Clemens, Jr., *Outer Space and Arms Control* (Center for Space Research,

Massachusetts Institute of Technology, 1966), p. 117.

61. Meyer, "Soviet Military Programmes and the 'New High Ground'", p. 212.

8. *Nixon and Ford: Continuity and Change*

1. "Nixon's Task Group Preparing a Vital Space Policy Report", *Space/Astronautics*, vol. 51, no. 6 (June 1969), pp. 28-32.

2. President's Space Task Group, *The Post Apollo Space Program: Directive for the Future*, Report to the President (September 1969), quoted in Lt. Col. Richard Bowers, *et al.*, "A Turning Point in Space: Has the Time Arrived for a National Review of Our Space Policy?" Study Project (US Army War College, Carlisle Barracks, 13 May 1977), p. 7.

3. Arthur Levine, *The Future of the U.S. Space Program* (Praeger, New York, 1975).

4. Aeronautics and Space Report of the President (Washington, DC, January 1970), p. 32.

5. Interview with a former Senior Air Force officer (Beta), Washington, DC, 10 October 1980.

6. "MOL Cancelled in Abrupt Decision", *Aviation Week & Space Technology*, vol. 90, no. 24 (16 June 1969), p. 28.

7. See, for further details, Curtis L. Peebles, "The Guardians: A History of the 'Big Bird' Reconnaissance Satellites", *Spaceflight*, vol. 20, no. 11 (November 1978), pp. 381-5.

8. US Congress, House, Subcommittee of the Committee on Appropriations, *Department of Defense Appropriations for 1971*, Hearings, 91st Congress, 2nd Session (1970), Part 6, p. 87.

9. See Paul B. Stares, "Space and U.S. National Security", *Journal of Strategic Studies*, vol. 6, no. 4 (December 1983), p. 45.

10. *The Next Decade in Space*, a Report of the Space Science and Technology Panel of the President's Science Advisory Committee, Executive Office of the President (Office of Science and Technology, March 1970), pp. 17-20.

11. Ibid., p. 20.

12. US Congress, House, *Department of Defense Appropriations for 1971*, Part 6, p. 88.

13. See Raymond L. Garthoff, "Banning the Bomb in Outer Space", *International Security*, vol. 5, no. 3 (Winter 1980/1), p. 26; and US Congress, House, Subcommittee on Space Science and Applications of the Committee on Science and Technology, *United States Civilian Space Programs, 1958-1978*, Staff Report, 97th Congress, 1st Session (January 1981), vol. 1, p. 869.

14. *Consolidated Space Operations Center Lacks Adequate DoD Planning*, Report MASAD-82-14 (General Accounting Office, 29 January 1982), p. 13.

15. US Congress, House, *Department of Defense Appropriations for 1971*, Part 6, p. 88.

16. US Congress, Senate, Committee on Aeronautical and Space Science, *NASA Authorization for Fiscal Year 1971*, Hearings, 91st Congress, 2nd Session (1970), Part 2, p. 876.

17. US Congress, House, Subcommittee of the Committee on Appropriations, *Department of Defense Appropriations for 1972*, Hearings, 92nd Congress, 1st Session (1971), Part 6, p. 22. Secretary of the Air Force Robert Seamans did provide, however,

a written description of the tests during congressional testimony in the previous year. But he did not categorically state that the Soviet Union had an ASAT system, only the potential. See US Congress, House, *Department of Defense Appropriations for 1971*, Part 1, pp. 662-3.

18. The information on this study and subsequent quotations are taken from an interview with a former senior member of the Department of Defense (Delta) Washington, DC, 21 May 1981.

19. US Congress, House, Subcommittee of the Committee on Appropriations, *Department of Defense Appropriations for 1973*, Hearings, 92nd Congress (1972), Part 4, p. 513.

20. *Arms Control and Disarmament Agreements: Texts and Histories of Negotiations* (US Arms Control and Disarmament Agency, Washington, DC, August 1980), pp. 141, 151.

21. Stuart Cohen, "The Evolution of Soviet Views on Verification" in W. Potter (ed.), *SALT and Verification* (Westview Press, Boulder, Colorado, 1980), pp. 54-5. See also "Arms Control Disarmament" (Novosti Press Agency Publishing House, Moscow, 1983).

22. See John Newhouse, *Cold Dawn: The Story of SALT* (Holt, Reinhart and Winston, New York, 1973); Gerard Smith, *Doubletalk: The Story of SALT I* (Doubleday, New York, 1980); and Thomas Wolfe, *The SALT Experience* (Ballinger, Cambridge, Ma., 1979).

23. Interview with former senior official of the State Department (Alpha), Washington, DC, December 1983.

24. Cohen, "The Evolution of Soviet Views on Verification", p. 63.

25. Smith, *Doubletalk*, p. 263.

26. Ibid.

27. Interview with former senior official of the State Department (Alpha), Washington, DC, December 1983.

28. Newhouse, *Cold Dawn*, p. 231.

29. Smith, *Doubletalk*, p. 265.

30. The above information comes from an interview with a former senior official of the National Security Council Staff (Beta), Washington, DC, 3 June 1981.

31. Interview with a former senior official of the National Security Council Staff (Alpha), Washington, DC, 4 June 1981.

32. Ibid.

33. This and the above information comes from an interview with a former senior official of the National Security Council Staff (Gamma), Los Angeles, 15 June 1981.

34. Ibid.

35. Interview with a former senior official of the National Security Council Staff (Alpha), Washington, DC, 4 June 1981.

36. Interview with a State Department official (Delta), Washington, DC, September 1983.

37. Donald Hafner, "Averting a Brobdingnagian Skeet Shoot: Arms Control Measures for Antisatellite Weapons", *International Security*, vol. 5, no. 3 (Winter 1980-1), pp. 50-1.

38. Ibid.

39. US Congress, House, Subcommittee of the Committee on Appropriations, *Department of Defense Appropriations for 1973*, Hearings, 92nd Congress, 2nd Session (1972), Part 4, p. 813.

40. US Congress, House, Subcomittee of the Committee on Appropriations, *Department of Defense Appropriations for 1977*, Hearings, 94th Congress, 2nd Session (1976), Part 3, p. 398.

41. See Chapter 10 for details.

42. US Congress, House, Committee on Armed Services, *Military Posture and Department of Defense Authorization for Appropriations for Fiscal Year 1977*, Hearings, 94th Congress, 2nd Session (1976), Part 1, pp. 125-6.

43. US Congress, Senate, Committee on Aeronautical and Space Sciences, *NASA Authorization for Fiscal Year 1977*, Hearings, 94th Congress, 2nd Session (1976), Part 3, p. 1968.

44. Ibid., p. 1982.

45. Quoted in "Warning to Soviets", *Aviation Week & Space Technology*, vol. 105, no. 19 (8 November 1976), p. 13 (my emphasis).

46. US Congress, House, Committee on Armed Services, *Military Posture and Department of Defense Authorization for Fiscal Year 1978*, Hearings, 95th Congress, 1st Session (1977), part 1, p. 1053.

47. US Congress, House, *Military Posture and Department of Defense Authorization for FY 1978*, p. 934.

48. US Congress, House, Subcommittee of the Committee on Appropriations, *Department of Defense Appropriations for 1978*, Hearings, 95th Congress, 1st Session (1977), part 3, pp. 229-320.

49. The report had been leaked to the *Los Angeles Times* after the Carter administration took office, possibly to ensure the continuity of the programme by that administration. See Norman Kempster, "New U.S.-Satellite Killer Project is Revealed", *Los Angeles Times* (31 March 1977).

50. Ibid.

51. Interview with a retired senior officer of the Air Force (Theta), Los Angeles, 10 June 1981.

52. For details of the reorganization of SAMSO see Barry Miller, "SAMSO Gears for Operational Challenge", *Aviation Week & Space Technology*, vol. 105, no. 3 (19 July 1976), pp. 59-67.

53. The only exception to this was voiced by Secretary of the Navy J. William Middendorf II during a meeting of the American Security Council: "There's no question that the Soviets have developed satellites that give them real time, mid-course guidance for their incoming missiles [and] that we've got to work like mad to get them down real fast." However, he later said that these satellites could be spoofed with chaff. See "Guidance Satellite", *Aviation Week & Space Technology*, vol. 104, no. 16 (4 October 1976), p. 13.

54. Interview with former senior official of the National Security Council Staff (Delta), Washington, DC (21 May 1981). This observer commented that in the last days of the Ford Administration the formal decision-making process broke down to the extent that "if you could get to Gerry Ford—you could get anything signed".

9. *Carter and the Two-Track Policy*

1. Interview with former official of the Department of State (Delta), Washington, DC, September 1983.

2. Press Conference of Jimmy Carter, Washington, DC, 9 March 1977, reproduced in Roger P. Labrie (ed.), *SALT Handbook: Key Documents and Issues* (American

Enterprise Institute, Washington, DC, 1979), p. 423.

3. Press Conference of Secretary of State Cyrus Vance, Moscow, 30 March 1977, in Labrie, *SALT Handbook*, p. 429. The other areas included a comprehensive test ban, limitations on civil defence preparations and chemical weapons, prior notification of missile test-firing, military limitations in the Indian Ocean, radiological weapons, conventional weapons and nuclear proliferation.

4. White House Press Release, "Description of a Presidential Directive on National Space Policy" (The White House, 20 June 1978).

5. Ibid.; and Craig Convault, "Unified Policy on Space", *Aviation Week & Space Technology*, vol. 108, no. 1 (2 January 1978), p. 14.

6. Interview with former official of the National Security Council staff (Epsilon), Washington DC, 19 May 1981.

7. Ibid.

8. Interview with former senior offical of the Department of Defense (Zeta), San Francisco, 17 June 1981.

9. US Congress, House, Subcommittee of the Committee on Appropriations, *Department of Defense Appropriations for 1979*, Hearings, 95th Congress, 2nd Session (1978), Part 3, pp. 726-7.

10. Ibid., Part 2, p. 826.

11. Covault, "Unified Policy on Space", p. 15.

12. Presidential Directive/NSC-37, *National Space Policy* (The White House, 11 May 1978). The national security recommendations in this document remain classified.

13. White House Press Release, "Description of a Presidential Directive on National Space Policy".

14. *Documents on Disarmament 1978*, Publication 107 (US Arms Control and Disarmament Agency, Washington, DC, October 1980), p. 586. This does not include President Johnson's off-the-record remarks on 15 March 1967 to a group in Nashville, Tennessee, at which he stated: "I wouldn't want to be quoted on this, but we've spent $35-40 billion on the space program and if nothing else has come out of it except the knowledge we've gained from space photographs, it would be worth ten times what the whole program has cost." Secretary of State William Rogers had also made an inadvertent reference to satellite reconnaissance just after signing of the SALT I treaty in 1972.

15. "Carter Confirms Recon Satellite Deployment", *Aviation Week & Space Technology*, vol. 109, no. 15 (9 October 1978), p. 22; and an interview with a former CIA official, Washington DC, 27 May 1981. See also Gerald M. Steinberg, *Satellite Reconnaissance: The Role of Informal Bargaining* (Praeger, New York, 1983), p. 177.

16. Presidential Directive/NSC-42, *Civil and Further National Space Policy* (The White House, 10 October 1978).

17. Presidential Directive/NSC-54, *Civil Operational Remote Sensing* (The White House, 16 November 1979). See also Craig Covault, "Space Policy Mandates U.S. Leadership", *Aviation Week & Space Technology*, vol. 109, no. 16 (16 October 1978), pp. 24-6; and *Civilian Space Policy and Options*, Office of Technology Assessment Report (Government Printing Office, Washington, DC, 1981), pp. 155-6.

18. "Brown States Russians Can Fell Satellites", *New York Times* (5 October 1977).

19. David Baker, *The Shape of Wars to Come* (Patrick Stephens, Cambridge, Mass., 1981), p. 119.

20. US Congress, Senate, Committee on Foreign Relations, *The SALT II Treaty*, Hearings, 96th Congress, 1st Session (1979), Part 1, pp. 423-4.

21. Clarence A. Robinson, Jr, "Soviets Push for Beam Weapon", *Aviation Week & Space Technology*, vol. 106, no. 18 (2 May 1977), pp. 16-23.

22. "Brown comments on Beam Weapons", *Aviation Week & Space Technology*, vol. 106, no. 22 (30 May 1977), p. 12.

23. Simon Kassel, "Pulsed Power Research and Development in the USSR", R-2212-ARPA (The Rand Corporation, Santa Monica, California, May 1978), p. 118. Kassel has also participated in other Rand studies of related Soviet developments. See Simon Kassel and Charles D. Hendricks, "Soviet Development of Needle-Tip Field Emissions Cathodes for High Current Election Beams", R-1311-ARPA (August 1973); Simon Kassel, "Soviet Development of Flash X-Ray Machines", R-1053-ARPA (October 1973); and Simon Kassel and Charles D. Hendricks, "High Current Particle Beam, 1: The Western USSR Research Groups", R-1552-ARPA (April 1974).

24. Clarence A. Robinson, Jr, "U.S. Pushes Development of Beam Weapons", *Aviation Week & Space Technology*, vol. 109, no. 14 (2 October 1978), p. 15; and Clarence A. Robinson, Jr, "Soviets Test Beam Technologies in Space", *Aviation Week & Space Technology*, vol. 109, no. 20 (13 November 1978), p. 14.

25. "Soviets Build Directed Energy Weapon", *Aviation Week & Space Technology*, vol. 113, no. 4 (28 July 1980), p. 47.

26. Interview with former senior offical of the Department of Defense (Zeta), San Francisco, 17 June 1981.

27. US Congress, Senate, Subcommittee of the Committee on Appropriations, *Department of Defense Appropriations for Fiscal Year 1979*, Hearings, 95th Congress, 2nd Session (1978), Part 5, p. 467.

28. See "Military Salyut Returns Data Quickly", *Aviation Week & Space Technology*, vol. 106, no. 10 (7 March 1977), p. 20; and "Soviet Launches of More Military Salyuts Expected", *Aviation Week & Space Technology*, vol. 109, no. 22 (4 December 1978), p. 17.

29. The PRM/NSC-23 Decision Paper was apparently only one page in length. Interview with former official of the Department of State (Delta), Washington, DC, September 1983.

30. Interview with former senior offical of the NSC Staff (Epsilon), Washington DC, 19 May 1981. See also "Killer Talks", *Aviation Week & Space Technology*, vol. 107, no. 22 (28 November 1977), p. 13. As a result of the leak of information on the respective departmental views in this article, the security surrounding the debate became even tighter, making these talks some of the most heavily classified ever conducted. See "Killer Anonymity", *Aviation Week & Space Technology*, vol. 108, no. 1 (2 January 1978), p. 13.

31. Robert Toth, "U.S.-Soviet Talks Seen Soon on Anti-Satellite Arms Ban", *Los Angeles Times* (5 November 1977).

32. "Killer Talks", *Aviation Week & Space Technology*, vol. 108, no. 13 (27 March 1978), p. 13.

33. "Soviets Said to Agree on Satellite Talks", *New York Times (1 April 1978).*

34. "U.S., Soviets to Begin Talks on Killer Satellites", *Baltimore Sun* (9 May 1978), p. 4.

35. Interview with former official of the State Department (Delta), Washington, DC, September 1983; and James Canan, *War in Space* (Harper and Row, New York, 1982), pp. 24-5.

36. Interview with former senior offical of ACDA (Alpha), Washington, DC, 29 May 1981.

37. "Antisatellite Move", *Aviation Week & Space Technology*, vol. 109, no. 8 (21 August 1978), p. 11. See also "U.S. Optimistic in Killer Satellite Ban", *Aviation Week*

& Space Technology, vol. 108, no. 26 (26 June 1978), p. 20.

38. "Soviets see Shuttle as Killer Satellite", *Aviation Week & Space Technology*, vol. 108, no. 16 (17 April 1978), p. 17.

39. Edgar Ulsamer, "Space Treaty Rift", *Air Force Magazine*, vol. 62, no. 1 (January 1979), p. 14.

40. See Donald Hafner, "Averting a Brobdignagian Skeet Shoot: Arms Control Measures for Antisatellite Weapons", *International Security*, vol. 5, no. 3 (Winter 1980-1), p 59.

41. "U.S. and Soviets Adjourn Meeting in Bern on Killer Satellite Output", *New York Times* (21 February 1979).

42. "Antisatellite Talks", *Aviation Week & Space Technology*, vol. 110, no. 17 (23 April 1979), p. 15.

43. See "Chinese Space Gains Hamper Antisatellite Limitation Treaty", *Aviation Week & Space Technology*, vol. 11, no. 2 (9 July 1979), p. 19.

44. Walter Slocombe, conference paper presented to the Stockholm International Peace Research Institute, July 1983, p. 9.

45. Robert Toth, "Progress is Slow on Discussion by U.S.-Kremlin on Satellites", *International Herald Tribune* (16 July 1979); and Zbigniew Brzezinski, *Power and Principle: Memoirs of the National Security Adviser 1977-1981* (Farrar, Straus, Giroux, New York, 1983), p. 341.

46. Interview with former Senior Official of the NSC staff (Epsilon), Washington DC, 19 May 1981.

10. *US Antisatellite Research and Development 1971-1981*

1. "History of Air Defense Command FY 1972", pp. 143-4.

2. Ibid., pp. 192-4. See also Final Report Program 437 Damage Assessment Hurricane Celeste (August 1972). The cost of repairing the damage was put at $24 million.

3. USAF Program Objective Memorandum (POM) for FY 1975-9.

4. USAF Program Management Directive for Program 437, Thor Missile Launch Support (10 August 1974).

5. Message from CINCONAD to JCS (6 March 1975).

6. Memorandum from Deputy Secretary of Defense Packard to Secretary of the Air Force, Subject: Anti-Satellite System (4 May 1970).

7. "History of Air Defense Command FY 1972", p. 145.

8. Letter from CINCONAD to Commander ADC, Subject: Project SPIKE (9 July 1971).

9. "History of Air Defense Command FY 1972", p. 147.

10. Memorandum From DCS/Operations, Subject: SPIKE (Briefings at HQ USAF, 21 September 1971).

11. Memorandum For Mr Hansen From Mr J. C. Jones, Subject: Project SPIKE (21 September 1971).

12. Ibid.

13. US Congress, House, Subcommittee of the Committee on Appropriations, *Department of Defense Appropriations For 1971*, Hearings, 91st Congress, 2nd Session (1970), Part 6, p. 710.

14. US Congress, House, Subcommittee of the Committee on Appropriations, *Department of Defense Appropriations For 1972*, Hearings, 92nd Congress, 1st Session (1971), Part 6, pp. 275-6.

15. US Congress, House, Subcommittee of the Committee on Appropriations, *Department of Defense Appropriations For 1973*, Hearings, 92nd Congress, 2nd Session (1972), Part 4, p. 676.

16. See "USAF Pushes Antisatellite Alerters", *Aviation Week & Space Technology*, vol. 95, no. 19 (8 November 1971), p. 18.

17. US Congress, Senate, Committee on Armed Services, *Fiscal Year 1974 Authorization for Military Procurement*, Hearings, 93rd Congress, 1st Session (1973), Part 3, p. 3502. Also, *Aviation Week & Space Technology*, vol. 98, no. 22 (28 May 1973), p. 11.

18. *Aviation Week & Space Technology*, vol. 102, no. 4 (3 March 1975), p. 9.

19. *Aviation Week & Space Technology*, vol. 102, no. 20 (19 May 1975), p. 11.

20. *Aviation Week & Space Technology*, vol. 103, no. 6 (11 August 1975), p. 13.

21. US Congress, House, Committee on Armed Services, *Military Posture and Department of Defense Authorizations for Appropriations for Fiscal Year 1977*, Hearings, 94th Congress, 2nd Session (1976), Part 5, p. 803.

22. US Congress, House, Committee on Armed Services, *Military Posture and Department of Defense Authorizations for Appropriations for Fiscal Year 1977*, Hearings, 95th Congress, 1st Session (1977), Part 3, p. 1758.

23. US Congress, House, *Hearings on Military Posture and Department of Defense Authorization for Appropriations for Fiscal Year 1977*, Part 5, p. 504. See also *Survivable Satellite*, DMS Market Intelligence Report (DMS Inc., Greenwich, Connecticut, January 1976).

24. *Fiscal Year 1979 Arms Control Impact Statements* (Government Printing Office, Washington, DC, June 1978), p. 102.

25. "U.S. to Build Satellite Killer in Case of War in Space", *Washington Post* (23 September 1977).

26. "U.S. Funds Killer Satellite Effort", *Aviation Week & Space Technology*, vol. 108, no. 6 (6 February 1978), pp. 18-19.

27. See US Congress, House, Subcommittee of the Committee on Appropriations, *Department of Defense Appropriations for 1980*, Hearings, 96th Congress, 1st Session (1979), Part 6, p. 692.

28. US Congress, Senate, Committee on Armed Services, *Department of Defense Authorization for Appropriations for Fiscal Year 1980*, Hearings, 96th Congress, 1st Session (1979), Part 6, p. 3016.

29. "U.S. Funds Killer Satellite Effort", p. 18; and Craig Covault, "US Pushes Antisatellite Effort", *Aviation Week & Space Technology*, vol. 109, no. 3 (17 July 1978), pp. 14-15.

30. For full details of the MHV system see Craig Covault, "Antisatellite Weapon Design Advances" *Aviation Week & Space Technology*, vol. 112, no. 24 (16 June 1980), pp. 243-7; and Bruce Smith, "Vought Tests Small Antisatellite System", *Aviation Week & Space Technology*, vol. 115, no. 19 (9 November 1981), pp. 24-5.

31. US Congress, House, *Department of Defense Appropriations for 1980*, p. 684.

32. Ibid., p. 682; and US Congress, Senate, *Department of Defense Authorization For Appropriations for Fiscal Year 1980*, pp. 3020 and 3037.

33. Jack Anderson, "Space Wars", *Washington Post* (16 August 1981). Interestingly, this report does not specifically identify the Soviet *radar* ocean reconnaissance satellites, only those that utilize ELINT techniques.

34. US Congress, Senate, Subcommittee of the Committee on Appropriations, *Department of Defense Appropriations for Fiscal Year 1981*, Hearings, 96th Congress, 2nd Session (1980), Part 1, p. 573.

35. US Congress, House, *Department of Defense Appropriations For 1980*, p. 692.

36. Ibid., p. 693.

37. "Avco Wins USAF Award For Antisatellite Targets", *Aviation Week & Space Technology*, vol. 110, no. 26 (25 June 1979), p. 23.

38. "Space Defense Review is Set for Dec. 18" *Aerospace Daily*, vol. 106, no. 3 (5 November 1980), p. 17.

39. *Fiscal Year 1979 Arms Control Impact Statements*, p. 102.

40. "Spacecraft Survivability Boost Sought", *Aviation Week & Space Technology*, vol. 112, no. 24 (16 June 1980), pp. 260-1.

41. "U.S. 'Nowhere Close' to Solving Satellite Survivability Problem", *Aerospace Daily*, vol. 106, no. 3 (5 November 1980), p. 35.

42. For further details see Barry Miller, "USAF Pushes Satellite Survivability", *Aviation Week & Space Technology*, vol. 106, no. 13 (28 March 1977), pp. 52-4; *Survivable Satellite*, DMS Market Intelligence Report (DMS Inc., Greenwich, Connecticut, 1980); "Spacecraft Survivability Boost Sought", pp. 260-1; and Desmond Ball, "Can Nuclear War Be Controlled?", *Adelphi Paper*, no. 169 (IISS, London, 1981), pp. 18-21.

43. This was reiterated in M. May, "War or Peace in Space", *The California Seminar on International Security and Foreign Policy*, Discussion Paper No. 93 (Santa Monica, Calif., March 1981).

44. See "USAF Considers Silo Launch of Satellite", *Aviation Week & Space Technology*, vol. 107, no. 1 (4 July 1977), p. 18; and "Industry Observer", *Aviation Week & Space Technology*, vol. 114, no. 15 (13 April 1981), p. 15.

45. See D. Ball, "Can Nuclear War be Controlled?", p. 21.

46. *Survivable Satellite*, DMS Market Intelligence Report, p. 2; and US Congress, Senate, *DoD Authorization for Appropriations for Fiscal Year 1980*, Part 6, p. 3026.

47. For further information see *Fiscal Year 1981 Arms Control Impact Statements* (Government Printing Office, Washington, DC, May 1980), pp. 197-9; Bruce Smith, "Ground-based Electro-Optical Deep Space System Passes Reviews", *Aviation Week & Space Technology*, vol. 111, no. 9 (27 August 1979), pp. 48-52; and "Space Surveillance Deemed Inadequate", *Aviation Week & Space Technology*, vol. 112, no. 24 (16 June 1980), pp. 249-55.

48. For further details see "New Colorado Operations Center Provides System Focal Point", *Aviation Week & Space Technology*, vol. 112, no. 24 (16 June 1980), pp. 250-1; and *Fiscal Year 1982 Arms Control Impact Statements* (Government Printing Office, Washington DC, February 1981), p. 166.

49. US Congress, House, *Military Posture and Department of Defense Authorizations for Appropriations for Fiscal Year 1977*, Part 3, p. 1758.

50. See "U.S. Group Structuring Five-Year Plan", *Aviation Week & Space Technology*, vol. 109, no. 20 (13 November 1978), p. 18; Clarence A. Robinson, Jr, "U.S. Pushes Development of Beam Weapons", *Aviation Week & Space Technology*, vol. 109, no. 14 (2 October 1978), pp. 14-21; and Clarence A. Robinson, Jr, "Beam Weapons Effort to Grow", *Aviation Week & Space Technology*, vol. 110, no. 14 (2 April 1979), p. 12.

51. "Beam Policy Reversal", *Aviation Week & Space Technology*, vol. 113, no. 5 (4 August 1980), pp. 44-68. Also, "Technology Eyed to Defend ICBMs, Spacecraft", *Aviation Week & Space Technology*, vol. 113, no. 4 (28 July 1980), pp. 32-42.

52. See Special Issue *Aviation Week & Space Technology*, vol. 113, no. 5 (4 August 1980), pp. 44-68. Also "Technology Eyed to Defend ICBMs, Spacecraft", pp. 32-42.

53. Source: Fiscal years 1981, 1982, 1983, *Arms Control Impact Statements*.

54. See "U.S. Nears Laser Weapon Decisions", *Aviation Week & Space Technology*, vol. 113, no. 5 (4 August 1980), pp. 43-54; and "Navy Schedules Laser Lethality Tests", *Aviation Week & Space Technology*, vol 113, no. 4 (28 July 1980), pp. 54-7.

55. See "Pentagon Studying Laser Battle Stations in Space", *Aviation Week & Space Technology*, vol. 113, no. 4 (28 July 1980), pp. 57-62; and "Laser Applications in Space Emphasised", *Aviation Week & Space Technology*, vol. 113, no. 4 (28 July 1980), pp. 62-3.

56. "U.S. Effort Redirected to High Energy Lasers", *Aviation Week & Space Technology*, vol. 113, no. 4 (28 July 1980), pp. 54-7.

57. Edgar Usamer, "The Long Leap Toward Space Laser Weapons", *Air Force Magazine*, vol. 64, no. 8 (August 1981), p. 63.

11. *The Reagan Presidency: Towards an Arms Race in Space, 1981–1984*

1. The following information is taken from "White House Fact Sheet Outlining United States Space Policy" (4 July 1982).

2. Ibid.

3. "Pentagon Space Policy: More of Same", *Aerospace Daily* (24 August 1982), p. 279.

4. Defense Secretary Weinberger's Testimony Before the Senate Armed Services Committee on US Strategic Capabilities, *Official Text*, ICA (US Embassy, London, 6 October 1981), p. 5; and "Pentagon Release on the U.S. Strategic Programme", *Official Text*, ICA (US Embassy, London, 5 October 1981).

5. See Jeffrey Richelson, "PD-59, NSDD-13 and the Reagan Strategic Modernization Program", *Journal of Strategic Studies*, vol. 6, no. 2 (June 1983), pp. 128-46.

6. Remarks on the completion of the Fourth Mission of the Space Shuttle Columbia, 4 July 1982. Interestingly, this was also the first time that the shuttle had been used by the Defense Department. See Thomas O'Toole, "Astronauts Operating Battery of Instruments in Secret Experiments", *Washington Post* (29 June 1982), p. 2.

7. US Congress, House, Committee on Science and Technology, *Hearing Before the Subcommittee on Space Science and Applications on National Space Policy*, 97th Congress, 2nd Session (4 August 1982), p. 13.

8. "White House Fact Sheet Outlining U.S. Space Policy".

9. Ibid. As a whole, the White House Fact Sheet was widely criticized for being too militarily oriented. See "A Space Policy Heat", *Aviation Week & Space Technology*, vol. 117, no. 3 (19 July 1982), p. 17; and "Space Debate", *Aviation Week & Space Technology*, vol. 117, no. 6 (9 August 1982), p. 17.

10. "White House Fact Sheet Outlining U.S. Space Policy".

11. "Pentagon Space Policy: More of Same".

12. To illustrate his argument Zeiberg stated that "the Soviets have a very redundant communication system to their forces. They have one of everything so taking out a communication satellite is not going to extract much of a price from them. If they took out one of our communication satellites it would be quite different". See US Congress, Senate, *Department of Defense Authorization for Appropriations for Fiscal Year 1980*, Hearings, 96th Congress, 1st Session (1979), Part 6, p. 3028. In the previous year, General Slay had made a similar comment and noted that the short lifetime and rapid replenishment rate of Soviet satellites meant that any one-for-one exchange would also favour the Soviet Union. See US Congress, Senate, Committee on Armed Services, *Fiscal Year 1978 Authorization for Military Procurement*, Hearings, 95th Congress, 1st Session (1977), Part 8, p. 5844.

13. "Pentagon Asks Arms Capability in Space", *Washington Times* (19 January 1983), p. 4 (my emphasis).

14. Charles Doe, "Use Technology to Offset Soviets, SecDef Advises", *Army Times* (11 April 1983), p. 32.

15. "Air Force Plan to Orr This Week", *Defense Week* (5 July 1983), p. 19.

16. US Congress, House, Investigations Subcommittee of the Committee on Armed Services, *Aerospace Force Act*, Hearing, 97th Congress, 2nd Session (1983), pp. 50-1.

17. Richard Halloran, "U.S. Military Operations in Space to Be Expanded Under Air Force", *New York Times* (22 June 1982). A Space Technology Center was also established at Kirtland AFB, New Mexico.

18. Edgar Ulsamer, "Space Command: Setting the Course for the Future", *Air Force Magazine* (August 1982), pp. 48, 51, 54-5.

19. "Navy Announces Establishment of Space Command", *Aerospace Daily* (16 June 1983), p. 260.

20. Robert Toth, "Joint Chiefs to Recommend Unified Military Space Command", *Los Angeles Times* (18 November 1983), p. 8.

21. "White House Fact Sheet Outlining U.S. Space Policy".

22. "Antisatellite Weapons Contract Awarded", *Aviation Week & Space Technology*, vol. 114, no. 5 (2 February 1981), p. 15.

23. Bruce Smith, "Vought Tests Small Antisatellite System", *Aviation Week & Space Technology*, vol. 115, no. 19 (9 November 1981).

24. Interview with retired Senior Air Force officer (Iota), Washington DC, February 1984.

25. *Aerospace Daily* (11 April 1983), p. 235.

26. "U.S. Antisatellite Program Needs a Fresh Look", Digest of Report By the Comptroller General of the United States, GAO/C-MASAD-83-5 (27 January 1983).

27. Ibid.

28. Clarence A. Robinson, Jr, "USAF Will Begin Antisatellite Testing", *Aviation Week & Space Technology*, vol. 119, no. 25 (19 December 1983), pp. 21-2.

29. Ibid.

30. US Congress, House, Subcommittee of the Committee on Appropriations, *Department of Defense Appropriations For 1984*, Hearings, 98th Congress, 1st Session (1983), Part 8, p. 501.

31. Fred Hiatt, "Anti-Satellite Weapon Research is Pressed", *Washington Post* (28 February 1984), p. A3.

32. C. A. Robinson, Jr, "U.S. Will Begin ASAT Testing", p. 20.

33. Fred Hiatt, "U.S. Tests Satellite Destroyer", *Washington Post* (22 January 1984), p. 1.

34. *Department of Defense Appropriations For 1984*, p. 499.

35. "Soviets Launch Second Satellite Interceptor in Nine Months", *Aviation Week & Space Technology* (9 February 1981), pp. 28-9.

36. Nicholas Johnson, *The Soviet Year in Space: 1981* (Teledyne Brown Engineering, 1982), p. 26.

37. "Soviets Test Another Killer Satellite", *Aviation Week & Space Technology*, vol. 114, no. 12 (23 March 1981), pp. 22-3.

38. Johnson, *The Soviet Year in Space: 1981*, p. 26.

39. "Cosmos Threat", *Aviation Week & Space Technology*, vol. 115, no. 22 (30 November 1981), p. 17.

40. "Pentagon Says It Has No Evidence of Soviet Space Battle Station", *Defense Daily* (28 October 1981), p. 279.

41. "Soviets Launch Antisatellite Target-Like Vehicle", *Aviation Week & Space Technology*, vol. 116, no. 24 (14 June 1982), p. 19; "Soviets Stage Integrated Test of Weapons", *Aviation Week & Space Technology*, vol. 116, no. 26 (28 June 1982), pp. 20-1.

42. Nicholas Johnson, *The Soviet Year in Space: 1982* (Teledyne Brown Engineering, 1983), p. 25.

43. *Department of Defense Appropriations For 1984*, p. 434. Interestingly, the spectre of the Soviets using this new booster to deploy "multiple orbiting bombardment systems" was also raised by DARPA Director Dr Cooper at these hearings.

44. George Wilson, "Soviets Reported Ready to Orbit Laser Weapons", *Washington Post* (3 March 1982), p. 1.

45. "Talk of Soviet Laser Weapon in Space By 1983 Called Nonsense", *Baltimore Sun* (4 March 1982), p. 15. For other critiques of the feasibility of laser weapons see Kosta Tsipis, "Laser Weapons", *Scientific American*, vol. 245, no. 5 (December 1981), pp. 51-7; and Jeff Hecht, "Lasers Make Ready for War", *New Scientist* (10 June 1982), pp. 714-17.

46. *Soviet Military Power*, 2nd edn (Government Printing Office, Washington DC, March 1983), p. 67.

47. Ibid., p. 68.

48. *Soviet Military Power*, 3rd edn (Government Printing Office, Washington, DC, April 1984), p. 36.

49. *Department of Defense Appropriations For 1984*, p. 485.

50. "DoD Space-Based Laser Program—Potential, Progress, and Problems", Unclassified Segment of GAO Report C-MASAD-82-10 (26 February 1982). See also Clarence Robinson, Jr, "GAO Printing Accelerated Laser Program", *Aviation Week & Space Technology*, vol. 116, no. 15 (12 April 1982), pp. 16-19.

51. Philip J. Klass, "House United Alters Laser Goals", *Aviation Week & Space Technology*, vol. 116, no. 17 (26 April 1982), pp. 18-19.

52. "Laser Critique", *Aviation Week & Space Technology*, vol. 116, no. 13 (15 March 1982), p. 13.

53. Michael Gordon, "Reagan's 'Star Wars' Proposals Prompt Debate Over Future Nuclear Strategy", *The National Journal*, vol. 16, no. 1 (1 January 1984), p. 16.

54. See D. Cannon and L. Cannon, "President Overruled Advisers On Announced Defense Plan", *Washington Post* (26 March 1983), p. 1; Leslie Gelb, "Aides Urge Reagan to Postpone Antimissile Ideas For More Study", *New York Times* (25 March 1983), p. 1; and Jeffrey Smith, "Reagan Plans New ABM Effort", *Science*, vol. 222 (8 April 1983), pp. 170-1.

55. J. Smith, "Reagan Plans New ABM Effort".

56. David Hoffman, "Futuristic Soviet Defenses Said Welcome Possibility", *Washington Post* (28 March 1983).

57. "Weinberger Softens Insistence 'On a Leak-Proof ABM System'", *Washington Post* (12 April 1983).

58. National Security Decision Directive No. 85, "Eliminating the Threat From Ballistic Missiles" (The White House, Washington, DC, 25 March 1983).

59. Ibid.

60. Memorandum From Secretary of Defense, Caspar Weinberger, Subject: Organization of Defensive Technologies Executive Committee—Action Memorandum (DoD, Washington, DC, 28 March 1983).

61. "Missile Defense Effort Indicates Directed-Energy, Other Means", *Aviation Week & Space Technology*, vol. 118, no. 22 (30 May 1983), p. 332.

62. "Energy Weapons", *Aviation Week & Space Technology*, vol. 119, no. 10 (5 September 1983), p. 17.

63. "Defensive Technolgies Study Sets Funding Profile Options", *Aviation Week & Space Technology*, vol. 119, no. 17 (24 October 1983), p. 50.

64. Clarence A. Robinson, Jr, "Study Urges Exploiting of Technologies", *Aviation Week & Space Technology*, vol. 119, no. 17 (24 October 1983), p. 50.

65. Fred S. Hoffman, *Ballistic Missile Defenses and U.S. National Security*, Summary Report (October 1983), p. 2.

66. Ibid.

67. Clarence A. Robinson Jr., "Panel Urges Defense Technology Advances", *Aviation Week & Space Technology*, vol. 119, no. 16 (17 October 1983), pp. 16-18; and Robert E. Tyler, "Reagan Set to Grapple with Space-Based Missile Defense Decision", *Washington Post* (27 November 1983), p. 24.

68. C. A. Robinson, Jr, "Panel Urges Defense Technology Advances", p. 17.

69. Ibid.

70. Gordon, "Reagan's 'Star Wars' Proposals Prompt Debate over Future Nuclear Strategy", p. 16.

71. Michael Getler, "Reagan Signs Anti-Missile Research Order", *Washington Post* (26 January 1984), p. 1. The Strategic Defense Initiative Justification Statement, however, identifies NSDD 116 as the decision document.

72. Charles Mohr, "Pentagon Backs Advanced Defense Despite Flaws", *New York Times* (9 March 1984), p. 9.

73. Quote from Rebecca V. Strode, "Commentary on the Soviet Draft Space Treaty of 1981" in Colin S. Gray (ed.), *American Military Space Policy: Information Systems, Weapon Systems and Arms Control* (Abt Books, Cambridge, Mass., 1982), p. 85.

74. "Draft Treaty on the Prohibition of the Stationing of Weapons of any Kind in Outer Space", Proposal to the 36th Session of the UN General Assembly, 20 August 1981.

75. Ibid.

76. Strode, "Commentary on the Soviet Draft Space Treaty of 1981", p. 88.

77. Ibid., p. 87.

78. US Congress, Senate, Subcommittee on Arms Control, Oceans, International Operations and Environment of the Committee on Foreign Relations, *Arms Control and the Militarization of Space*, Hearings, 97th Congress, Second Session (20 September 1982), p. 11.

79. Ibid.,

80. Ibid., p. 27.

81. US Congress, Senate, Committee on Foreign Relations, *Controlling Space Weapons*, Hearings, 98th Congress, First Session (14 April and 18 May 1983), pp. 77-81.

82. See, for example, "Communique on the Meetings of the Political Consultative Committee of the Warsaw Treaty Member States", *Tass* (5 January 1983); John F. Burns, "Andropov Offers Ban on Space Weapons", *New York Times* (28 April 1983), p. 1; and "Excerpts from Gromyko Speech Reversing Soviet Union's Foreign Policy", *New York Times* (17 June 1983), p. 5.

83. Dusko Doder, "Andropov Urges Ban on Weapons to Attack Satellites", *Washington Post* (19 August 1983), p. 1.

84. See John Pike, "Limits on Space Weapons: The Soviet Initiative and the American Response", *FAS Staff Study* (12 September 1983), p. 9.

85. "Test of Soviet Draft Treaty on Banning the Use of Force in Space and from

Space with Respect to the Earth".

86. US Department of State News Briefing, 26 August 1983 (transcript prepared by John Pike).

87. Pike, "Limits on Space Weapons", pp. 11-12.

88. "Peace in Space?" *The Economist* (27 August 1983), pp. 11-12.

89. See Flora Lewis, "A Lock for Pandora", *New York Times* (30 August 1983).

90. Establishing Criteria Governing the Test of Antisatellite Warheads, Section 1235, Public Law 98-94 (24 September 1983), pp. 695-6.

91. "Antisatellite Ban Called Unsound", *New York Times* (16 March 1984), p. 8.

92. Walter Pincus, "High Pentagon Officials Opposing U.S. Actions on Arms Control", *Washington Post* (20 March 1984), p. A17.

93. Covering Letter from Ronald Reagan to Speaker of the House of Representatives, *Report to the Congress on U.S. Policy on ASAT Arms Control* (The White House, 31 March 1984).

94. Ibid.

95. Helen Dewar, "Antisatellite Tests Backed, With Condition", *Washington Post* (13 June 1984).

96. Dusko Doder, "Chernenko Suggests Antisatellite Ban Before It's 'Too Late'", *Washington Post* (12 June 1984).

97. Helen Dewar, "Antisatellite Tests Backed, With Condition".

98. Fred Hiatt and Rick Atkinson, "U.S. Considering Controls on Use of Arms in Space", *Washington Post* (16 June 1984), p. 1.

99. Dusko Doder, "U.S.-Soviet Talks Set on Space Arms", *Washington Post* (30 June 1984), p. 1.

100. Bernard Gwertzman, "U.S. Says It Weighs Kremlin's Motives in New Arms Offer", *New York Times* (1 July 1984), p. 1.

101. Leslie H. Gelb, "U.S. Agreed to Talk About Space Arms Without Clear Plan", *New York Times* (3 July 1984), p. 1.

102. Dusko Doder, "Soviets Oppose Link of Talks on Missiles, Antisatellite Weapons", *Washington Post* (2 July 1984), p. 1.

103. Walter Pincus and David Hoffman, "No Conditions Set on Space Talks, U.S. Tells Soviets", *Washington Post* (3 July 1984), p. 1.

104. Seth Mydans, "Soviet Says Talks on Space Weapons are 'impossible'", *New York Times* (28 July 1984).

12. *Conclusion*

1. Arthur M. Schlesinger, Jr, *A Thousand Days* (Houghton Mifflin, Boston, 1965), p. 290.

2. Philip Klass, *Secret Sentries in Space* (Random House, New York, 1971), p. xiii.

3. Gerald M. Steinberg, *Satellite Reconnaissance: The Role of Informal Bargaining* (Praeger, New York, 1983).

4. Ibid., p. 95.

5. Ibid., p. 129.

6. SIPRI, *Outer Space—Battlefield of the Future*, pp. 60-1; and Klass, *Secret Sentries in Space*, p. 126.

7. Robert P. Berman and John C. Baker, *Soviet Strategic Forces: Requirements and Responses* (The Brookings Institution, Washington, DC, 1982), pp. 115-17.

8. Alton Frye, "U.S. Space Policy: An Example of Political Analysis" in E. S. Quade and W. I. Boucher (eds), *Systems Analysis and Policy Planning: Applications in Defense* (American Elsevier Publishing Company, New York, 1968), p. 316.

9. Quoted in R. F. Futrell, *Ideas, Concepts, Doctrine: A History of Basic Thinking in the United States Air Force 1907-1964* (Air University, Maxwell AFB, Alabama, 1971), p. 435.

10. See M. Mihalka, "Soviet Strategic Deception", *Journal of Strategic Studies*, vol. 5, no. 1 (March 1982), pp. 40-93; and *Khrushchev Remembers Vol. 2: The Last Testament*, Strobe Talbott, trans. and ed. (Penguin, London, 1977), p. 79.

11. Herbert F. York, *Race to Oblivion*, p. 124. McNamara described it as "nothing more than a paper study of a very esoteric system". See Futrell, *Ideas, Concepts, Doctrine*, p. 434.

12. Leitenberg, *The History of U.S. Antisatellite Weapons Systems*, p. 33.

13. Raymond Cohen, *International Politics: The Rules of the Game* (Longman, London, 1981), p. 50. Cohen uses the Nazi-Soviet tacit understanding on aerial surveillance between 1940-1 to illustrate this point. Here:

> Since neither side was officially supposed to suspect the good intentions of the other, and since the permissibility of overflights by military of aircraft of each other's territory could hardly be given formal sanction, an understanding could not be openly stated in written or even verbal form. Nevertheless, its advantages in removing suspicions and thereby avoiding preemptive and untimely attack could not be denied. (That Nazi Germany saw it as a temporary and cynical expedient did not obviate its utility in the short term.)

14. Even here, in an atmosphere free of public posturing, misperceptions of intent can still occur. For example, it was only Garthoff's personal contact with the Soviet diplomat Usachev that cleared the original Soviet misperception of Fisher's *private* approach to Gromyko in New York in 1962, and led to the UN resolution. For a theoretical exposition of the problems of misperceiving arms control offers, see Thomas C. Schelling, "A Framework for the Evaluation of Arms Control Proposals" in Long and Rathjens (eds), *Arms, Defense Policy, and Arms Control* (Norton, New York, 1976), pp. 187-200.

15. Christoph Bertram, "The Future of Arms Control: Part II. Arms Control and Technological Change: Elements of a New Approahc", *Adelphi Paper*, no. 141 (Institute for International Strategic Studies, London, 1978), p. 16.

16. See Raymond Garthoff, "ASAT Arms Control: Still Possible", *Bulletin of the Atomic Scientists*, vol. 40, no. 7 (August-September 1984), pp. 29-31.

17. See Donald L. Hafner, "Outer Space Arms Control: Unverified Practices, Unnatural Acts?" *Survival*, vol. xxv, no. 6 (November/December 1983), pp. 242-8.

18. Ashton Carter, *Directed Energy Missile Defense in Space*, Background Paper (Office of Technology Assessment, Washington, DC, 1984), p. 97.

19. A string of unsuccessful tests, however, would diminish US bargaining leverage. Either way it will become progressively more difficult for the United States to give up its ASAT programme.

20. See Hafner, "Outer Space Arms Control".

21. Article III of the Outer Space Treaty states that the exploration and use of outer space shall be carried out in accordance with the UN charter. Article II: 4 of the latter prohibits the use of force or threat of use of force.

22. There is the contrary argument that states that ASAT weapons are stabilizing as

certain satellites, such as those used for counterforce targeting or for ASW purposes, undermine deterrence.

23. In a prepared statement before Congress in March 1979, General Stafford, Deputy Chief of Staff of the Air Force for Research Development (and a former astronaut), stated:

> Under certain circumstances space may be viewed as an attractive arena for a show of force. Conflict in space does not violate national boundaries, does not kill people, and can provide a very visible show of determination at relatively modest cost. Because of this possibility, it is desirable to provide the National Command Authorities with the additional option of ordering a response in kind to a space attack. The nation is presently limited to doing nothing, diplomatic protest or economic and military sanctions which may be escalatory.

See US Congress, Senate, Committee on Armed Services, *Department of Defense Authorization for Appropriations for Fiscal year 1980,* Hearings, 96th Congress, 1st Session (1979), Part 6, p. 3019.

Index

For Product Safety Concerns and Information please contact our EU representative GPSR@taylorandfrancis.com
Taylor & Francis Verlag GmbH, Kaufingerstraße 24, 80331 München, Germany

www.ingramcontent.com/pod-product-compliance
Lightning Source LLC
LaVergne TN
LVHW011112300926
849160LV00008B/363

* 9 7 8 0 3 6 7 5 5 7 9 9 7 *